RESTORING CLIMATE

RESTORING CLIMATE

DORIN PREDA

QSOLAR

To my parents,

Nicolina and Constantin Preda

Title: RESTORING CLIMATE

Author: Dorin Preda

First published in January 2020 "Home: Calculations for Hope"; Q Solar, ISBN: 978-0-6487764-1-3.

Second version of Oct. 2021 "Earth: Calculations for Climate"; Q Solar, ISBN: 978-0-6487764-2-0.

This is the third version, of July 2024 "Restoring Climate"; QSolar, ISBN: 978-0-6487764-4-4 .

Illustrations or additions by Dorin Preda, except where attributions were made to the original authors. The author thanks the copyright holders for the illustrations quoted in the book.

Subjects (keywords): global warming, deforestation, climate change, environmental protection, drought abatement, built environment, transportation, land and climate rehabilitation.

Target audience: this book is addressed to all high school graduates interested in climate change and climate repair who want to assess and check by themselves new notions and basic calculations.

Publisher: QSolar, Queensland, Australia, ABN: 26857431550 https://dpreda.net/qsolar/restoring/

TABLE OF CONTENTS

Preface

I am an Earth Sciences engineer specialised in coastal studies (1982-1986), that allowed the understanding of the border areas between the evaporating sea and the mainly condensing land. That understanding was enhanced with the interaction of atmosphere and Earth surface while doing scientific research for a meteorology and hydrology institute (1986-1990). Needing a very efficient heating system for my house, I invented and in 1992 I was granted a patent of invention for a heat pipe system that uses latent heat transfers, similarly to the atmosphere. From 2001 to 2024 retirement, I was the director of Q Solar Pty Ltd, conducting solar air heating research, for which I was granted several patents of invention. Like my previous fields of experience, solar air heating is also relevant for climate, representing the sensible heat transfers in the atmosphere.

The reasons for my sudden interest in climatology since 2019 were the forest mega-fires which confirmed major climate changes and the necessity of multidisciplinary studies to allow proper understanding. Unsettled present climatology issues like the role of forests and water in climate, and whether reforestation is suitable for climate repair, showed that more analysis is needed.

In 2020, the first version of this book [52] explained new mechanisms through which water and vegetation cool the climate, and suggested revegetation as a solution. This 2024 version confirms the theoretical basis, explains the damages of intertropical deforestation, and suggests solutions.

The reason of the inaction on climate is the divergence of opinions, which is due on one hand to the diverging interests of different parties, and on the other hand to most people not understanding climate change basics. Therefore, in this book I explain the main problems and find realistic solutions within a unifying framework, that would result in benefits and opportunities for all parties.

I think that the international commitment should be focused simultaneously on the transition to renewable energy to reduce fossil fuel emissions, and on the urgent revegetation of the intertropical areas with useful trees to reduce atmospheric carbon dioxide and to restore climate. It is extremely important to restore the climate through Nature, not damage it with geoengineering.

This personal study has not been externally funded or influenced.

I am grateful to the late Professor E. Pora, a former member of the Romanian Academy of Sciences, for setting me on the path of accurate science many years ago.

Dorin Preda

Introduction

This book argues that the current understanding of climate science underestimates the critical role of water and vegetation in regulating the Earth's climate. Immediate action is needed to address these scientific oversights and start restoring the climate by using water and vegetation.

Improvements in the understanding of water vapours, clouds and vegetation resulted in identifying a large imbalance in the global water cycle that is more powerful than other climate causes, while all working together to create a greater than known existential threat for people and Nature. The severity of the situation requires that the problem is assessed and understood by all affected people; thus, it will be plainly explained here, considering the fact that climate change is disputed.

Intensely disputed issues like climate change [120] can achieve progress only through open discussion, rational debate and identification of mutually beneficial solutions, like in this book. Many novel suggestions for agriculture, architecture, transport, environment protection, climate restoration, are presented here for the benefit of individuals, entrepreneurs and organisations.

This study revolves around the climate role of water and areas with dense trees (forests, orchards and plantations), which we shall just call "tree areas". They were studied more in the past with land use and water cycle research, through direct observations and measurements. However, the weight of climatology studies changed in the last 20 years to remote sensing and computer simulations, which brought benefits, but also incertitude due to their great leap forward.

Climate science was focused on high latitude areas, more familiar to most climatologists [121]. Extended research has been conducted on glaciers, Milankovitch cycles and albedo variations in frozen areas, which I argue are important at high latitudes, but less significant in the more energetic lower latitudes. Despite the intertropical area is the driver of climate, it received less attention, and it does not have enough meteorological stations to provide reliable statistical data. As result, the climatic simulations run by high latitude climatologists through the understanding of their local climate, did not have proper qualitative and quantitative input from the lower latitudes, resulting in their decreased capacity to properly represent global climate. Therefore, tropical climatic changes such as drought through deforestation, and desertification through land mismanagement have received less attention [122], leading to less efficient solutions like Africa's Great Green Wall.

The observed extreme climate events have been correlated with a rise in air temperature that in turn was observed to be directly proportional to carbon dioxide variation. Consequently, climate change has been linked to global warming, which associates with the atmospheric carbon dioxide, while the above-mentioned imperfections of the computer simulations excluded deforestation and water phenomena as climate drivers, leading to imbalances in the water cycle.

In the past, we have altered Earth's surface by clearing forests for settlements, agriculture, and mining, creating widespread pollution, excessive carbon dioxide emissions, and generally disregarding consequences for benefits. However, the changes we have made especially by burning fossil fuels and vegetation and through mechanical deforestation now make our lives more difficult rather than better, manifesting in warming, extreme weather events, desertification, sea level raise, melting of perennial ice, soil, air, and ocean degradation called "climate change".

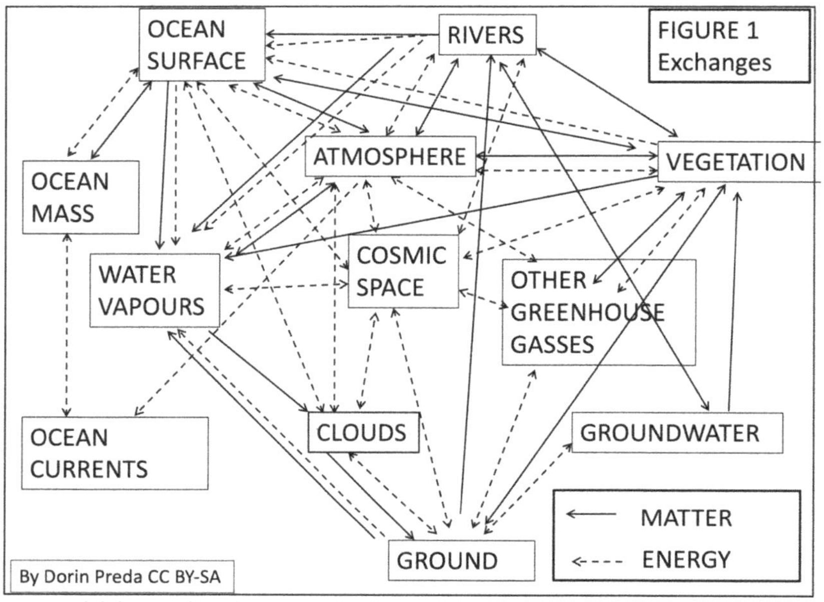

There is a psychological problem that reflects in our present way of dealing with the planet. Despite being our cradle, unique in the Universe, and our only asset, we call our planet "earth", thus it is easier to disregard and destroy. I suggest to call it Earth, and to call nature "Nature", because we did not find life on other planets. It would be ridiculous and absurd to destroy the entire planet (as we transformed the beautiful Garden into the Middle East desert), while we lay down plastic grass in air-conditioned containers, and pretend that we are happy. But this is exactly what we now do to the Earth: devise all sorts of geoengineering contraptions that pretend to "repair" the world, while all they can do is ruin it more. Increasing the artificial changes to "repair" our past mistakes has always put us in a worst situation. Some want to exhaust Earth resources preparing to leave for Mars because Earth would be "unrepairable". More logical and safer is to withdraw on the same path, repairing our past mistakes.

The climate, as depicted in Figure 1, is a complex system with many interconnected parameters, resulting in rather unpredictable outcomes across space and time. To this complexity, different human interests add, so speculations arise, leading to the climate change debate. The interplay of economic, social, financial and psychological factors is the major obstacle in front of solving climate change. Section 7.8 shows how the interests of the involved entities can be appeased, providing large benefits to those entities and to the wider society, while restoring climate.

Now that there is widespread agreement that humans are responsible for climate change and action must be taken to address it, the focus shifts to determining the most effective mitigation strategies. Most climate-conscious people agree that carbon gases in the atmosphere must be reduced, but a specialist group paradoxically opposes sequestration through reforestation; therefore, new alternatives that can sequester carbon are investigated and justified in this study.

Intertropical deforestation decreased local precipitation, overheating and flooding higher latitudes [90]. Supposing that we do not stop deforestation and its solar air heating as demonstrated in the next chapters, it is very probable that we shall soon lose to desertification in an avalanche effect (positive feedback) not only Amazonia (Section 7.3.6.5), but Central Africa also, leading to formation of a hot, dry low latitude belt and a dark, warm, too wet and stormy high latitude belt.

Therefore, it is urgently necessary to promote rational debate, collaborate in devising correct, efficient climate restoration measures, and oppose further damage through geoengineering [123]. Climate was intensely studied especially in the past 50 years, but lately elements of natural cooling like water and vegetation are presented by insufficient computer simulations as unwanted climate warmers (Section 2.1), while supporting instead damaging deforestation [97] and geoengineering.

In Figure 2, it is shown that we create in our minds a model of the much more complex real world.

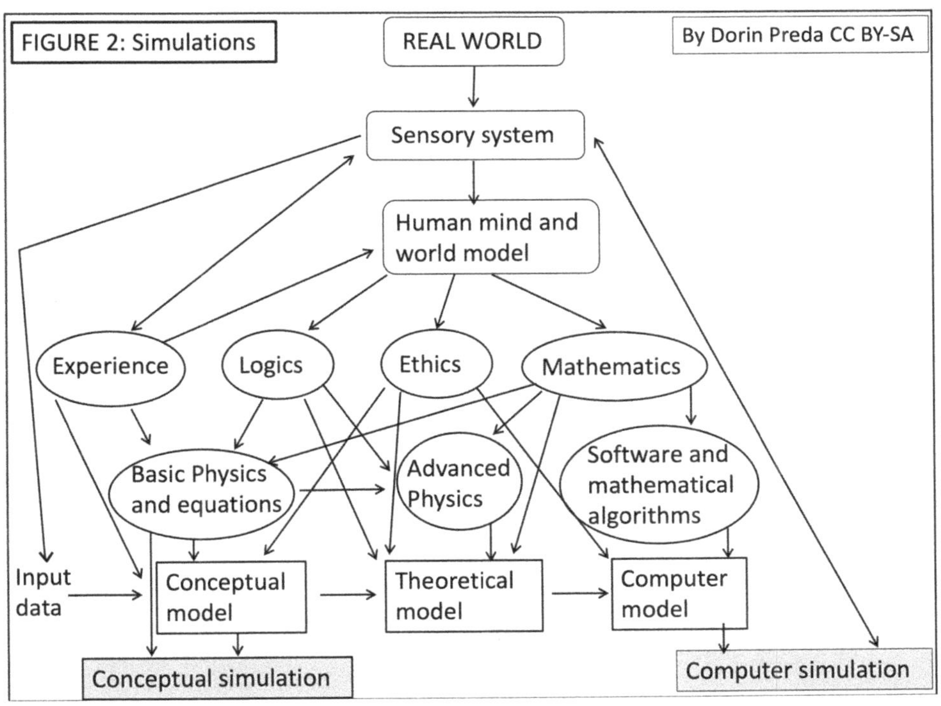

I believe that science is an ever-changing (and generally improving) collection of models.

The model that anthropogenic CO_2 causes global warming was debated and accepted, while this book demonstrates the more obvious fact that deforestation is also a driver of climate change.

Since our world model relies on our sensory system, we can sometimes be wrong (we cannot feel CO_2, but it exists). Also, our mind's reasoning system is not perfect, but has developed sophisticated logic, ethics, and mathematics as models for knowledge. People over 20 years old have gained experience, a model that informs of what works and what doesn't in the real world. An even more abstract level of models was created by the cleverest people when they developed physics, mathematical algorithms, and computer software systems. We can interpret Figure 2 as a graph where operational performance increases downwards, while certainty increases to the left.

Computer models and simulations have great potential when based on solid conceptual models, while validated remote sensing can provide valuable data and operational capacity. The safest area is in the lower left corner of Figure 2, where our experience, basic or empirical equations, and direct observations allow us to evaluate more speculative models on the right, which in turn can lead to significant advancements when matured.

We hope that one day computer simulations will provide accurate global climate predictions. The biggest problem is the difficulty to assess computer simulations combined with the impatient desire of using them for real world climatic management. This is all part of scientific improvement, and we should proceed with caution, stepping back when unsure, rather than stating uncertain assumptions as truths that can conduct to irreparable damage like loss of soils or of living species.

The fundamental principle should be that advanced physics should derive from basic physics, and both should use ethical logic. The urgency of the present existential threat asks for cautious action.

Therefore, for the time being, complex and very important issues like climate change should be validated through a framework of conceptual simulations based on ground data measurements.

Special attention will be paid to the powerful intertropical phenomena, as they are the engine and major cause of climate change, that have been overlooked until 2020 in [52], and ignored since.

This book aims to provide observational data, arguments and accessible calculations focused on low-latitude regions to enhance our understanding of the major climate drivers. We must start urgent climate restoration because the already produced changes make it more difficult by the day.

1. Chapter One: Known Basic Notions

<u>This chapter contains basic notions collected from literature; specialists can skip it.</u>

Earth's climate is a complex and dynamic system composed of many interlinked parameters. Its state at any time is a reflection of both its past and current conditions. Throughout Earth's long history, conditions have varied significantly, leading to natural changes in the climate, typically occurring at a slow pace. However, the changes we now witnessing are more significant and occurring at a faster rate, indicating a human-induced cause.

1.1 Cosmic Solar Radiation

The Sun emits solar radiation (irradiance) as electromagnetic waves and particles (the solar wind), with the latter being mainly deflected by Earth's magnetosphere. Unlike the charged particles, electromagnetic radiations (ultraviolet, visible and infrared) are not influenced by the magnetosphere and reach the top of the atmosphere at an almost constant rate of 1,361W/m^2.

FIGURE 3 The Milankovitch Cycles and the glaciations

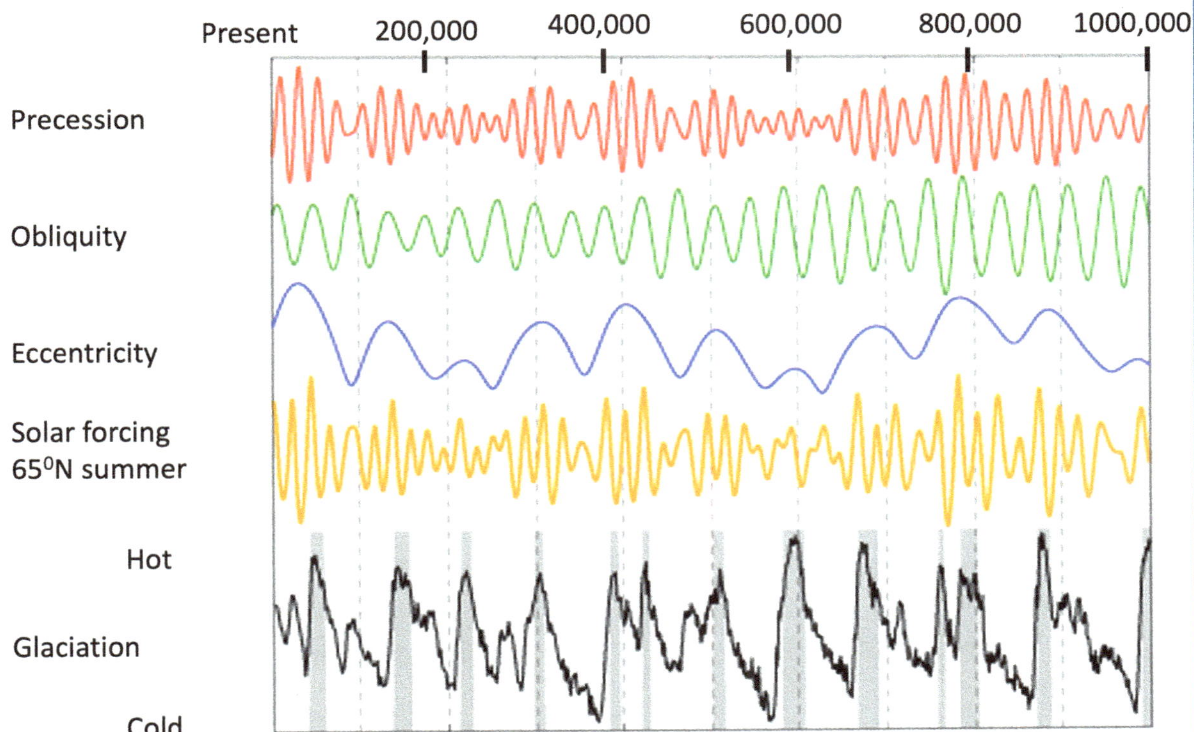

Modified image from that of Robert A. Rohde from publicly available data, and is incorporated into the Global Warming Art project. - Own work, CC BY-SA 3.0, https://commons.wikimedia.org/w/index.php?curid=567537

There is less than 5% variation in the solar radiation received at the closest annual position (perihelion) versus the most distant one (aphelion) because the Earth's orbit is an almost circular ellipse (see Figure 3). The Milankovitch cycles produce periodic variations in Earth's orbit eccentricity, axial tilt, and precession movement [5]. These orbit variations could result in up to a20% cyclical variation in the solar energy reaching the 65^{0}N areas (the Polar Regions), which is important because most land is in the Northern Hemisphere. I argue that the influence of these cycles is much smaller in temperate areas and almost negligible closer to the Equator. The Milankovitch cycles are widely accepted by most climatologists as a cause of the inferred climate cycles of 100,000 years, 41,000 years, and the glaciations (see the lowest graph in Figure 3). Similar fluctuations observed in geological records have been attributed to the Milankovitch cycles.

1.2 The Atmosphere

The atmosphere is approximately 10,000km thick, with 3/4 of its mass concentrated in the lowest 10km.There are significant variations in gas components and temperature, as illustrated in Figure 4 [The volume fraction of the main constituents of the Earth's atmosphere as a function of height, based on the MSIS-E-90 atmospheric model. Figure 4 is an adaptation by the author of (https://ccmc.gsfc.nasa.gov/modelweb/models/msis_vitmo.php].

Please note the exaggerated vertical profile of the diagram, highlighting essential features at lower altitudes, while disregarding the vast and featureless spaces above.

1.2.1 The Troposphere is the densest layer, and contains most of the water, leading to intense weather patterns. Air is transparent to solar radiation, which cannot directly heat it. Instead, all troposphere dynamics is driven by solar radiation heating the ground and the ocean. This heat warms the air in contact with these surfaces, and also causes water evaporation. The result is an inversion in air density (where lower layers are warmer and lighter than those above). This produces the vertical air movements, the wind and the high turbulence characteristic of the troposphere.

By comparing the density of a 30^{0}C air layer in contact with the ground to a 20^{0}C air just above it, we can observe why heated air slabs rise, transporting heat to the upper troposphere. Evapotranspiration (water vapours from land, water bodies and vegetation) also rises and condenses primarily in the middle troposphere, releasing heat in the process. Unfortunately, the present theoretical physics of the atmosphere has incomplete explanations for these processes.

1.2.2 The Stratosphere is heated through the absorption of ultraviolet radiation by ozone molecules (most of the ozone is in this area).

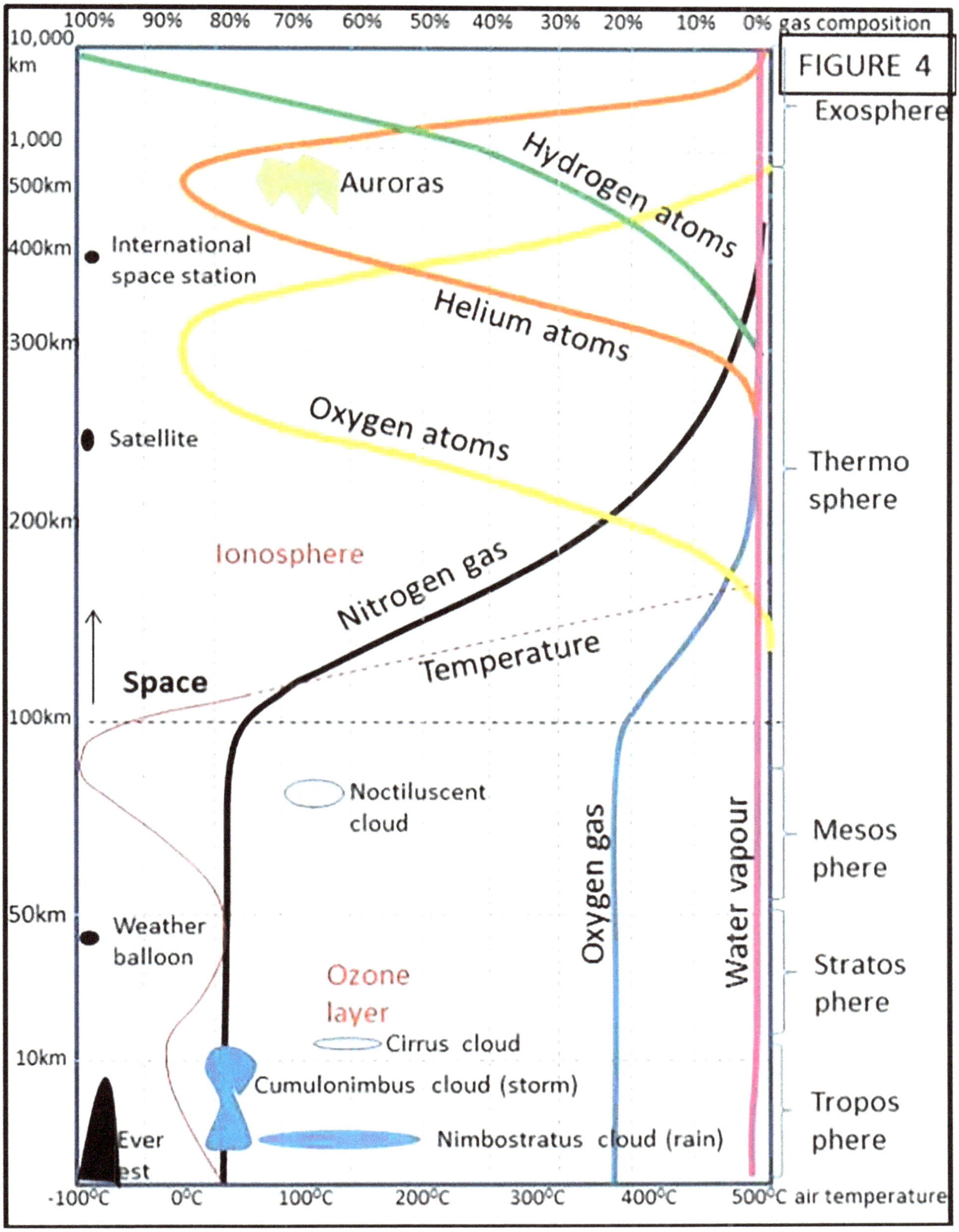

FIGURE 4

Ozone is formed by ultraviolet radiation from oxygen molecules, although there are no reasons for this not to also happen in the mesosphere. Because of this heating from above, the stratosphere has a normal temperature stratification (cold lower, hot upper), making it more stable than the troposphere and preventing the formation of clouds in its middle-high layers, despite still having some water vapours. Clouds form only in the lowest stratosphere, which is colder.

1.2.3 The Mesosphere is too rarefied for airplanes and balloons, and too dense for satellites, making it the least studied area of the atmosphere. It is the coldest part of the atmosphere, at less than minus 100^{0}C; noctiluscent clouds form here from ice crystals. Although it is freezing, the air here moves a lot with strong East-West winds, atmospheric tides and pressure pulses coming from the lower atmosphere.

1.2.4 The Thermosphere is so rarefied that atoms and molecules rarely interact. As they stop behaving as a gas, but as individual particles, at an altitude of about 100km cosmic space starts. The molecules and atoms stratify according to their specific gravity: lower oxygen, nitrogen, helium and hydrogen at the top. The ionosphere forms here in the daytime due to solar radiation and the solar wind which bombard the particles of the thermosphere, ionise them and impart high kinetic energy equivalent to more than 1,000^{0}C. However, this is not actual temperature (heat) but particle speed, therefore the dotted temperature line in Figure 4.

1.2.5 The Exosphere is the topmost layer of the atmosphere, at its base consisting of helium, hydrogen, and oxygen atoms. Its top part is formed only of hydrogen atoms.

Therefore, there is an alternation of atmospheric layers: the stratosphere, the thermosphere and the exosphere which are heated from above by solar radiation have normal thermal stratification and are calm, while the troposphere and the mesosphere are heated from below, have inverse thermal stratification and are turbulent.

Carbon dioxide is 0.04% of the atmosphere, while methane is only 0.00018% because it is quickly broken down by solar radiation in carbon dioxide and water.

Different components of the atmosphere absorb radiation differently, as seen in Figure 5.

Part of the solar radiation is reflected by the atmosphere and clouds, while air impurities absorb part of the remainder. From the received total solar radiation shown under the yellow line, the yellow area is the proportion that passes through the atmosphere, with the difference being absorbed by the atmosphere as shown in the grey areas of the Percent graph underneath. The depiction of the Major Components below shows which wavelengths are absorbed (grey colour) by different greenhouse gases (only greenhouse gases can absorb infrared radiation).

Air and water are very different fluids that interact in the atmosphere and oceans. About 0.25% of the atmosphere (4% by volume) is water vapours, but its distribution varies a lot in the atmosphere and in time. Due to its high affinity for heat and the fact that it continuously changes its state between gas, liquid and solid, water is the main agent in Earth's climate.

FIGURE 5

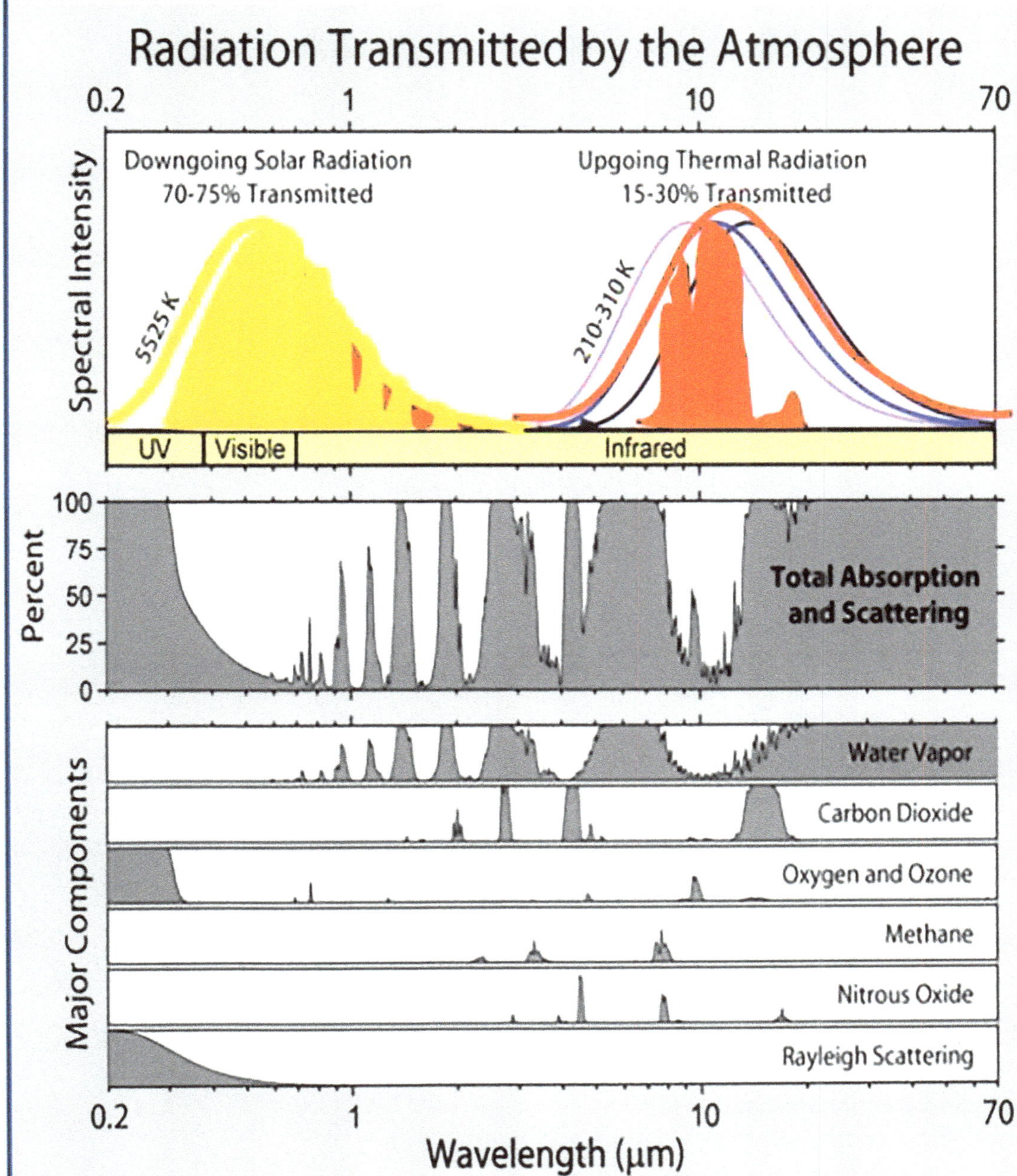

MODIFIED from Robert A. Rohde for the Global Warming Art project

1.3 Solar Radiation at Ground Level

Figure 6 displays the ground level direct and diffuse Global Horizontal Irradiation.

FIGURE 6

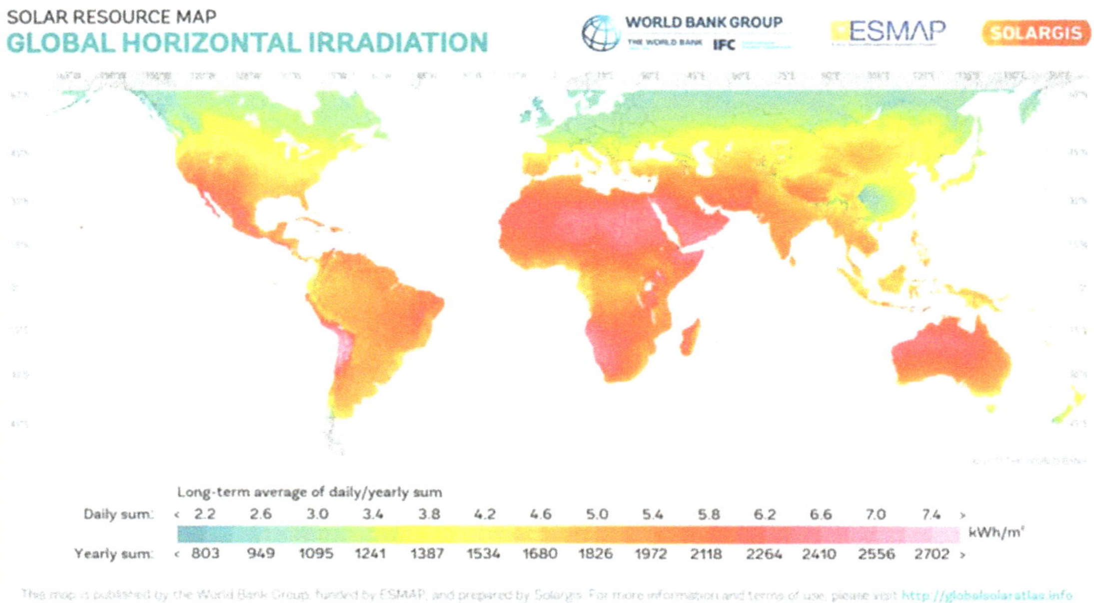

FIGURE 7 Outgoing Radiation at the Top Of Atmosphere (TOA).

Both images represent the same place at the same time.

The shortwave TOA Flux shows the large reflection of solar radiation by the white clouds and the smaller reflection by the blue atmosphere (blue because of the humidity).

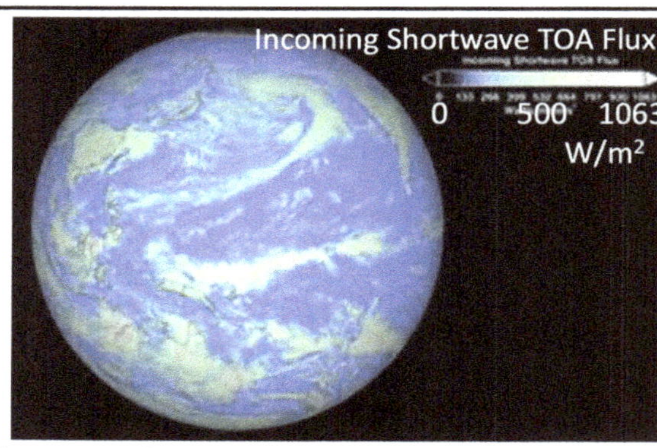

Title: NPP Ceres Shortwave Radiation.ogv Author: NASA/Goddard Space Flight Center

The intense infrared emitted by the hot ground and oceans is displayed in yellow and red, while the clouds emit less because their tops are colder than the land or the ocean. There is some incertitude upon the sensitivity of the sensors (the 5 – 8 microns gap). https://en.wikipedia.org/wiki/Earth%27s_energy_budget#Earth's_energy_imbalance

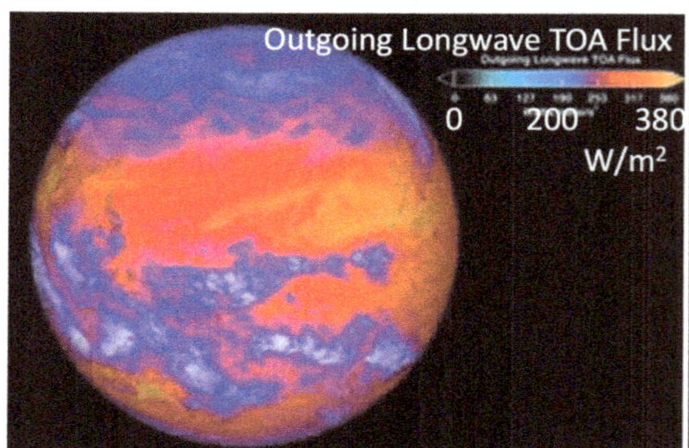

Title: NPP Ceres Longwave Radiation.ogv Author: NASA/Goddard Space Flight Center

A notable negative anomaly is observed in Central China, which receives half the solar radiation of Iraq, despite being at the same latitude. The reason for this is the continuous local smog and pollution, that block the solar radiation.

All bodies emit infrared radiation exponentially with their temperature. After being absorbed by the Earth, solar radiation is converted into heat and infrared radiation which then move upwards in the atmosphere as shown on the right side of the diagram in Figure 5.

Most of the infrared radiation is absorbed and re-emitted by greenhouse gases, causing a delay (cushioning), that is the greenhouse effect of global warming. The Earth emits the entire infrared spectrum, but only the radiation in the red-coloured area escapes without significant bouncing back and forth in the greenhouse effect.

It is important to note that water vapours absorb most of the infrared radiation (the widest grey area in Figure 5), making it the most potent greenhouse gas from a qualitative standpoint. Considering that water vapours concentration is also tens of times larger than other greenhouse gases, it is evident that the greenhouse effect in the troposphere is primarily a water effect.

1.4 Earth's Energy Budget

The heat that raises the temperature of the ground and water bodies is called sensible heat. It stores energy in the molecules of those bodies as vibration and produces infrared radiation. When water molecules vibrate strongly, they break away from the liquid as water vapours which take a lot of latent heat with them.

The vapours transported by wind may condense as rain or snow thousands of kilometres away. This way, water disrupts the local energy budget and makes it global.

Figure 7 shows that 50% of solar radiation reflects from the clouds, which cover about 50% of the Earth's area; thus, decreasing the incoming solar radiation clouds by about 25% globally.

About 10% more is reflected by the atmosphere and the land or oceans (the blue areas).

The lowest picture of Figure 7 shows that the infrared energy is dissipated towards cosmic space by the land and ocean. The clouds besides emitting some infrared have also reflected a lot of solar energy as seen in the top image. We observe that the maximum infrared emission is only about 400W/m^2, while the clouds reflection could be as much as 600W/m^2.

FIGURE 8

Net Flux July

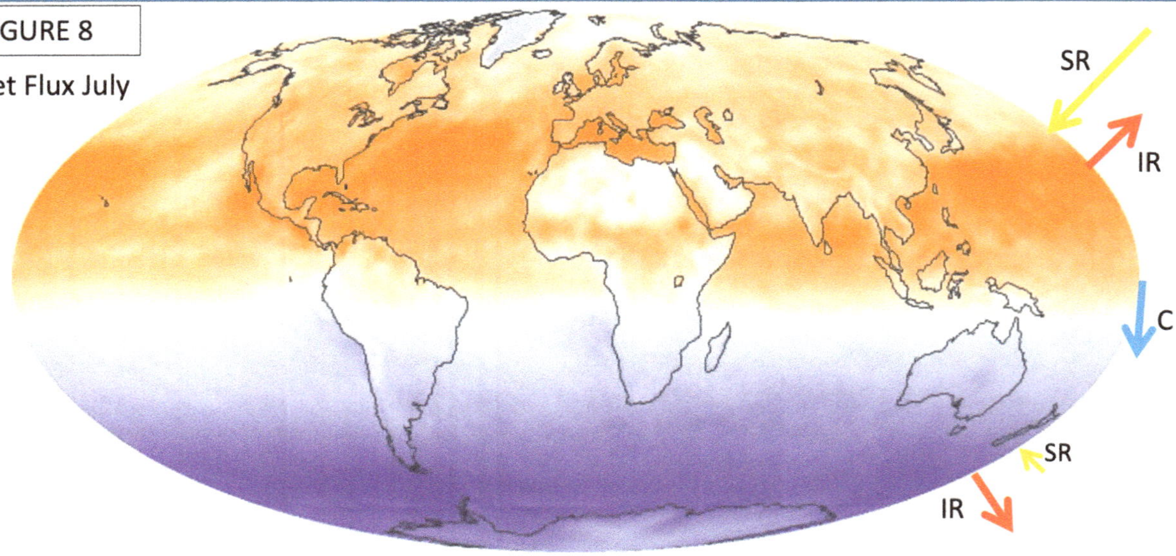

From: https://earthobservatory.nasa.gov/global-maps/CERES_NETFLUX_M

FIGURE 9

Net Flux January

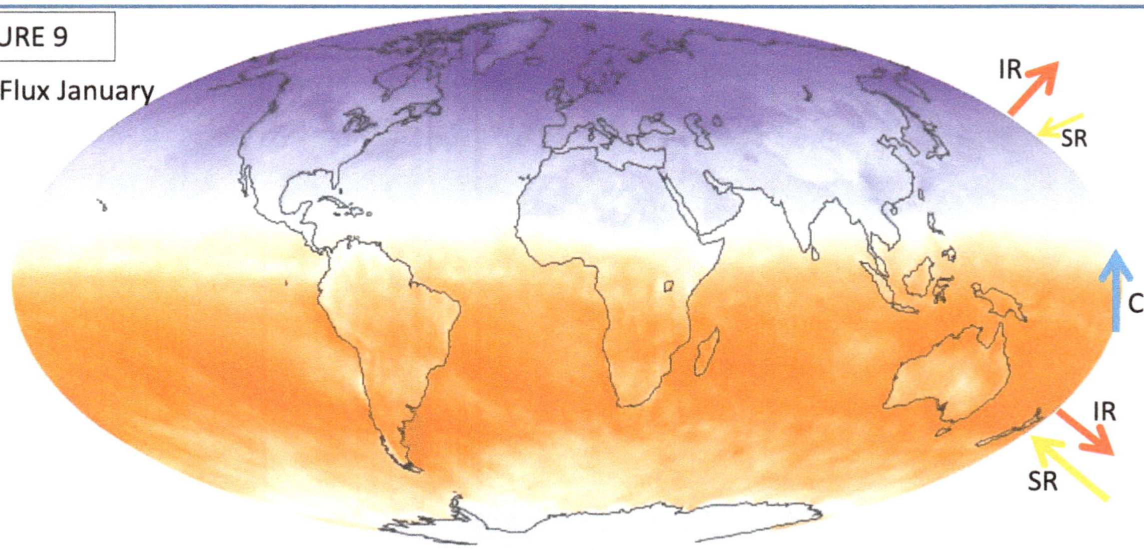

From: https://earthobservatory.nasa.gov/global-maps/CERES_NETFLUX_M

The net flux of Figures 8 and 9 show the difference between all radiation received from the Sun and all radiation emitted towards cosmic space by each place over 24 hours. In July more solar radiation SR is received (orange difference) in the Northern Hemisphere and more infrared IR is emitted (blue difference) from the Southern Hemisphere, while in January the situation reverses. "C" is the convective heat transported by the huge heat conveyor belts of atmosphere and oceans.

Figure 10 shows that since Earth is in the cosmic vacuum, the quantity of radiation that enters the Earth's atmosphere must also come out, otherwise our planet would quickly heat up. Thus, the sum of the downwards arrows largely equal the sum of the upwards ones (there are small imbalances though).

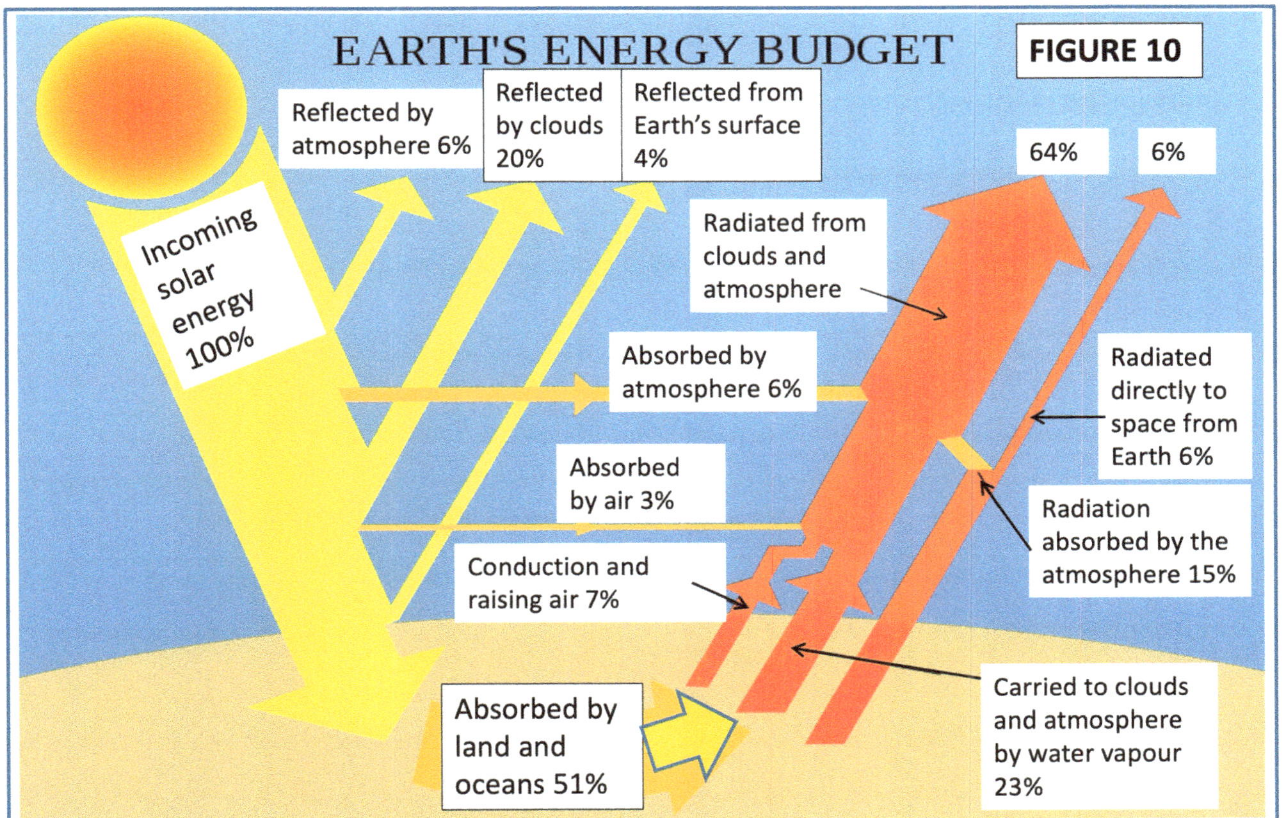

By NASA - https://web.archive.org/web/20140421050855/http://science-edu.larc.nasa.gov/energy_budget/ quoting Loeb et al., J. Clim 2009 & Trenberth et al, BAMS 2009, Public Domain, https://commons.wikimedia.org/w/index.php?curid=32285340

As an example of imbalance, in the past, the Earth has emitted less, as some solar energy has been stored in fossil fuels. Now that we burn those fuels, the planet has to emit more heat than it receives. However, that fossil heat cannot be represented in Figure 10 because it is much smaller than the other components.

As seen, three times more heat is transported back up by latent heat in water vapours (23%) than by conduction and convective air currents (7%). Thus, water not only has the main role in the radiative heat transfer among greenhouse gases as shown in Figure 5, but it is the largest latent and sensible heat transporter at the surface of the Earth, and the main heat storage of the oceans, as detailed further.

1.5 Air Temperature

The atmosphere is transparent to solar radiation, which cannot directly heat the air. When solar radiation touches water or objects on the ground, it is transformed mainly into heat which raises the temperature of those objects.

The average annual air temperature has slightly varied in the past as we know from:

- measurements with thermometers in the past 300 years,
- climate records in old living beings, like tree rings (correlations)
- historical records in the past 5,000 years that allow interpretations
- pictorial representations of the past 30,000 years allowing interpretations
- old geological records (climate proxies) that enable correlations

Since air temperature varies a lot in time and space, we must use the long-time records from 7,280 old and present meteorological stations all over the world (Figure 11).

The longest reliable record is since 1701 in Berlin [3].

The Intergovernmental Panel on Climate Change (IPCC) stated: "The global average and combined land and ocean surface temperature, show a warming of 0.85 [0.65 to 1.06] °C, in the period 1880 to 2012, based on multiple independently produced datasets. ...The trend is faster for land than ocean, faster for Arctic regions, and faster since the 1970s than the longer period."[4]

As seen in Figure 12, there is a consistent trend of temperature increase in the past 150 years.

The measurements of air temperature above land (as have been made for hundreds of years and shown in Figure 12) represent the temperature of the air inside a standard meteorological box. The IPCC uses the term Land Surface Temperature (LST) for most of their studies which refer to the temperature of the ground, and is usually different from the temperature of the air.

The increase trend continues and appears to accentuate, as almost each year sets a new record.

Although the average increase is small, individual events can be very significant and justify the attention paid to them. For example, in Figure 13, we can see temperatures many degrees above average during the 2006 heatwave in Europe have made many victims through excessively hot air.

The amplitude of the variations is also essential, as we would not like to be confronted each summer with +40^{0}C and each winter with -10^{0}C (although their average is a pleasant +15^{0}C). While Central Europe was excessively hot in 2006 (Figure 13), Eastern Europe was excessively cold. Thus, the large temperature fluctuations in time and space are another undesired factor.

FIGURE 11 From: https://en.wikipedia.org/wiki/Global_Historical_Climatology_Network

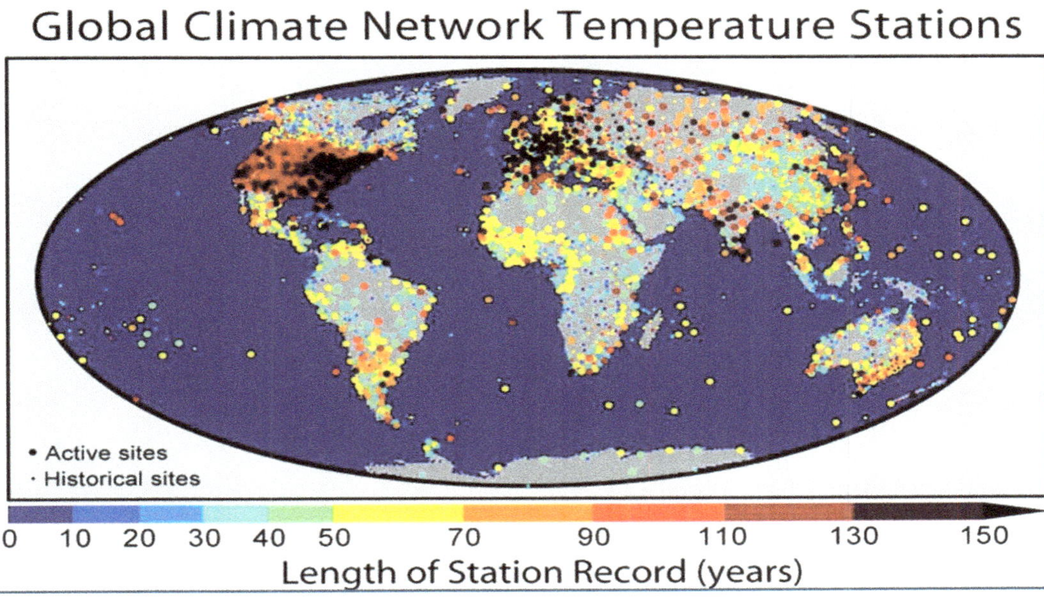

FIGURE 12

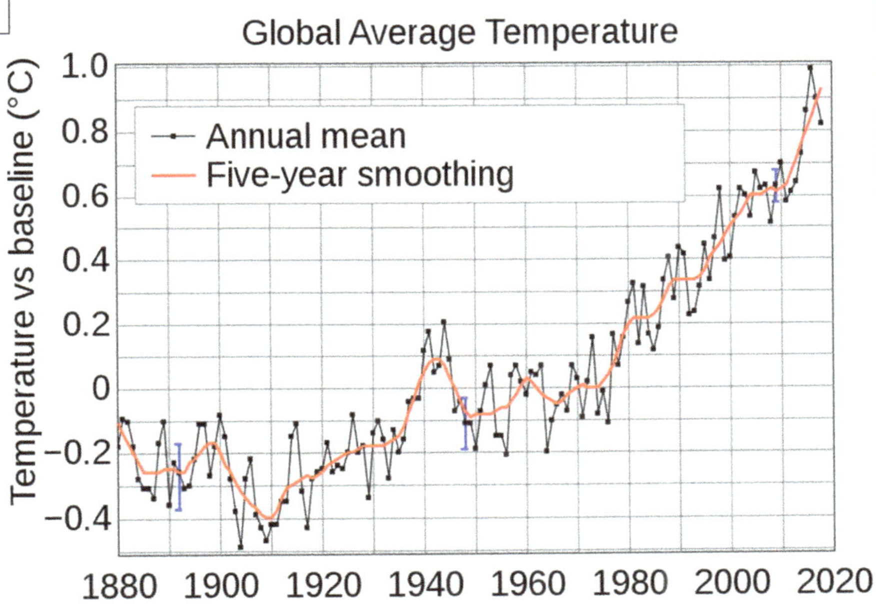

By NASA Goddard Institute for Space Studies - http://data.giss.nasa.gov/gistemp/graphs/, Public Domain, https://commons.wikimedia.org/w/index.php?curid=24363898

Solar radiation heats both the ground and the ocean. Still, the latter warms less (see Figure 14) because of the greater water evaporation from the oceans (evaporative cooling) and because the sea has a huge mass which takes time (and a lot of heat) to warm up. Unfortunately, once the ocean heats up, it takes a lot of time to cool down.

The diagram in Figure 14 shows that the land heat is the real driver of global warming, while the ocean is the follower. In a cooling situation, the ocean air would be warmer than the land air.

FIGURE 13
The 2006 Heatwave in Europe

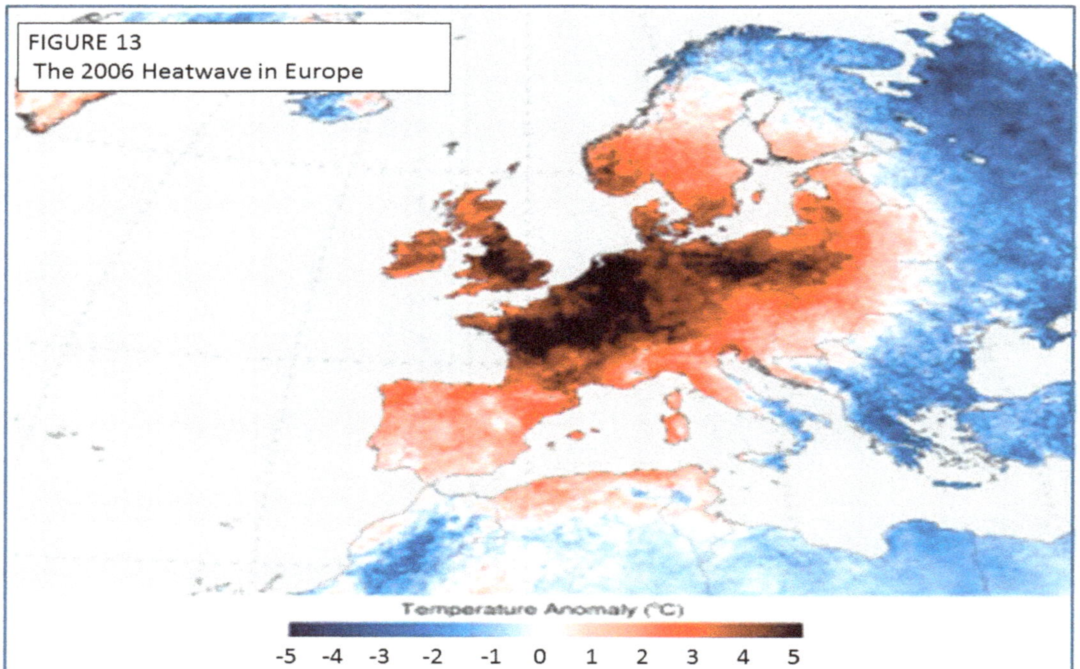

By Giorgiogp2 - Own work, CC BY-SA 3.0, https://commons.wikimedia.org/w/index.php?curid=21231713

FIGURE 14

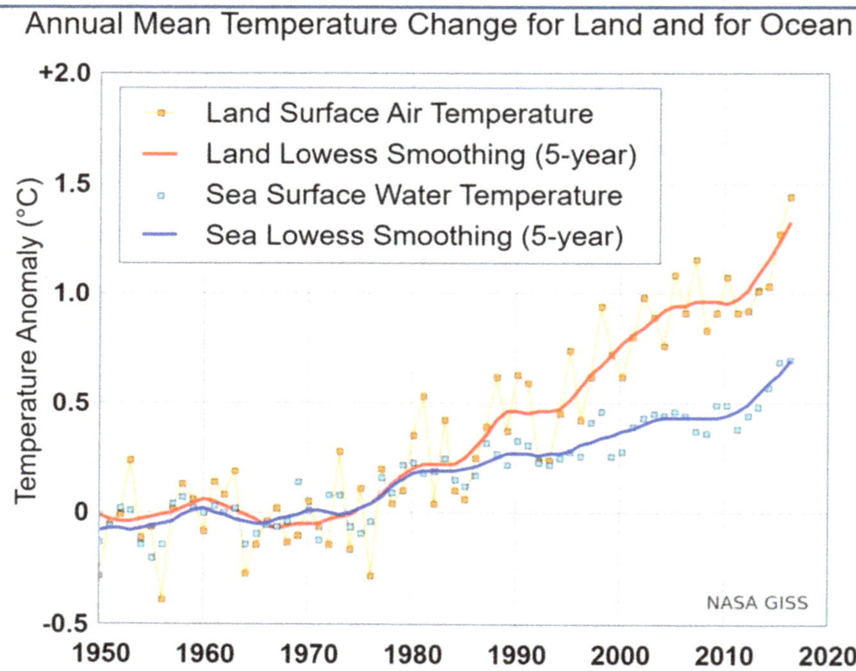

By NASA - https://data.giss.nasa.gov/gistemp/graphs/, Public Domain, https://commons.wikimedia.org/w/index.php?curid=63276564

The current sharp increase in surface air temperature is evident when compared to the evolution of temperature over the past 2,000 years, which marks the time of the largest changes we have seen on Earth (see Figure 15). There was a slight warming trend before the year 1000 AD, followed by a slight cooling period.

FIGURE 15

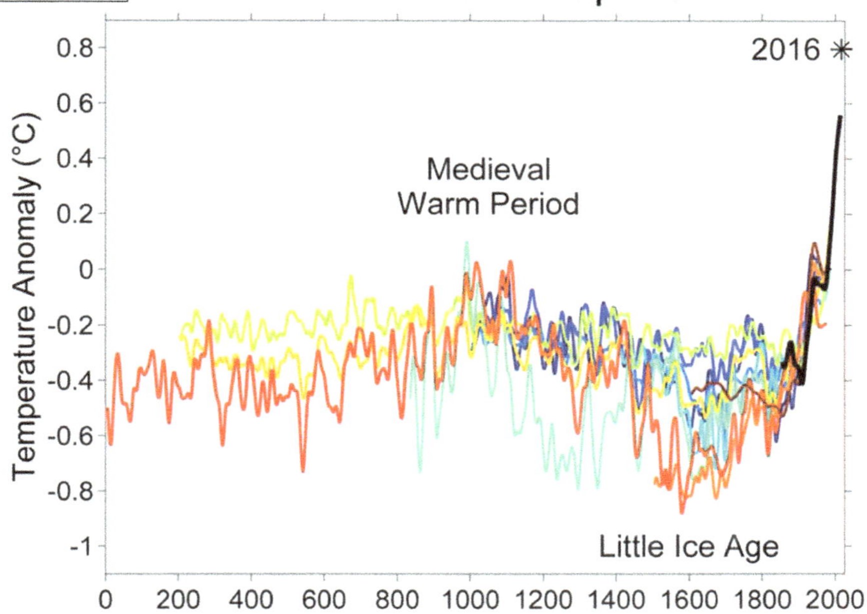

However, between about 1600 and 1850, the weather was significantly colder. Despite the average temperature dropping by only about 0.5^{0}C, the winters were very harsh. Rivers like the Thames and other water bodies, along with some shorelines froze. Crops were lost, leading to famine and social hardship across the Northern Hemisphere, especially in Western Europe. There were various inconclusive explanations for the so-called "Little Ice Age". The cooling effect during that time in the Southern Hemisphere was far smaller, or even warming in the Pacific Islands.

FIGURE 16: Temperature in the past 10,000 years

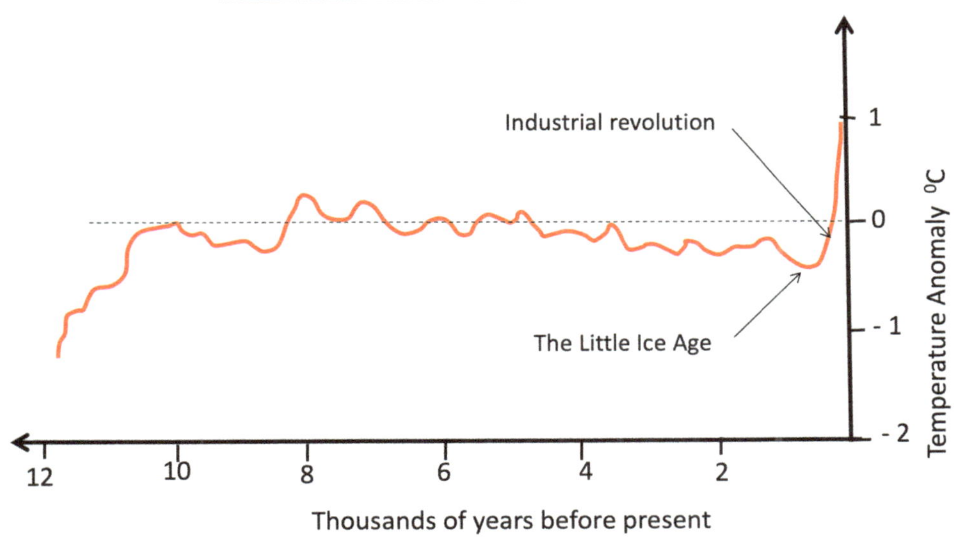

From: https://commons.wikimedia.org/w/index.php?curid=466269

Looking further back into the past, to the beginning of human civilisation, we can see in Figure 16 a reconstruction by climatologists of the temperature in the past 12,000 years. The red line shows the warming after the last glaciation, followed by a small maximum at 10,000 years, a larger maximum at 7800 years, a plateau until about 3,500BP, and then a gradual cooling until 2000BP.

At the beginning of the 19th century, the temperature started to increase sharply and continues to do so today. Finally, Figure 17 shows the previous four glaciation cycles, including the times before humans had an impact on the natural environment.

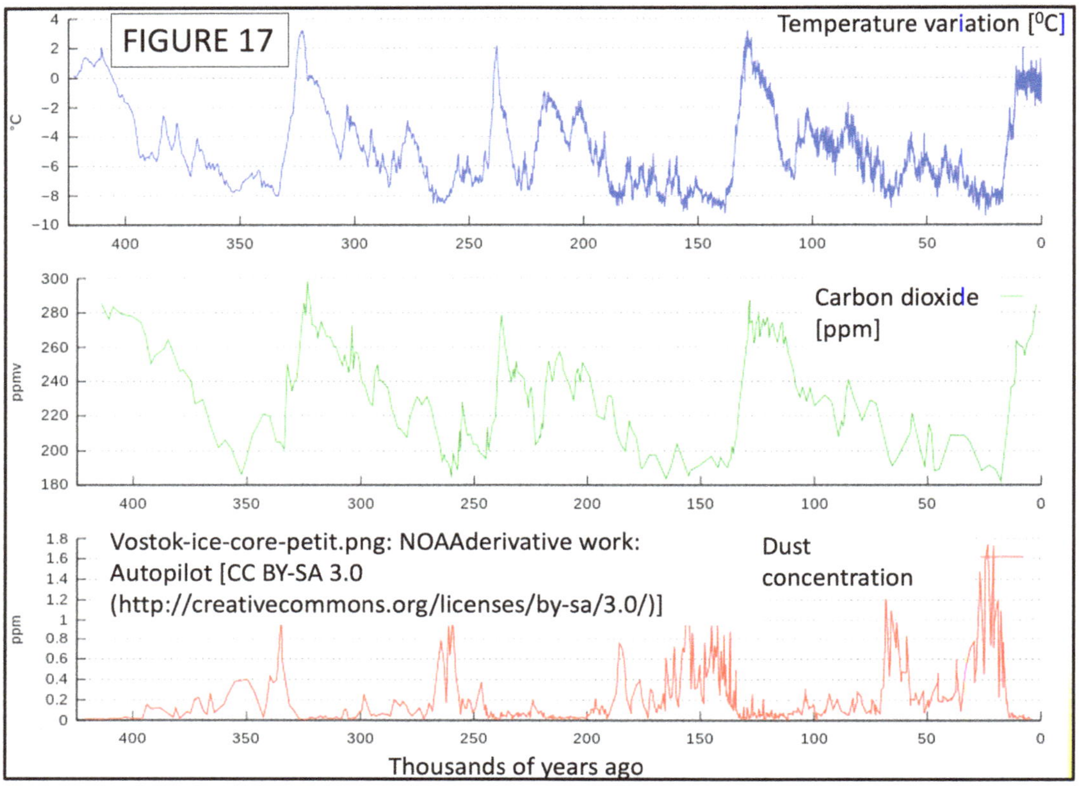

These measurements are from ice cores from Antarctica, providing objective data interpreted and correlated with the parameters of interest. Temperature is inferred from the proportion of hydrogen isotopes, CO_2 content is measured from the air bubbles trapped in the ice, and the dust remains after melting the ice.

The low points of the curves represent the cold glaciation periods, while the peaks represent the deglaciations (the warm periods). All of written human history (about 5,000 years) is contained in the peak on the right end of the graphs, along with the times from the oldest settlements (about 10,000 years ago) or Neolithic period. The rest of the graph includes the Middle Stone Age, when humans were only in the Equatorial and tropical areas, thus not affected by the glacial temperatures further North.

These glaciation cycles correlate well with the Milankovitch cycles of Figure 3, which stem from calculated variations in received solar energy, due to astronomic fluctuations. The Milankovitch cycles theory has been further improved by others and more recently associated with similar variations observed in geological records (ice core, oceanic sediments, etc.) as shown in Figure 3. Small variations in solar radiation could have large effects at high latitudes and have created significant climate variations in Northern Europe and North America.

Woolly mammoths were large herbivores adapted to the freezing climate, but there were no mammoths or polar bears in the Southern Hemisphere, proving that tropical and equatorial areas of the Earth were never cold.

Conversely, the warm periods were never too long, otherwise polar ice would have disappeared, changing the arctic animals. For example, arctic land mammals have white fur because they need it for camouflage which would be detrimental to them in a coloured forest. In conclusion, as seen in the graphs in Figure 17, the past warm periods were short, just spikes in the temperature curve.

Unlike the past warm periods, the present one (the top right peak) is a much broader temperature peak on the graph, lasting about 7,000 years longer than the others. From figures 12 and 15 we see that the temperature is not decreasing now but increasing rapidly. Therefore, this current warm period is projected to be even hotter and longer lasting than before due to the human impact.

The industrial revolution only occurred in the past 250 years. However, in Figure 17 the top right peak, which includes human history, is broader by about 7,000 years compared to previous peaks. Could it be possible that humans have had a significant influence on the global climate not just for 250 years, but for over 7,000 years? As we will discuss further, some researchers have shown that this is possible, and we will present new arguments.

1.6 The Greenhouse Effect

We are interested in the greenhouse effect because of its potential to cause global warming, which is one of the significant impacts of human activity on the Earth's climate. As seen above at point 1.5, the increase in air temperature was measured mainly at the ground level, but greenhouse gases also influence temperature high in the atmosphere. Traditional greenhouse gases are compounds of carbon, nitrogen and a few other minor gases. A particular position is held by water, which will be demonstrated has both a global warming and global cooling effect.

All air molecules emit infrared radiation, but only greenhouse gases (the red dots in Figure 18) can absorb it. The greenhouse gases are immersed in the much larger air mass (not shown).

When solar electromagnetic radiation hits opaque objects (not air), it is transformed into sensible heat of those objects, which is the source for latent heat and infrared radiation. As temperature of those objects rises, their water starts to evaporate and locks a lot of latent heat in the water vapours. During the evaporation process the temperature remains constant, although a lot of heat is absorbed as long as water is available. By not allowing the temperature of wet objects rise, water prevents the increase in their infrared radiation (the source of heat for the greenhouse effect).

FIGURE 18: The greenhouse effect

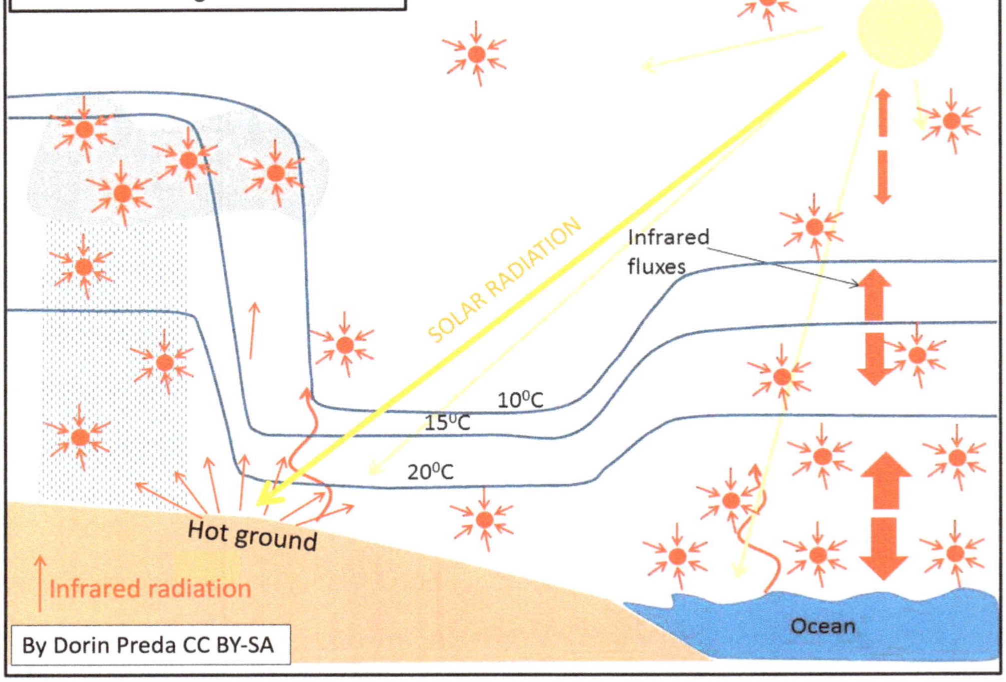

After available water evaporates, the incoming solar radiation increases the sensible heat of those opaque objects, their temperature and therefore the infrared radiation they emit. That infrared radiation is absorbed by the greenhouse gases which in turn heat the air around them. While the CO_2 content of the atmosphere is small and rather constant, water vapours are in much larger proportion and their concentration varies a lot in time and space (see Figure 18 where only water vapours are shown because CO_2 is in too little proportion). Thus, the layers of air containing more greenhouse gases (like the clouds) absorb more infrared heat than layers of air with lower concentration of water. The greenhouse gases in the atmosphere induce multiple levels of absorption-emission of the infrared radiation with the ground (the fluxes in Figure 18), and this way they create a warmer air layer near the ground, producing the classical greenhouse effect.

Thus, greenhouse gases passively warm the lower atmosphere by delaying heat escape to space.

In Figure 18 there are 22 molecules of water represented as greenhouse gases which absorb infrared radiation; we could not represent any CO_2 there, which is about 50 times less than water in the troposphere. Moreover, considering the difference in their infrared radiation absorbance (Figure 5), it results that the greenhouse effect of water vapours is much larger than that of CO_2.

Carbon dioxide and the other non-condensing greenhouse gases become relevant in the upper troposphere where there is less water and because they trap the infrared radiation emitted by the clouds below, and delay its escape to cosmic space.

1.7 Carbon and its Gases

Organisms produce a significant amount of methane (CH_4), which plays a crucial role as a greenhouse gas. However, since methane is lighter than air, it rises high into the atmosphere where it is broken down (oxidized) by solar radiation into CO_2, releasing that energy as heat at high altitude. Thus, methane serves as a chemical transporter of heat to the upper atmosphere.

The role of CO_2 in global warming and climate change has been widely debated, as its fluctuations coincide with changes in air temperature, and because it is well known as a greenhouse gas.

Figure 19 indicates that CO_2 levels in the atmosphere increase until May, then decreases until October due to the activity of land vegetation in the Northern Hemisphere, which has more landmass than the Southern Hemisphere [6].

In 2019, the concentration of CO_2 in the atmosphere reached 415ppm (parts per million) and increases by approximately 2ppm annually. About 7Gt (gigatons) of CO_2 or roughly 2Gt of carbon correspond to each 1ppm in the atmosphere. 1Gt is equivalent to the weight of 1 Km^3 (cubic kilometre) of water. Over the past 2.5 million years CO_2 has fluctuated as shown in Figure 17 due to the glaciations of the Quaternary Ice Age (the last ice age). Naturally, atmospheric CO_2 concentration in warm periods should decrease because of increased plant and coral growth, as well as higher rock weathering, as illustrated in the seasonal variations of Figure 19. However, in the glaciation cycles, the correlation is paradoxically inverse (more CO_2 during warm periods). A convincing explanation for this deglaciation effect is still pending.

FIGURE 19

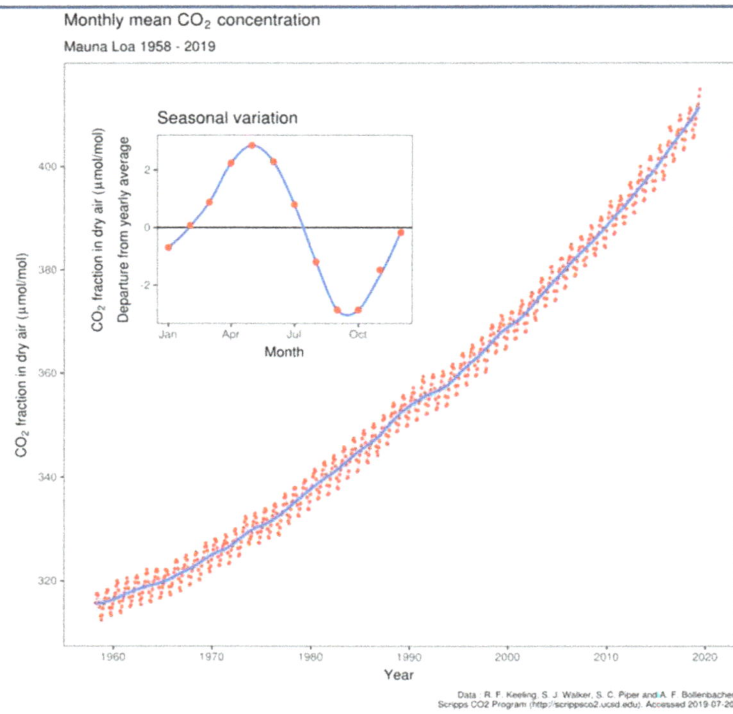

By Delorme - Own work. Data from Dr. Pieter Tans, NOAA/ESRL and Dr. Ralph Keeling, Scripps Institution of Oceanography., CC BY-SA 4.0, https://commons.wikimedia.org/w/index.php?curid=40636957

The concentration of CO_2 has significantly decreased since the first plants appeared (Figure 20).

FIGURE 20 Carbon Dioxide Variation Since Life Begun on Earth

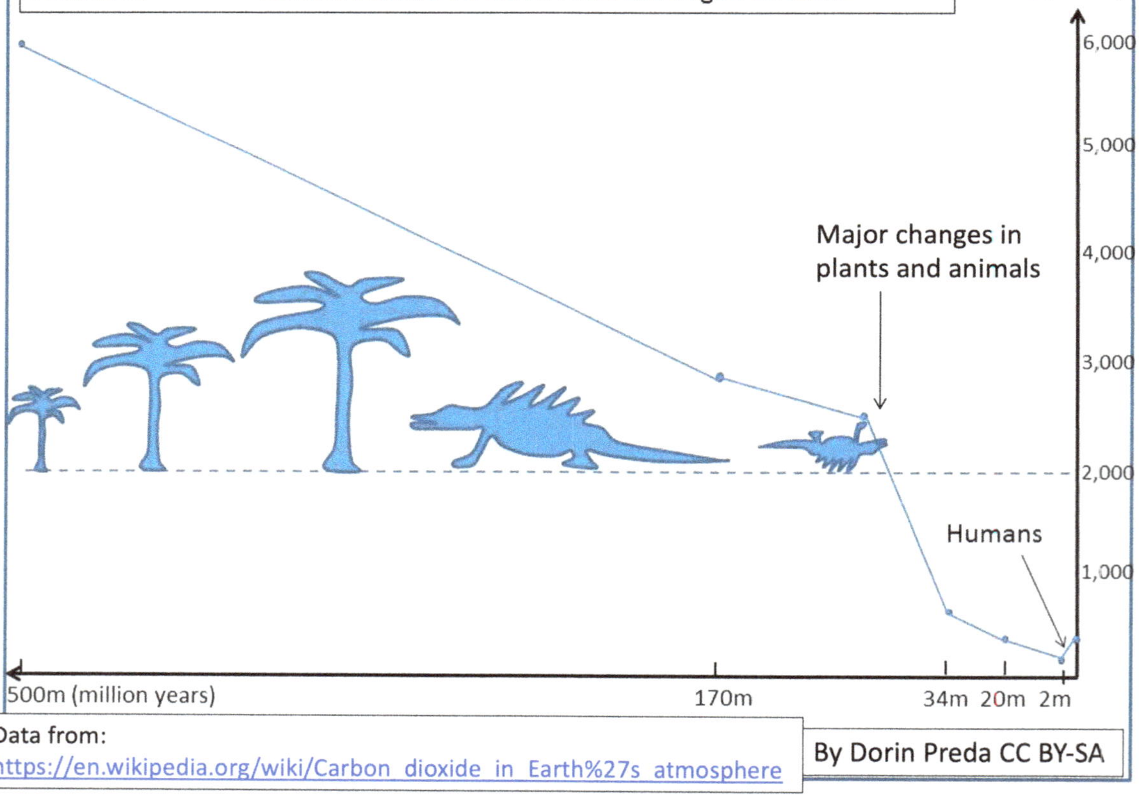

Data from:
https://en.wikipedia.org/wiki/Carbon_dioxide_in_Earth%27s_atmosphere

By Dorin Preda CC BY-SA

Life on Earth is carbon-based, so higher concentrations of carbon dioxide enhance biological activity (agriculturalists pump carbon dioxide into their greenhouses to increase plant productivity).

As humans learned to heat themselves by burning whale fat, they were able to venture further from the tree line, the forested areas. They crossed the Bering Sea on the Aleutian Isles which were dry land at that time due to the low sea level. This led to the discovery and colonisation of North America and eventually South America. The retreat of glaciers 20,000 years ago allowed humans to settle at higher latitudes (Figure 21).

FIGURE 21 Carbon Dioxide Variations in the Past 40,000 Years

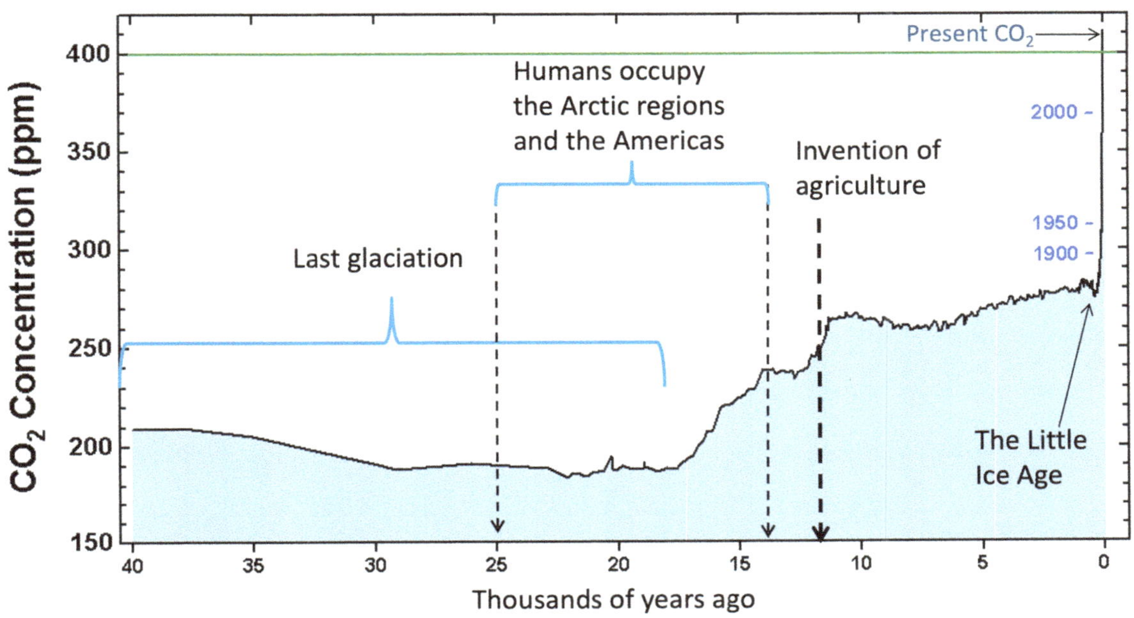

Historical information added by Dorin Preda to original graph By Renerpho - Own work, CC BY-SA 4.0, https://commons.wikimedia.org/w/index.php?curid=78909796

A significant increase in CO_2 content was observed after the glaciers and snow retreated. The major peak around 12,000 years ago coincided with the invention of agriculture. Historians hypothesize that the rise in atmospheric CO_2 at that time stimulated plant growth and aided in the development of agriculture, a topic we will discuss further.

Water has a strong affinity for CO_2. Ocean water stores about 60 times more CO_2 than the atmosphere, with that quantity continuing to increase. There is constant exchange based on concentrations, temperatures and pressures. If anthropogenic CO_2 emissions were to cease now, their atmospheric concentration would decrease slightly, until reaching the ocean's equilibrium concentration. However, a temperature increase could lead to the ocean releasing large quantities of CO_2 into the atmosphere.

CO_2 dissolved in ocean water leads to acidification, potentially harmful to corals and other organisms with a carbonate skeleton. Yet, these organisms rely on CO_2 to build their $CaCO_3$ skeletons. Corals emerged about 500million years ago, when there was much more CO_2 in the atmosphere and therefore, in oceans (see Figure 20). Corals have played a crucial role in reducing the atmospheric CO_2 by sequestering it in large rock formations.

There also is a mechanism of natural inorganic precipitation of CO_2 from water as carbonates, that has locked away even greater amounts of carbon as limestone and dolomite, forming large mountain ranges.

Chemical weathering of basalt releases calcium and magnesium ions in the ocean water, that react with CO_2 forming carbonates, and permanently locking carbon. Basalt is the main ocean basement rock; it serves as an acidity buffer against CO_2 where it is not covered by sediments.

Plants also play a role in removing CO_2 from the atmosphere through photosynthesis, utilizing solar radiation to build their structures from carbon, and releasing excess oxygen to the atmosphere. Animals consume these plants, and the buried remains of both plants and animals transform into fossil fuels. Currently, we extract and burn fossil fuels for energy, releasing CO_2 into the atmosphere at a rate thousands of times faster than their formation. This process involves consuming oxygen from the air to produce CO_2 and water. Fortunately, there is an abundance of oxygen in the atmosphere, so there is no risk of dying through asphyxiation from burning hydrocarbons.

Another significant way to reduce the CO_2 in the atmosphere is through the weathering of certain rocks by atmospheric agents which changes silicates to carbonates.

These pathways form a partially closed cycle of carbon between the atmosphere and underground or underwater carbon sinks (see Figure 22).

About 3.5 billion tonnes (Gt) of CO2 correspond to 1 billion tonnes (Gt) of carbon and 2.5 billion tonnes (Gt) of oxygen. 1 Gt is equivalent to the weight of 1 cubic kilometre of water. Some of these carbon sinks are naturally closed, like the fossil pool (although we have opened it up). Others, such as soil carbon, are partially closed, as soil erosion can release carbon and convert it to CO_2.

There is a continuous transformation of CO_2 and other forms of carbon in the soil and water. Therefore, CO_2 is interconnected with past and present Earth processes, from the deepest depths to the upper atmosphere.

FIGURE 22 The Carbon Cycle (Gt/year)

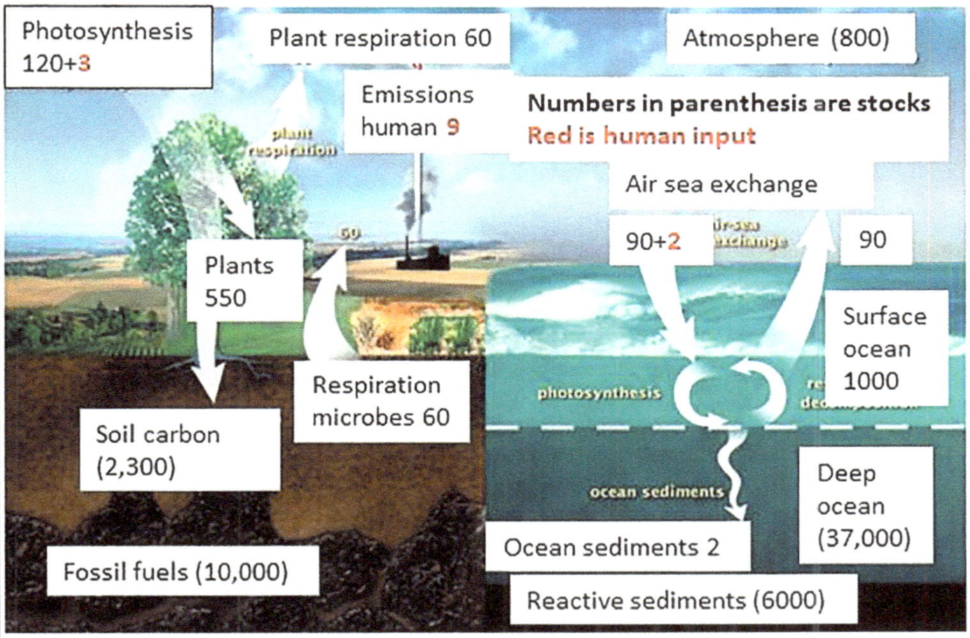

Diagram adapted from U.S. DOE, Biological and Environmental Research Information System. - http://earthobservatory.nasa.gov/Features/CarbonCycle/

1.8 Deforestation

If prehistoric man could have flown into cosmic space and looked back, he would have called Earth "The Blue-Green Planet". Unfortunately, we can now only refer to it as "The Blue Planet" because there is not much green left.

FIGURE 23A Deforestation in Mato Grosso – Brazil (the same location)

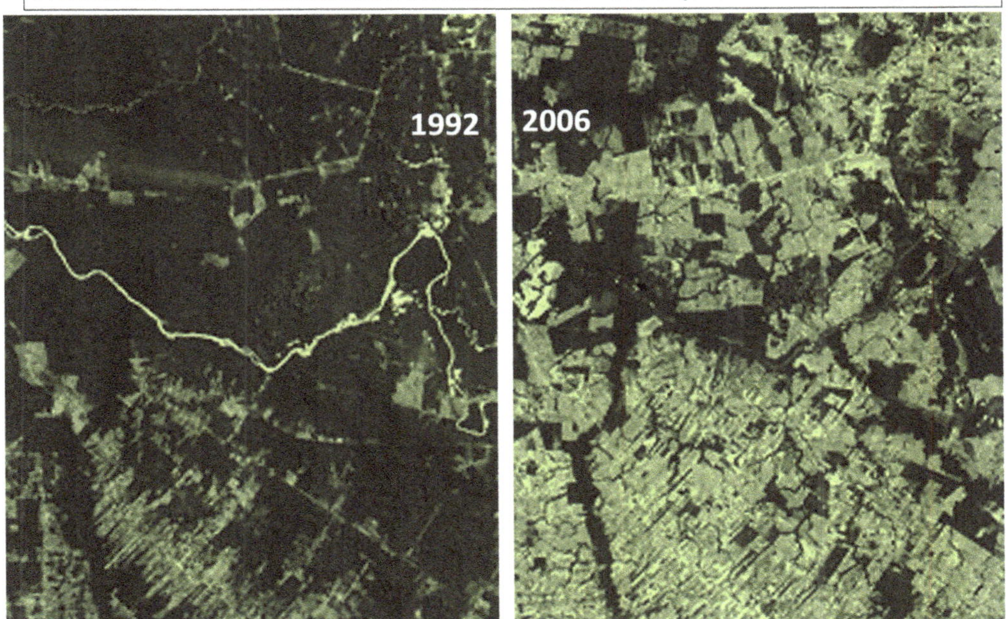

By The Advanced Spaceborne Thermal Emission and Reflection Radiometer (ASTER) on NASA's Terra satellite - http://earthobservatory.nasa.gov/IOTD/view.php?id=35891, Public Domain, https://commons.wikimedia.org/w/index.php?curid=5198881

However, it is crucial that Earth remains blue, as it indicates the presence of water, which sustains life and creates a climate. Before humans, almost all of Earth was covered in forests. Currently, only about one-third of land has forests, with approximately 200,000km^2 (the size of Syria) being lost every year, resulting in the annual emission of 5Gt CO_2.

An important aspect, often ignored or forgotten, is that not only do we have fewer forests in terms of total area, but they are also less dense, with smaller trees (younger or less healthy) compared to the original forests. Lush, tall, cold rainforests with trees of more than 1m diameter have disappeared. While we retain one-third of the initial forested areas, we likely only have one-tenth of the original forest mass.

"About 8% of global CO_2 emissions currently come from tree cover loss in tropical forests, but these same forests could provide 23% of the cost-effective climate mitigation needed before 2030." [11]

Tropical forests hold 230Gt of carbon, yet deforestation is more prevalent in the tropics, with less than 1/10 of the original tropical forests remaining.

Deforestation is more pronounced in the under developed countries and has stagnated in developed countries, especially in China, where reforestation has begun.

As shown further, due to the computer inability to simulate water and forest processes, deforestation (land use) is considered in IPCC reports as having a global cooling (therefore good effect) because it would increase Earth' albedo, despite it is also mentioned as producing CO_2. Therefore, at the Climate Conference COP26 of 2021, deforestation extension was pushed to 2030.

Causes of deforestation include agriculture, logging, pastoralism (especially goat herding), firefighting, roads, railways, mining, natural gas drilling, human settlements, war, mismanagement, arson, biofuels, and global warming. Deforestation generally occurs because of interests other than the forest, or disinterest; it is never accidental. Fire is the most common method of deforestation, with natural fires not leading to deforestation, otherwise we would not have found very old trees. Conditions leading to fires can be intensified through maltreatment of the land, years before the fire (for example slash and burn, traditional fires, forest thinning, grass fires), as detailed further.

Effects of deforestation include increased CO_2 emissions, aridity (reduced precipitation), soil erosion leading to irreversible soil loss to the ocean, loss of biomass and biodiversity, and desertification. Soil loss is the most significant consequence of deforestation, resulting in desertification. By removing trees, the topsoil breaks down as there are no more roots or organic matter to hold it together. The sun scorches the ground and creates deep cracks where air and solar radiation penetrate, further degrading the soil and causing it to break down and be carried away by storms.

Reforestation is difficult after soil erosion of a deforested area. Forested areas may have an average of two to three metres of porous ground that can hold a lot of water, essential for forest resilience. As soil erodes and is carried away, less water can be retained in the bedrock, making it difficult for replanted trees to survive. Eroded lands cannot recover without the thick soil.

Statistics show that despite their importance, forests receive only 3% of climate mitigation finance.

1.9 Air Circulation

The troposphere is a thick gaseous layer with continuous and substantial movements of air in all directions, resulting in equal energies in its different regions. Due to the rotation of the Earth, solar radiation heats only one side of the Earth, while thermal infrared radiation escapes continuously in all directions with varying intensities. The marked uneven distribution of energy in the troposphere leads to continuous air movement.

While local weather systems vary and move frequently, there are larger air currents with dominant circulation as seen in Figure 23 B due to the large-scale physiography of the Earth.

The highest annual solar radiation falls on the thermal equator, a region around the geographical Equator with maximum solar radiation. The largest air currents rise from the Equator carrying heat and water vapours to high altitude, where heat is partially lost upwards and partially transported horizontally towards areas of descending air (high barometric pressure around the 30^0 latitude). The air flows then along the ground towards the Equator, completing the Hadley cell.

Less intense but stable are the polar cells and the Ferrel cells, which help balance the airflow differences between the Hadley and polar cells.

Due to Earth's rotation, the Coriolis force causes air currents to deflect to the right in the Northern Hemisphere and to the left in the Southern Hemisphere, resulting in dominant Easterlies Trade winds near the Equator and Westerlies winds at mid-latitudes. Other major wind patterns are associated to El Nino, displaying multi-annual oscillation patterns. On a smaller scale, coastal winds on the sides of continents and local sea breezes exhibit diurnal changes in direction as the land and sea alternate in temperature.

FIGURE 23 B Global Air Circulation

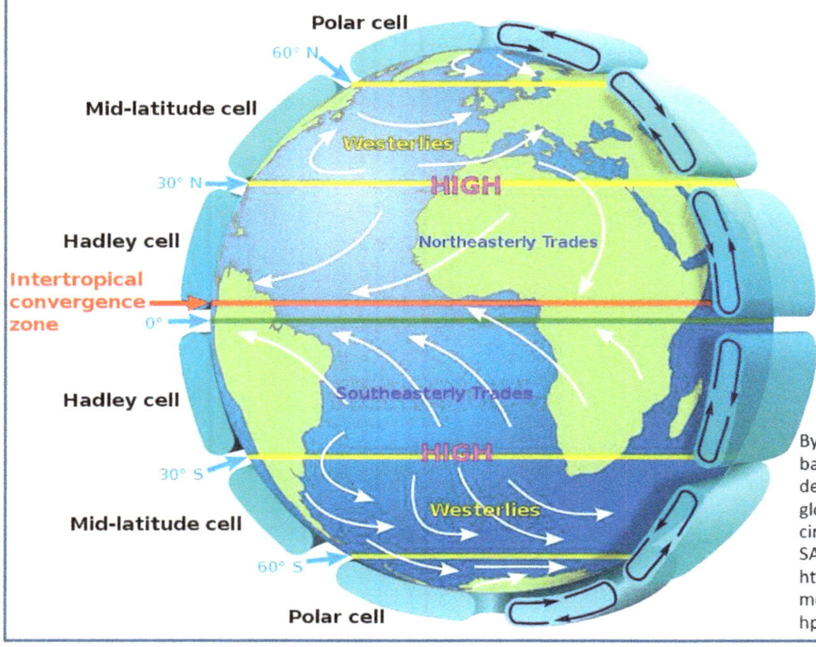

By Kaidor - Own work based on File:NASA depiction of earth global atmospheric circulation.jpg, CC BY-SA 3.0, https://commons.wikimedia.org/w/index.php?curid=23902538

The study of three-dimensional air circulation has been challenging due to apparatus limitations. Historically, air circulation has been measured only horizontally through wind measurements.

FIGURE 24 Anemometers

A

By photographer: Archival Photograph by Mr. Sean Linehan, NOS, NGS - The Aims and Methods of Meteorological Work by Cleveland Abbe, in Maryland Weather Service, Johns Hopkins Press, Baltimore, 1899. Volume I, page 316.NOAA's National Weather Service (NWS) Collection, Image ID: wea00920.Source link: http://www.photolib.noaa.gov/htmls/wea00920.htmDirect link to image: http://www.photolib.noaa.gov/bigs/wea00920.jpg, Public Domain, https://commons.wikimedia.org/w/index.php?curid=9682749

B

Public Domain, https://commons.wikimedia.org/w/index.php?curid=417624

C

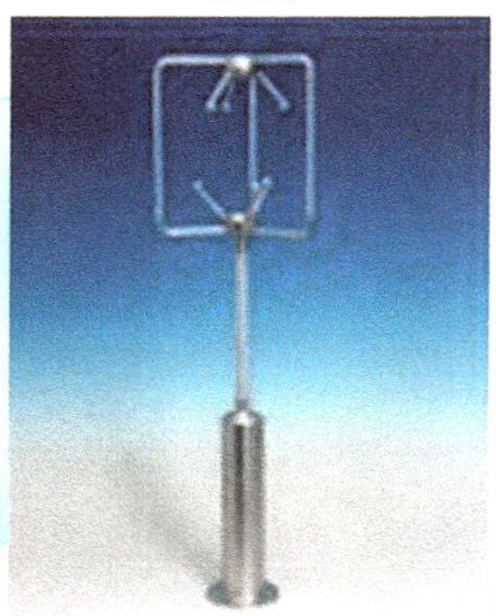

By Gillinstruments at English Wikipedia - Transferred from en.wikipedia to Commons., Public Domain, https://commons.wikimedia.org/w/index.php?curid=4668794

Air circulation has been historically measured systematically only through its horizontal component (wind) because people live close to the ground, which flattens the air movement. Thus, anemometers were designed to only measure horizontal air currents (see Figure 24, A and B).

Recently a few 3D anemometers (see Figure 24, C), have been developed to study 3D air circulation. However, a very large number of such instruments would be necessary to provide a reasonable spatial picture. Satellite data has been useful in studying large scale air dynamics, but is less effective at medium-small scale. The lack of proper instrumentation has hindered our understanding of air movements and heat exchange. Meteorologists infer vertical air circulation from barometric pressure variations, but it does not work properly at medium-small scales.

Despite its importance in climatic studies, vertical air circulation remains largely unmeasured.

1.10 The Water Cycle

Long ago, people observed that water circulates between the oceans and the land. Since then, our understanding of the water cycle has gradually improved and continues to improve even for the most fundamental processes.

Most water (Fig 4) stays in the lower troposphere (under 5,000m altitude), where all people live and meteorological stations are located. Water evaporates from ocean and land, rises in the atmosphere to usually less than 5,000m altitude (rarely to the top of the troposphere) where it forms most clouds and usually condenses and falls as precipitation.

Water is an extremely dynamic component, that generates much of the movements of the troposphere and transports most of the heat due to its very high latent heat and the fact that its vapours are lighter than air. Humid parcels of air rise in the atmosphere, and wind can transport them long distances (Figure 25).

Along with the water-laden air currents of the atmosphere, ocean currents are the working means of the vast thermal engine of Earth, which moves solar heat from the low latitudes to Polar Regions.

Of the 1,400,000,000km^3 of water on Earth, 97% is ocean water, about 30,000,000 km^3 is freshwater in glaciers and ice caps, approximately 23,000,000 km^3 is fresh and saline groundwater, 100,000 km^3 is freshwater in rivers, and lakes, around 10,000 km^3 is in the atmosphere, and about 2,000 km^3 is in the biosphere [2].

As liquid water only flows downwards, there would be no water on continents if it did not evaporate from ocean, condense and fall as rain or snow on the land. This simple truth is central to this book.

FIGURE 25 The Water Cycle

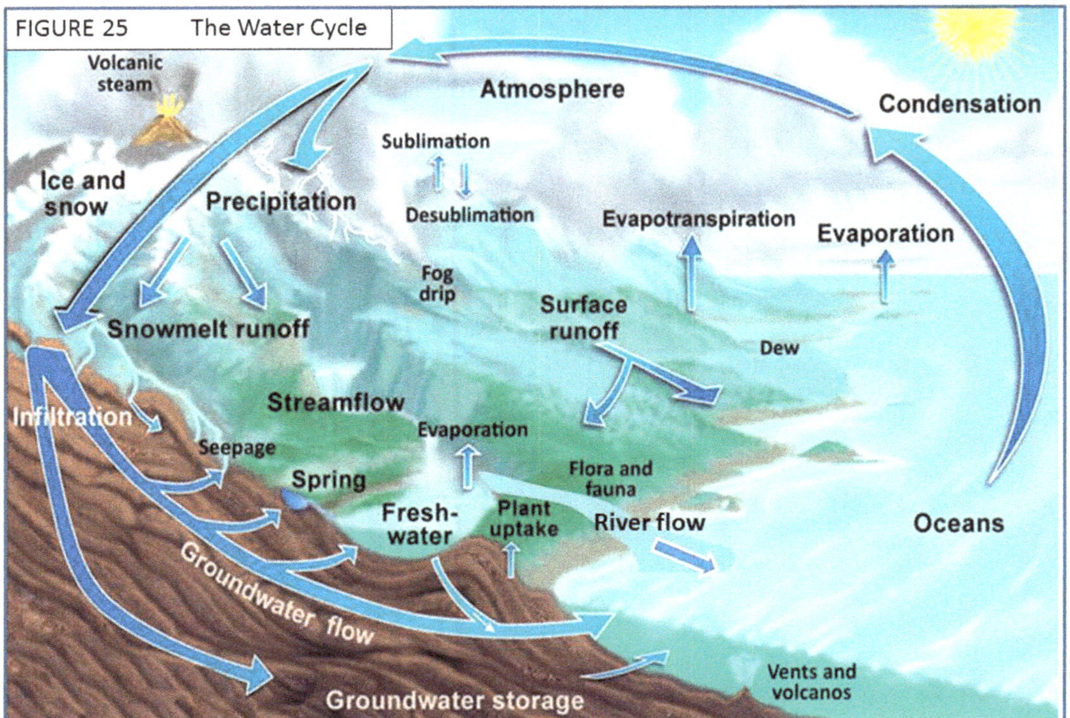

By John Evans and Howard Periman, USGS - http://ga.water.usgs.gov/edu/watercycle.html, Public Domain, https://commons.wikimedia.org/w/index.php?curid=26818355

FIGURE 26 Annually Exchanged Water Globally

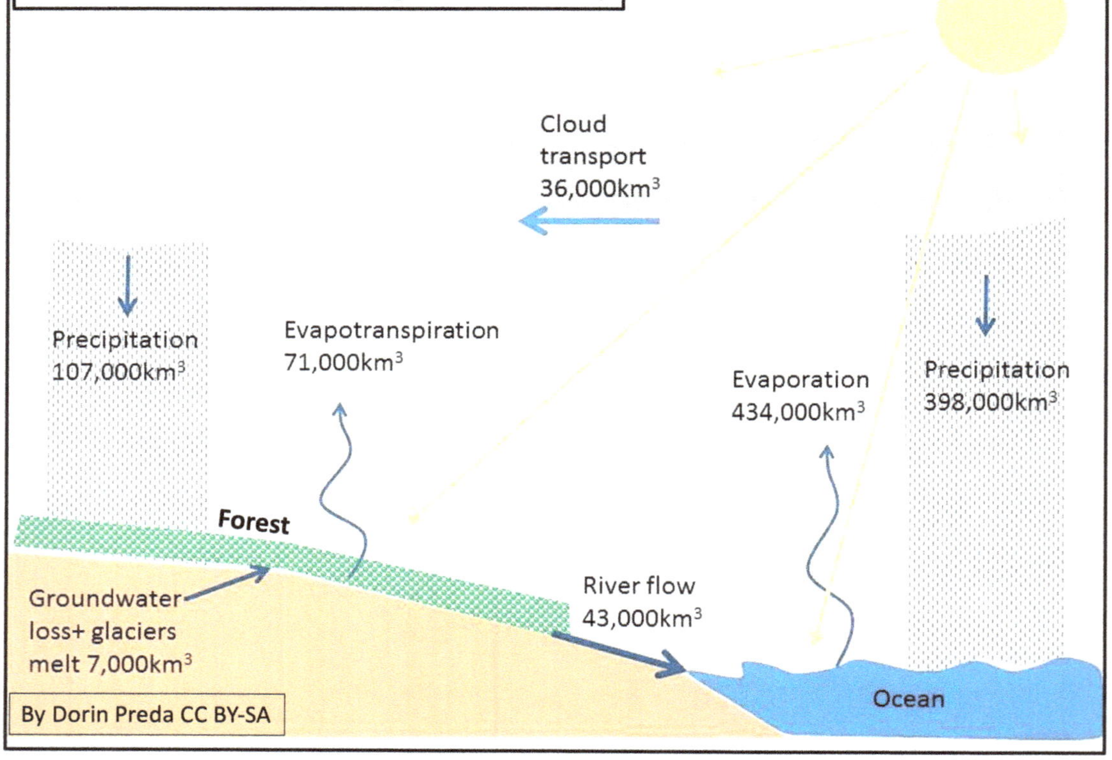

By Dorin Preda CC BY-SA

The evaporation, precipitation and river flow data in Figure 26 are from [7] and [8]. From that data it can be inferred a net annual cloud transport from the ocean to the land of 36,000km^3. There is very little storage of water in the atmosphere, so if the rate of evaporation/precipitation from the ocean decreases, it will result in less rain on land. If the quoted water volumes are correct, there would be a deficit of water on continents of about 7,000km^3 per annum, which could be explained as groundwater loss and melting losses from the glaciers. Despite variations, this tendency of dryness of continents is already evident through increased droughts and forest fires.

From Figure 26 and the known areas of the oceans (361,000,000km^2) and land (150,000,000km^2), we infer the following (expressed as columns of water):

- The rate of evaporation from the oceans is 1200mm/annum, while evapotranspiration from the land is only 480mm/annum.
- Precipitation on the ocean is 1100mm/annum, while on the land is 720mm/annum
- There is a slight evaporation/precipitation excess on the ocean, but a strong evaporation/precipitation deficit on the land.

Figure 27 shows the large evaporation/precipitation rate between different zones of the Earth.

FIGURE 27

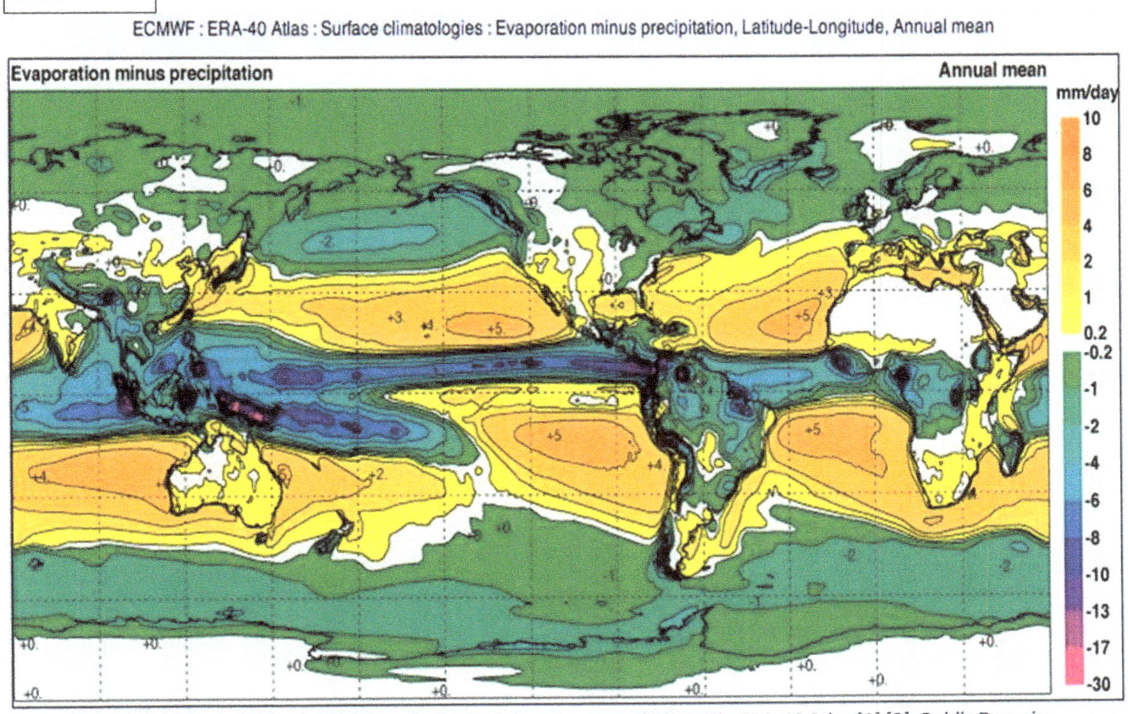

By NASA & European Centre for Medium-Range Weather Forecasts (ECMWF) - ERA-40 Atlas [1] [2], Public Domain, https://commons.wikimedia.org/w/index.php?curid=52851991

Generally, evaporation (orange and yellow) is dominant in tropical areas, while precipitation (green and blue) dominates the smallest and largest latitudes. Deserts are white because they evaporate as much water as they receive when there is no surface water input. The difference between the tropical Sahara (bare desert) and tropical South-East Asia (lush vegetation) is due to the mountainous South Asia being surrounded by seas, and much better water management in those countries.

There is large evaporation in the equatorial areas of the Earth, but precipitation is even larger.

As water evaporates from salty oceans or marshes, it rains as pure, freshwater which is of particular interest to us, but not evenly distributed. Human settlements started mostly on the shores of rivers. Except for the North American Great Lakes, the other are in less accessible areas. In Australia, for example, most renewable freshwater is in the Cape York Peninsula, which is almost uninhabited.

In some European, South American and Asian areas, freshwater is from the summer melting of mountain glaciers, coinciding with high water demand. In contrast, in regions like Australia, water availability fluctuates greatly, and droughts are the norm, as storages like Lake Eyre are unreliable.

Fresh groundwater was greatly over-exploited (by extracting more than the natural recharge).

The level of the ocean varied greatly, depending on the quantity of ice stored in glaciers and ice sheets and due to variations in the stored groundwater. It is estimated that at the peak of the last ice age, about 1/3 of the land was covered by glaciers, and the ocean level was about 100m lower than now. Conversely, in the past warm period, the ocean level might have been about 50m higher than now. Local sea-level fluctuations due to plate tectonics are also significant.

The IPCC estimated that global warming would intensify the water cycle, making dry areas drier, and wet regions wetter [7].

The freshwater monitoring GRACE satellite of NASA operated only until 2017; there are US and German plans for joint deployment of another satellite, essential for diagnosing water cycle problems. However, local water management observations should also show us the imbalances.

Ultraviolet radiation dissociates thermosphere water into oxygen and hydrogen. The hydrogen curve in Figure 4 is open at the top of the atmosphere (the concentration does not drop back to zero as for the other gases), confirming that hydrogen escapes from Earth into cosmic space at a rate of about 3kg/second [9]. Therefore, there is a slow loss of water from Earth.

1.11 Energy Transfer Processes

We shall use basic Physics to discuss some atmospheric processes of Earth, aiming for representativeness, and to infer relative sizes and effects of the main phenomena through basic calculations and estimations.

1.11.1 Radiative Energy Transfer

The Earth is surrounded by vacuum; thus, it only can exchange energy with the cosmic space through radiation (we ignore the solar wind and cosmic particles which have negligible energy).

Firstly, we need to define the following:

- **Solar radiation (SR)** or shortwave radiation having a wavelength smaller than 0.78 microns
- **Infrared radiation (IR)** or longwave radiation having a wavelength between 0.78 – 1000 microns
- **Planetary equilibrium temperature T_P**: the theoretical temperature of a planet with black body surface, without atmosphere, that emits through Stefan-Boltzmann law the energy it receives from its sun. It is a theoretical, unreal temperature, because the black body is only a concept.
- **Above-ground air temperature**: the air temperature measured by meteorological stations (at 1.2m height), serving as the benchmark for global warming and climate change studies.
- **Radiative flux**: the radiation emitted by a square metre of a body [J/m^2*s] or [W/m^2]
- **Total radiation flux**: the energy from one or more sources passing through a square metre of space in one second [J/m^2*s] or [W/m^2]. Multiple radiative fluxes with the same or different wavelengths emitted by different bodies may pass through the same area of space, resulting in a total radiation flux that is the sum of all those radiative fluxes.
- **Radiative forcing (RF):** the difference between incoming and outgoing total radiation fluxes passing through the top of the stratosphere; it is used to assess global warming/cooling. [W/m^2]
- **Greenhouse gases (GHG)** like CO_2 (carbon dioxide), water vapours, GH_4 (methane), NO_2, etc, absorb infrared radiation (IR), unlike O_2, N_2, and other atmospheric gases.
- **Greenhouse effect (GHE)** is produced by GHG, and by the clouds, when they absorb infrared radiation and heat the surrounding atmosphere instead of letting it escape to space.

Infrared radiation (IR), has different wavelengths, inversely proportional to the temperature of the body that emitted it.

Different substances absorb and emit radiation differently:

- Clean dry air without carbon dioxide cannot absorb solar radiation or infrared radiation, but it emits infrared radiation.
- Carbon dioxide and water vapours do not absorb solar radiation, but absorb and emit infrared radiation
- Clouds is said not to absorb solar radiation, but they absorb and emit infrared radiation.
- Methane absorbs infrared radiation and is broken down by the ultraviolet radiation in the stratosphere

Secondly, we should clarify the systems:

- **The inner system** consists of Earth and its atmosphere with GHG, water, and parameters including radiative, sensible and latent energy fluxes. Earth's temperature in the inner system is the above-ground air temperature as measured by the meteorological stations.
- **The outer system** above Earth's atmosphere is the cosmic space including the Sun. Its parameters are solar radiation, RF, Earth's shortwave reflection (albedo), and the emitted infrared radiation. Earth's temperature in the outer system is the planetary equilibrium temperature, a theoretical temperature, different of the above-ground air temperature.

Albedo (A) indicates the fraction of the incident light that is reflected by a celestial body. It is a dimensionless parameter between 1 (when all light is reflected) and 0 (when all is absorbed). Typical examples of albedo include ocean 0.06, fresh asphalt 0.1, conifer forest 0.12, deciduous forest 0.16, bare ground 0.2, grass 0.25, desert 0.3, clouds 0.7, snow 0.6, fresh snow 0.8.

The equilibrium energy equation for a planet without variation of internal energy is: solar radiation = reflected solar radiation + infrared emission. Thus, if reflected solar radiation increases, infrared radiation of that body (hence its temperature) should decrease. This is why it is claimed that an increase in reflected solar radiation (albedo) through deforestation would mean a decrease in Earth temperature.

The law of conservation of energy states that a system in equilibrium emits as much energy as it absorbs, meaning that solar radiation (SR) received by Earth must be balanced by reflected solar radiation (R) plus infrared radiation (IR): $SR = R + IR$

Astronomers introduced albedo (A) to infer properties of remote planets from the light they reflect, and assumed it would be consistent enough to characterise a surface: R = SR x A.

The energy equation becomes: SR = SR x A + IR

The Stefan-Boltzmann law defines the relationship between the temperature T of a body and the flux of infrared radiation it can emit: **IR = σ x T^4**, where σ is the Stefan-Boltzmann constant = 5.7 x 10^-8 W/m^2K.

Therefore, the law of conservation of radiative energy becomes: SR (1 – A) = σ x T^4

The equation shows that if the albedo of a body increases, its temperature should decrease, creating an inverse proportionality between albedo and temperature.

The single case in which the albedo interpretation holds is a planet without an atmosphere, with a very dark surface, and which quickly spins. It is said that if Earth did not have an atmosphere, it would be a rocky desert with an albedo A = 0.3. We divide SR = 1,361W/m^2 by 4 because Earth's total area is 4 times larger than its cross-section. The Earth's resultant planetary equilibrium temperature without an atmosphere becomes: T = 255^0K (- 18^0C).

1.11.2 Sensible Heat Transfer

In atmosphere heat is less transferred through conduction between two layers of air than through convection and mixing of those layers. The most important conduction takes place between ground and air or between solid particles in the atmosphere and air. For example, air cannot absorb solar shortwave radiation, but dust particles can, and they transfer that energy to the surrounding air.

Greenhouse gases (GHG) in the atmosphere absorb infrared radiation and transfer it through collisions (conduction) to the air.

Air has a very low thermal conductivity; it is a thermal insulator.

1.11.3 Latent Heat Transfer

It involves a change of state of a substance from liquid to solid or gas, or from gas to solid, and it requires a large quantity of energy. The most important fluid undergoing latent heat changes or transfers in the atmosphere is water. Very important processes that are central to this study are those between vegetation, water, air, and ground, where solar radiation is the source of energy.

The basic three formulas of sensible and latent heat transfer:

$Q = m \times c_p \times \Delta T$ is for the sensible heat,

$Q = m \times \Upsilon$ is for the latent heat of vaporisation/condensation

$Q = m \times \lambda$ is for the latent heat of fusion/melting Where:

Q – quantity of heat [J], [kWh] ; m – mass [kg]

c_p – specific heat capacity at constant pressure [J/kg^{0}K]

ΔT – difference in temperature [^{0}C], [^{0}K]

Υ – heat of vaporisation [J/kg]; λ – heat of melting [J/kg]

TABLE 1 Property	Air	Water vapours	Liquid water	Ice	Carbon dioxide
Specific gravity [kg/m^3]	1.2	0.6 (at 100^{0}C)	1,000	917	2
Specific heat capacity [J/kg K]	1,040	1,870	4,200	2,050	1,080
Heat of vaporization [J/kg]	-	-	2,260,000	2,838,000	574,000
Heat of melting [J/kg]	-	-	-	334,000	184,000
Colour	invisible	invisible	light blue	light blue	invisible
Light transmissivity	V. good	Very good	good	poor	Very good
Infrared absorption	no	Very good	Very good	Very good	good

This book verifies the result of the complicated climatology methods, models and equations with basic physics and simple calculations that use multiple angle observations, measurements and reasonable estimations, to achieve a higher level of certitude than computer simulations.

Readers are encouraged to check the principles and the basic calculations by themselves. This will allow them to make an informed opinion on climate phenomena and the best measures to protect themselves. It would also allow for choosing the proper method of climate restoration.

2 Chapter TWO: Processes, Models, Calculations

In addition to new explanations for the heat transfer mechanisms, this chapter addresses three classically unresolved problems: albedo and the climatic role of water and vegetation.

2.1 The Problems of Transition to Computer Simulations

More than 50 years ago computers were totally unable to simulate the complex climatic processes, therefore Earth science (meteorology and other sciences) were used to understand the current phenomena. They were correlated with past indications to infer the paleoclimate, and extrapolate that knowhow to forecast the future as logical continuation of the present measurements data.

Land use and land cover change (LULCC) and water cycle studies dominated at that time the climatic understanding, and clearly implied not only that weather acts upon the surface of the Earth under the clouds (at the meteorological stations), but that the land also influences the atmosphere [70].

Computations were done mostly with empirical equations, or with basic theoretical formulas in iterations running on the early electronic computers. At that time, there was a tight connection between input and output information through the reasoning of the scientists, who used proper reality feedbacks to make the necessary adjustments and keep the results representative.

Lately, remote sensing has significantly increased through the use of instruments on aircraft and satellites. It has evolved into a distinct branch of Earth's physics which has greatly benefited from the increasing processing power of computers. Data acquisition systems provide satellite data to land data processing centers which in turn deliver the information to computer modelling scientists.

Computer power has increased greatly, in parallel with advanced modelling software that allowed modelling local weather such that good approximations are forecast for a week in advance. Forecasts are improved by correcting with newer information as that event approaches.

One week forecasting of local weather led many to believe that we would now be able to forecast the climate of the entire Earth for years in the future. Advanced and continuously improving theoretical models use the most powerful computers to produce simulations that are adjusted and improved continuously by many teams around the world, in a race for climate change modelling.

The above is on the normal path of scientific progress, and we all believe that correct climate modelling will be gradually achieved. However, the task is enormous, not only because of the size of the Earth system, but because of the complexity of the many interacting phenomena, and especially because of the complex feedbacks that interlink everything and decrease certitude. The imperfect computer simulations of the past 20 years have resulted in the problems shown below.

2.1.1 The Water and Vegetation Computer Problem

Only the "well mixed GHG" are considered climate drivers, and CO_2 is considered a well-mixed gas, while water is not, thus it is only considered through its feedback in the CO_2 cycle, despite studies like [125]. Water is considered a follower, that would exist in the atmosphere only because of CO_2.

Climatology considers that forests form where there is suitable precipitation from abiotic causes; vegetation is not considered now a driver of climate, but a follower of water. Many say that evapotranspiration water is carried away by wind and does not necessarily create clouds above forests. Drought is said to be mostly caused by increased evapotranspiration demand [106].

Formation of Sahara and other deserts is explained through a global climate change towards dryness due to the Milankovitch cycles, without and before human intervention. Sahara heat and dryness are explained as dry cold air of the Hadley cell warming up as it descends onto the desert.

Only irrigation is included in the radiative forcing, as seen in Figure SPM2 of [102], not deforestation water variations, that are larger.

<u>Computer simulations result in mistakenly indicating that the overall effect of water vapours and condensed water (clouds) would be of increasing the global warming:</u>

- "The net feedback from the combined effect of changes in water vapour, and differences between atmospheric and surface warming is extremely likely positive and therefore amplifies changes in climate." [4]

- "The net radiative feedback due to all cloud types combined is likely positive". [4]

- "With a fixed temperature profile, increasing the greenhouse gas content makes the higher parts of the atmosphere opaquer to infrared radiation upwelling from below, replacing this radiation with OLR emitted from the colder regions" [22]

- "If water vapour at such places increases as the climate warms, then the additional reduction in OLR requires the new equilibrium to be warmer than it would have been if water vapour content had remained fixed. This is referred to as a positive water vapour feedback." [22]

- "Water vapour feedback continues to be the most consistently important feedback accounting for the large warming predicted by general circulation models in response to a doubling of CO2. Water vapour feedback acting alone approximately doubles the warming from what it would be for fixed water vapour (Cess et al., 1990; Hall and Manabe, 1999; Schneider et al., 1999; Held and Soden, 2000). Furthermore, water vapour feedback acts to amplify other feedbacks in models, such as cloud feedback and ice albedo feedback. If cloud feedback is strongly positive, the water vapour feedback can lead to 3.5 times as much warming as would be the case if water vapour concentration were held fixed (Hall and Manabe, 1999)" [22]

- There is high confidence from observations and climate models that the overall cloud feedback on climate change is positive [106]

Computers could not correctly infer whether clouds emit more infrared downwards (global warming) or upwards (global cooling), and its proportion with the reflective capacity of the clouds.

Additionally, the climate effects of vegetation are not properly represented, because:

- it is usually stated that vegetation depends on water and not the other way around,
- the vegetation-water feedback is usually ignored, certainly in conclusions
- the below influence of the albedo issue.

The result of the improper dealing with vegetation results in the idea that deforestation would cool the Earth, stated in many scientific papers like [94] and reflected in the IPCC conclusions upon land use, illustrated in Figure39 of this book and the more recent Figure SPM2 of [102].

The above views will be challenged in this study using fundamental physics, which discerns the planetary temperature from the above ground air temperature, because we know that clouds provide reflective cooling, and that evapotranspiration cools the entire Earth, including the meteorological stations. Obviously, the difference between "global" and "local" is well understood.

2.1.2 The Albedo Computer Problem

We must thoroughly investigate the albedo issue because practically all scientific papers on climate indicate that through its slightly lower albedo, vegetation would warm global climate, resulting in IPCC reports stating that deforestation (land use) would cool the Earth. For example, Figure SPM2 of [102] and [94] among many others show that land use (deforestation) increases reflectance of

Earth (albedo) and would cool the planet. This, unfortunately, is a minor truth, as it ignores the rest of the linked phenomena that are opposite to albedo and outdo it, as suggested in [126].

Albedo problems include:

- Albedo of water bodies increases a lot in the evening, morning or usually due to waves
- Albedo of cities with tall, dense buildings is not higher, but lower than deciduous forests, thus they radiatively warm the Earth more than forests [35]
- Bare soil has very variable albedo, depending on humidity [53].

There are also other problems with complex albedo like in [127] which result in its large variability:

- Its definition is not unique; different things are called “albedo”, including brightness
- Correct albedo measurement of a remote planet is impossible
- Correct albedo measurement on Earth is very difficult, thus it is just approximated
- Is measured in different ways, with different instruments, giving different results.

Albedo of forests is included in simplistic computer models which do not consider all interplaying parameters, resulting in errors with significant global consequences when they are used in land and even Earth management, necessitating the detailed albedo analyses in this chapter.

2.1.3 The Opposition to Reforestation Problem

There were many proposals in the past 20 years for reforestation as an easy, natural way of sequestering carbon from the atmosphere, like [55], but they were intensely opposed. Moreover, in the past 4 years, there was a flood of papers in the scientific journals that unsoundly criticize reforestation and afforestation, risking to dominate the IPCC reports and silence the calls for Earth rehabilitation through revegetation [126]. Despite being known that plants naturally absorb a lot of carbon dioxide, the opposite is stated and disproved with minor or defective arguments.

One argument is that carbon absorption would diminish as carbon dioxide increases to levels that become saturating for photosynthesis when in fact photosynthesis increases to more than double present CO_2. Another argument is that afforestation should not be used in grasslands and savannas, ignoring that most of them are results of prehistoric deforestations. The proof that they were forests is that, if allowed, forests naturally take over nearby grasslands.

Some appear to fear that if storage in trees would fail, the release of carbon back to the atmosphere would be worse than before, when in fact trees cannot release more than they received.

Another argument invokes biodiversity conservation like in [128], ignoring that forests are biodiverse in themselves, while other claim incompatibility of species.

A defective argument is that grasslands would store more of their carbon in the ground than above, and when replaced by forests that carbon would be lost; obviously, it would not be lost, and forests will add their underground carbon, that is much larger, not smaller than that of grasslands.

A recurring argument which is frequently invoked by different authors is the supposed disadvantage of forest albedo; based on it, [97] suggests deforestation as method of climate change mitigation. This argument will be disproved by calculation in Section 2.6 and discussed in 2.5.2 and 2.3.1.

There are, though, good arguments against some trees, like eucalyptus and other species which should not be used because they are highly flammable.

The opposition to reforestation is supported by well-known forestry specialists asking unsound questions like "How much carbon would be released to the atmosphere by restoring forests?" https://science.nasa.gov/earth/climate-change/examining-the-viability-of-planting-trees-to-help-mitigate-climate-change/ Therefore, there are complicated human issues involved.

<u>The above discussion addressed reforestation and afforestation of grasslands/savannas, while the method suggested in this book is not reforestation/afforestation.</u>

The main reason for the opposition of climatologists to reforestation and afforestation appears to be that it would distract from reducing the carbon emissions. However, these are two distinct issues. Supposing that we would be able to halve our annual carbon emissions of 36GT to 18GT, the remaining 18GT emissions would still be much more than the 2GT sequestration capacity of the ocean, thus the atmospheric carbon concentration will continue to grow; the greenhouse effect would continue to increase despite drastic reduction in emissions. <u>Emissions reduction is not a method of reducing the carbon concentration in the atmosphere,</u> but only reduces the rate of increase of that concentration. As known, one cannot get poorer by diminishing his income, if it is still larger than his expenses.

On the other hand, absorbing carbon dioxide from the atmosphere (with trees), effectively reduces the carbon concentration from the atmosphere, therefore the greenhouse effect. <u>Carbon</u>

sequestration (in trees) is the real method of reduction of the greenhouse effect. This is similar to drawing money from the bank account, which really makes someone poorer.

Obviously, if both reduction in emissions and sequestration in trees take place, the faster and easier will be to reduce the greenhouse effect. Therefore, we should do both as much as possible.

In conclusion, there should be no opposition to revegetation with trees, that should be started immediately to reduce carbon in the atmosphere, in parallel with emissions reduction.

2.1.4 Limitations of Computer Modelling

Real phenomena are too vast in size and complex in structure; thus, Figure 28 depicts the approximation utilised to simplify Earth models and is the foundation for the solar forcing concept.

As solar radiation passes through the atmosphere, it is distributed in a pattern due to statistical cloud cover, which is heavy in South America, Central Africa and China. It assumes that all areas of the Earth would receive continuous solar radiation over 24 hours, but in a smaller measure.

This way, computer models simplify the natural processes by eliminating the day-night variation, then add other approximations [124], [70], which result in the problems requiring this study.

Satellite radiometers are currently unable to measure long-term variations in energy fluxes on a large scale, hence the human impact on climate cannot be measured, but is instead simulated using computer models. Unfortunately, the magnitude and high complexity of water phenomena, which are essential for climate, prevents their accurate computer simulation.

Simulations mainly stop water from reacting to changes in other elements, because otherwise the model would show that Earth climate is a water climate. This and other restrictions of the computer models lead to the problems analysed further. Two fundamental changes were made: firstly, the ground point of view was changed with the astronomical one (albedo, RF, planetary temperature), and secondly there was high formalization, which decreased the possibility of human assessment. As result, computer simulations cannot yet properly represent natural processes, as justified above.

Significant advancements have been made, as stated in literature [129], and we hope that in the future computer models will be able to model global climate, but at present they can model only non-condensable greenhouse gases (CO_2, CH_4, other hydrocarbons, nitrous oxide, ozone), but not water and its vapours, which is the main working fluid of the Earth.

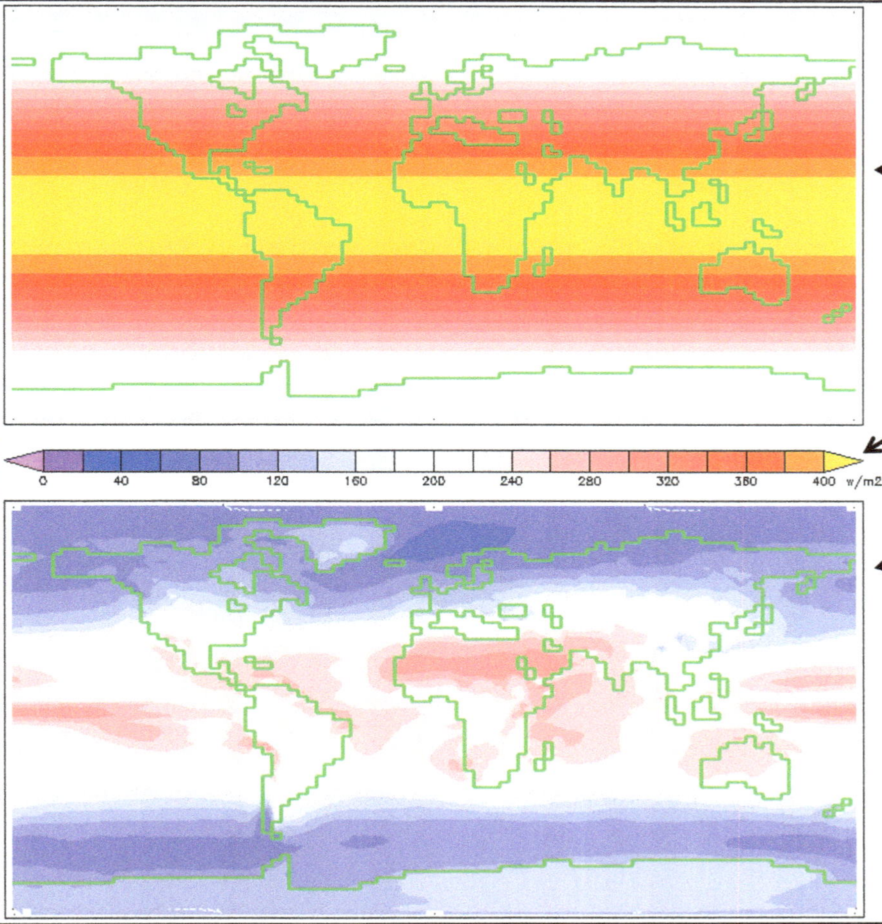

FIGURE 28

Annual solar irradiance at the top of the atmosphere

Common legend in W/m²

Annual solar irradiance at ground level after passing through the atmosphere

By William M. Connolley using HadCM3 data. - Own work, CC BY-SA 3.0, https://commons.wikimedia.org/w/index.php?curid=1335849

Most if not all papers on cloud dynamics [109] and latent heat at ground level [25] show that the models do not behave realistically. As a result, the concluding IPCC reports like AR4 Physical Science Basis [105] in Chapter 8, sections 8.2 and 8.2.1.3 signal important modelling issues.

Problems are gradually resolved, but significant issues remain, thus the IPCC Sixth Assessment Report (AR6 Physical Science Basis) shows that significant challenges remain in accurately representing cloud and water vapor processes (citing or quoting from it is forbidden). The persisting deficiencies suggest that the conceptual models should be improved.

As shown in the Introduction section, Physics is a collection of models in continuous development. Undoubtedly, a lot has been achieved, but there still are problems as shown above, thus computer simulations should not yet be used in real Earth management. Instead, policy makers should use for land management the realistic assessments of land use specialists, hydrologists, meteorologists.

Conceptual models relying on basic physics and supported by simple calculations will be used in this book to make correct conceptual simulations confirmed by direct measurements.

2.2 Direct Measurements, Conceptual Models and Conceptual Simulations

From this point onwards, unsettled issues will be analysed, and improvements suggested.

The first necessary step is to eliminate any bias, because Earth's unique value does not allow any, and the urgency of the climate crisis requires to quickly analyze problems and act to resolve them.

Real climatic processes are highly interdependent, full of feedback loops and apparent paradoxes; thus, the second step is to relax the cartesian system of thinking whenever it is necessary to accept a paradoxical reality. Subsequently, a more complex cartesian understanding will be achieved.

2.2.1 Water and the Greenhouse Effect (GHE)

Climatology refers to "well mixed GHG", but neither CO_2 or water are not well mixed, but their concentration is larger near their sources and sinks. It is impossible to characterize correctly the climate only from the anthropogenic variation of the other parameters while ignoring the anthropogenic variation of water for three reasons. Firstly, water is ignored as a driver because it is not well mixed, too variable in space and time. However, its wide variations are due to the powerful evaporation-condensation cycle, which in turn produces the large energy flux that makes water dominant in climate. Secondly, water is prevented from reacting in the computer simulations and some partial conclusions are drawn, then it is allowed to vary as a follower of the drivers for other partial conclusions. Despite this, anthropogenic variations of water have the largest effects.

As shown in section 2.1.1, from computer simulation assessment, water vapours would be the largest contributor to global warming through its GHE, and therefore would be undesirable for its feedback to CO_2 [130]. However, water also has multiple cooling effects, es explained further.

I argue that the greenhouse effect does not depend on relative humidity, but on absolute humidity.

As known, when air temperature rises, evaporation and absolute humidity increase, even if locally the air seems dry (low relative humidity due to higher temperature). In a desert the hotter air may have low relative humidity but high absolute humidity, while in cool tree areas relative humidity might be high, but the absolute humidity and therefore the greenhouse effect could be lower than in the desert. This is only the radiative part of the GHE.

Estimation: A desert at 40^0C and 25% relative humidity (like Sahara) holds 10g/kg of water, while a tree area like a forest at 20^0C and 50% relative humidity holds 8g/kg of water, hence less than the desert. A conceptual equation of GHE would be: **GHE = Infrared Radiation x GHG**, where desert infrared emission (through Stefan-Boltzmann law) is 550W/m^2 and the tree area one is 420W/m^2, resulting in the combined desert GHE being 1.6 times larger than that of the tree area.

Numerical experiments [67] and satellite measurements [68] claim that around 90% of the entire greenhouse effect takes place in the upper troposphere. However, as seen in Figure 4, naturally there was very little water vapours in the upper troposphere. This does not account for the effect of clouds above tree areas, or clouds condensing the high-altitude absolute humidity, further decreasing their GHE, while hot deserts propel more water in their tall convection cells, further increasing their GHE. Thus, deserts produce larger GHE than tree areas, infirming that trees and water would increase GHE. Satellite observations show that in hot dry areas the infrared emission, therefore the GHE of GHG is larger than in cloudy tree areas, as seen in Figure 35A.

The greenhouse effect (GHE) may be smaller in vegetated areas than in devegetated areas.

2.2.2 Vegetation and Water

Literature addresses this important issue in an analogue way [126] and by computer models [131].

The climatic effects of water are large, and on land they mostly revolve around vegetation, which vehiculates large amounts of water. The leading climatologists believe that plants only benefit of abiotic humid conditions, do not create those conditions, thus plants would not be drivers of climate. A brief analysis is thus necessary for the role of vegetation and especially of trees in climate.

The question would be if vegetation (and especially trees) significantly influences the above land air temperature (the benchmark of global warming) and the precipitations, and by how much?

2.2.2.1 Attraction of Precipitation

We start from the observation that at global level and over long-time, evapotranspiration must equal precipitation. However, on land there are rivers which only flow into the ocean, this water having to come from evaporated ocean water which precipitates over land. If the precipitation and evaporation rates over the ocean would equal those on land, the land would dry out because of river runoff. There would be no water on land if it would not evaporate from the ocean and preferentially precipitate on land. Therefore, the land must attract more precipitation than the ocean, and land precipitation must exceed land evapotranspiration by river runoff.

There is no precipitation in deserts, which occupy a large part of the land, thus precipitation must be much larger in vegetated areas to compensate for the lack of precipitation in deserts, where just abiotic conditions are active. Hence, it results that vegetation has the capacity of attracting much larger precipitation than abiotic deserts, as was also inferred by other researchers [82].

There is very sparse rain in an abiotic desert; water seeps through the degraded soil and the high heat rapidly evaporate surficial water without any river formation. Conversely, perennial rivers are known to flow out from forests at all geographic latitudes. Moreover, it is well known that if the forest of a wet region is cut down, soon the regional climate and the river will dry out [126].

As shown in the above Section 1.10:

- Average evapotranspiration from oceans is 1200mm/year
- Average evapotranspiration from land is 480mm/year
- Average precipitation on the ocean is 1100mm/year
- Average precipitation on land is 720mm/year
- There is a slight evaporation/precipitation excess on ocean
- There is a strong evaporation/precipitation deficit on land.

The known 2000mm/year precipitation in rainforests is 3 times more the land average, proving that forests just do not start growing in land areas with anomalously high precipitation that cannot be justified by abiotic factors, but they attract and increase local precipitation.

2.2.2.2 Water Extraction from Air

For example, the Amazonian Forest attracts a net flow of humid air from the ocean equalling the Amazon River flow, of 200,000m^3/s. Since the front of wind entrance in Amazonia is around 4,000km wide and 5km high (half of the troposphere), and if the average wind speed would be 2m/s, it means that 40km^3/s of air are carrying the 200,000m^3/s of water, resulting in a significant average humidity extraction of 5g/m^3. In fact, since humidity is not extracted evenly from the entire air thickness of 5km, the water extraction efficiency of the Amazonian Forest would be even larger. It appears that not only the Atlantic air is drained of its humidity, but any air part of the global wind circulation coming above Amazonia. It is revelatory that this largest performance of humidity extraction from air corresponds to the largest rainforest on Earth.

In Amazonia precipitation is more than 10 times larger than in Somalia, despite both Amazonia and Somalia are in the equatorial area and have eastern coastlines. The difference is that Somalia was deforested in the past 2000 years for the traditional charcoal export, and turned into a desert through the traditional goat herding. There was no global shift in climate responsible for the African drought, as Amazonia and Southeast Asia were not affected. Somalia was left with the same abiotic precipitation factors like Amazonia: similar latitude, coastline orientation, coastal wind and ocean humidity, but it has lost its precipitation when people deforested and turned it into a desert.

It results that large, compact forests have high capacity to attract and extract precipitation.

2.2.2.3 Air Conditioning

In Somalia, daytime temperatures are around 10^0C higher than in the Amazonian Forest, while night time temperature is around 10^0C lower than in the forest, making the tree area a more life-friendly environment, despite both the forest and the desert have the same abiotic conditions. Tree areas create environments suitable for their needs because they are anti-entropic living systems.

Global warming is the increase in air temperature measured over the past 200 years at the meteorological stations in settlements around which the land was gradually developed during those 200 years, losing both density and areal extent of vegetation. This is part of the "urban island heat" effect which is known to rise temperatures by a few degrees Celsius and was discussed in [86], then officially denied, for example in [87].

As less water is evaporated due to areal development from less vegetation in and around meteorological stations, air temperature increases by much more than the 1.5^0C of official global warming. Corrections are applied to the recorded air temperature to exclude this urban heat island effect to account only for the greenhouse effect, as stated in the 2007 report of IPCC. Unfortunately, such corrections do not go back for all the historic recordings.

Therefore, it is confirmed that vegetation has a powerful regional cooling effect through its evapotranspiration of water. Conversely, air temperature significantly increases all over the Earth because of intertropical deforestations, as explained further.

2.2.2.4 The Trees Humidity Pump

The result of the above 3 Sections is the trees humidity pump, which attracts and promotes precipitation through trees' anti-entropic mechanisms above forests, plantations and orchards.

This "trees humidity pump" happens with all dense trees, not only with forests, especially if the ground is thick, porous and humid. This tree humidity pump is the main mechanism of water gain on intertropical land, since there is almost no rain in un-vegetated tropical land (desert).

At higher latitudes, where air temperature is lower and closer to the dew point, this humidity pump is eased, and takes place with plants that do not develop a tall canopy (usual trees or even grasses and crops), while in the arctic areas it is produced even by the moss beds.

People with local knowledge of forests state that tree areas "make rain", while scientifically was proven [75] that around half of forest precipitation can be attributed to local evapotranspiration. Therefore, deforestation decreases precipitation and sooner or later leads to desertification. However, due to officially misunderstood phenomena, deforestation is allowed and sometimes encouraged. This is helped by statements that drought is caused by evapotranspiration demand.

An outstanding but disputed theory [83] described a forest mechanism of attracting ocean humidity, while in 2020 in [52] is presented an improved theory showing that:

- Pumping is due to precipitation vacuum, not to tree area evapotranspiration rising.
- Vapour condensation vacuum and rising of the resulting dry air are the main engine of continental scale humidity pumps.
- Pumping works better when trees evapotranspiration is smaller than ocean evaporation
- The pump does not stop with evaporation or flooded ground, but it increases.
- It is less dependent on the lapse rate of temperature in the forest.
- A closed canopy is not a condition for the pump to work successfully.

Through this tree humidity pump, vegetation produces larger precipitation conditions above tree areas than above the ocean, attracting marine humidity, and usually creating a river (Figure 29). Once the humidity pump starts, the vegetation effect becomes dominant over abiotic parameters, and increases qualitatively and quantitatively with the increase in vegetation area, height and density because it receives more humidity.

Water evaporating from the ocean increases regional atmospheric pressure and is mixed with air by wind shear, remaining invisible at low altitude and traveling with the wind. When it rises in lighter-than-air large parcels, it creates clouds through partial condensation. Total cloud condensation (precipitation) is usually prevented above the ocean due to high vapour pressure and to heat and infrared radiation received from the surface.

FIGURE 29: The Trees Humidity Pump

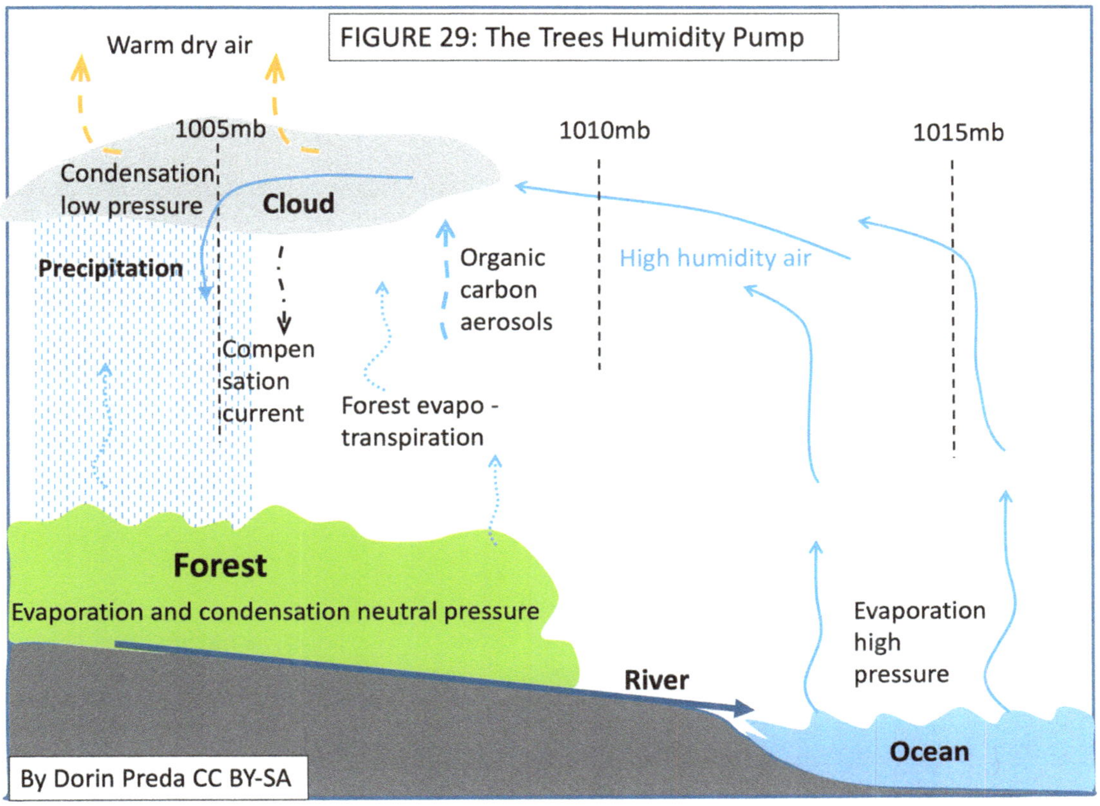

This mass of humid air can only produce rain when meeting a cold air front, when pushed to high altitudes as shown in Figure 51, when receiving more moisture as illustrated in Figure 74, or when transported above forests, dense tree plantations or large, dense orchards.

The tree area canopy consists of layers of leaves alternating with layers of air, creating a thermally insulated solar radiation reflector/absorber and evaporative cooler. It may produce daytime rainfall when its air mixes with the incoming humid ocean breeze if the sky is cloudy, thus it is low solar heating. This process is especially intense (and is more frequent) at night, when solar radiation does not reduce relative humidity. At night, the top layers of leaves emit some infrared radiation and quickly cool down due to their low mass, preventing the tree area from radiatively warming the clouds above. There is still evapotranspiration from the tree area at night, that partially is recaptured by the underneath surfaces of the cool top leaves, and partially escapes in the air above the tree area which is humid oceanic air. From this mixing results a mass of cool and very humid air, to which organic carbon aerosols of the tree area add up and trigger the rainfall of the clouds. Rainwater supersaturates tree area humid air, raising its dew point, triggering also condensation and rain in the tree area. Condensation above the tree area generates a barometric low, which is amplified by the rising warm air after losing its water, these phenomena attracting more humid air from the high-pressure evaporative regions above the ocean, to produce more condensation on land, and create an avalanche effect, or positive feedback. This way the trees extract water from the ocean over large distances.

The humidity pump closes the water cycle with the rivers that flow out from the tree area.

Condensation of water vapours in the clouds and usual dry warm air rising above the clouds can induce air pressure drops of up to 20 mbar, which is enough to attract ocean humidity on any continental scale, as shown in Section 2.2.2.2. Exceptional condensation and warm air rising in cumulonimbus clouds can produce significantly larger pressure differences with ocean areas [83].

Trees humidity pumps are due to the outstanding capacity of plants, which cool and humidify air, actively and anti-entropically producing rain, which they need. This is in line with the Gaia hypothesis.

Tree areas evaporate water to create clouds that reflect most solar radiation and through their condensation reduce barometric pressure to initiate the attraction of humid air from elsewhere. Tree areas then spend organic carbon aerosols and water to cool and saturate the incoming oceanic air and trigger the rain, which will bring down into the tree area both the ocean and tree area water. At night, the cooling effect is enhanced radiatively with the cold night sky, increasing precipitation of the ocean humidity, in a feedback loop. The condensation air is dried and warmed up, thus it rises above the clouds and readily emits its energy to the sky through the dried-up top troposphere.

The “atmospheric rivers” in literature are different in their mechanism from tree humidity pumps.

2.2.2.5. The Avalanche Greening

Vegetation comprises many diverse species of plants that both co-operate and compete in their strive for survival and reproduction as individuals, and for spreading in territory as species.

Vegetation has the capacity to slowly take over new land, even if it is very inhospitable initially, for example hot, dry and drought-prone; this is how Central Australia is at present. Small drought-resistant plants would take roots initially in the sand or through the cracks of rocky terrain. They would take most of their necessary water from the night dew that is precipitated from air by the cold desert strongly radiating to the clear night sky. The small plants gradually grow in density, covering the ground and cooling it down through shadowing, slightly cooling and humidifying the local climate. This way, they create cooler conditions for larger plants to creep in and produce more litter that enhances the soil. The larger plants just co-exist with the small plants, further cooling the climate with the brought-in shadowing clouds and with more evapotranspiration cooling.

The greater biological production of the area from solar radiation, water and some ground minerals mean more organic matter as plant residue that start create a substantial soil with bacteria and invertebrates that attract insects. This soil in turn allows better root growth for the plants, that develop larger foliage, in positive biological feedback. The larger plants have deeper roots that tap into the water table and assure a perennial source of water to the plants, which evaporate more water and increase the average cloud cover and precipitation, creating a local climate feedback.

In these favourable conditions, if the wind or birds or animals bring seeds of trees, they will have good chances of survival, that would not have been possible in the initial hot and barren desert. The tree saplings quickly grow and overtake in size the other plants on the ground, which served as start-up to the green takeover of the land. The trees grow and their shadow might decrease the solar radiation to the point that the other plants are greatly diminished, changing the vegetation cover to a much stronger and more permanent structure. Extreme events like storms are unlikely to wipe out completely the new biotope, making it more resilient and self-supported.

The new forest creates its soil and its local climate, and hydrological cycle. As it grows in size to hundreds of kilometres or more, it starts to influence the regional climate and to interact with other hydrological entities, like large bodies of water, which produce large masses of water vapours. Since forests, as anti-entropic living beings and potent entities at the ground-atmosphere interface need water, they evolved mechanisms to attract humidity (Section 2.2.2.1). The condensation in the clouds of the water evaporated by the forest creates a vacuum which attracts the high partial pressure of the water body evaporation, and starts a humidity pump (Section 2.2.2.4) that enhances the conditions in which the forest thrives. This mechanism allowed plants to take over the Earth.

This natural mechanism is different of the artificial Avalanche Greening and Cooling (AGC) below.

2.2.2.6 The Green Island

When (new) vegetation takes over a stretch of land, it is usually as a front like shown above, but if the available land conditions are restricted, the newly vegetated area would take an isometric land shape, like an island. This is also important for a patch of baren land in the middle of a desert, replaced with green development by people, in an opposite effect to the Urban Heat Island (UHI), where vegetated land is deforested to create an urban island surrounded by larger green areas.

A green island is a natural or artificial (AGC of Section 7.3) system with very different abiotic and biotic conditions versus its surroundings, these conditions being anti-entropically maintained by its vegetation. The features of this system separate it, but also integrate it with the surroundings:

- Tall trees create a canopy protecting from solar radiation the smaller plants underneath.
- Vegetation on the sides protect from wind the humid and temperate inner environment.
- Very high plant density and low temperature means plants benefit of each other's humidity
- Green islands (like an oasis) cool and humidify their area, oppositely to our UHI.
- Depending on their surroundings, large green islands may generate excess water (river).
- If surrounded by desert, green islands cannot setup a tree humidity pump with the ocean.
- Solar radiation, rain, minerals and CO_2 initially generate accumulating organic matter.
- Initially, the plants and soil increase, then only the soil increases up to equilibrium.
- In green islands that develop a river, soils stop increasing, thus they do not store carbon.
- Green islands generate through photosynthesis oxygen from air CO_2 and water
- Water and carbon usually also exit the mature green islands, mostly equalling the inputs.
- The animals in green islands recycle carbon with the plants, thus much input is not needed
- Excess rainwater flows out as a river, carrying the excess organic carbon to the ocean.
- Oceans absorb CO_2 from the air and receive organic carbon from the land and store both.
- The organic carbon is buried as hydrocarbons, dissolved CO_2 is fixed chemically; all is lost.
- Biosphere 2 (https://en.wikipedia.org/wiki/Biosphere_2) was a fully enclosed green island.
- Earth is an evolving green island, most of its life being limited by the available CO_2.
- There are bacterial forms of life that do not need CO_2; thus, life will continue after CO_2.

2.2.2.7 Vegetation and the Radiative Forcing

Anthropogenic-induced variation of vegetation cover is a driver of the water cycle, thus of the climate, and should be included in the global radiative forcing (as it used to be [70]) because:

- millenary observations and studies like [100] show that forests make rain, changing climate
- studies like [82], [83], [92] show that forests actively attract ocean precipitation
- section 2.2.2.2 proves the high water precipitation capacity of trees (55% of water at 25^0C)
- evapotranspiration clouds strongly influence the planetary temperature through reflection
- evapotranspiration is indistinguishable from irrigation water (that is already considered in Figure SPM2 of [102]), thus devegetation lost evapotranspiration should also be included
- vegetation is a driver, acts anti-entropically for its own interest against abiotic factors
- people act upon vegetation, thus indirectly driving the climate through the water cycle.
- further calculations will show its major role in the global energy budget and radiative forcing

Hence, anthropogenic vegetation variations should be included in the global radiative forcing, and its value calculated through the imbalances in the water cycle it produces.

Human land use is included in the global radiative forcing, but it is poorly defined, as it can mean a change from savannah to urban, which is totally different as water processes (thus temperature) from the change savannah to orchard or plantation. It would be appropriate to replace "land use" with "vegetation change", that would reflect the change in water handling capacity, resulting in a few categories, in the style shown forward in Section 4.1.

2.2.3 Cloud Effects

Clouds are essential for Earth's climate, as they influence all spatial and terrestrial parameters and are a part of the powerful water cycle [132]. Only the radiative clouds effects are discussed here.

Classically, it is believed that clouds stay at high altitude because of their warmer-than-surroundings air between their fine water droplets, because they are pushed upwards by the thermals that created them, and that they are not floating, but slightly falling. However, the assumptions of [52] indicate that a lighter than air parcel of humid air rises and cools adiabatically, resulting in condensation of a proportion of its vapours, such that cloud's density equals that of the surrounding air, although not all the cloud's vapours have condensed. Any rising would condense more vapours making the cloud too heavy, while any sinking would evaporate more droplets resulting in increased buoyancy. In this state the cloud is integrated in the mass and energy transfer system that surrounds it, capturing any vapours, heat and infrared radiation from the ground and emitting heat and infrared radiation upwards and radiation downwards. The evolution of the cloud depends on its mass and especially energy budget. I assume that the uncondensed vapours would be attached to the micro-droplets with adhesion forces and provide the necessary equilibrium buoyancy to support the weight of the condensed water.

An additional explanation for the cloud's equilibrium altitude [52] would be that ascending clouds are heated through radiation, convection conduction by the ground, and that they rise until the density of greenhouse gases above them decreases enough to allow their vapours and micro-droplets dissipate more energy than they receive from the greenhouse gases above, so the vapours condense and stop cloud ascension. The remaining vapours of large parcels of air at the top of the cloud continue to cool radiatively and condense and increase the weight of their air parcel which descends to the bottom of the cloud where it receives more heat from the ground, re-evaporate part of the droplets and repeat the cycle while transporting heat from the bottom of the cloud to its top, generating the observed cloud convections.

Climatologists also decided about 20 years ago that clouds would not absorb solar energy, but only scatter it around, which is contradicted by measurements of solar radiation above and under the clouds. It is stated in climatology that average stratocumulus clouds would have an albedo of 0.65, while the densest clouds (cumulonimbus) would have an albedo of 0.8. However, our measurements of solar radiation under compact stratocumulus clouds show only around 15% of the solar radiation above clouds, while under the densest cumulonimbus clouds solar energy is only 5% of that above cloud. This means that stratocumulus would absorb 20% of solar radiation, while cumulonimbus clouds would absorb around 15% of solar radiation. This is opposite than expected, but suggests correlation with high thin clouds. More consistent measurements would be needed.

Another argument in favour of cloud absorption of solar radiation is the imperfection of anything; we cannot assume billions of perfect reflections of each ray when passing through a cloud.

Additionally, our cloud observations in windless early mornings show that a stratocumulus cloud would rise and disappear through evaporation as the solar radiation increases but sometimes does not even reach the ground. We should also note that a thermal cannot form so early morning and evaporate a cloud in just a few minutes; conduction anyway is very slow.

I assume that clouds absorb some solar radiation and then use that energy internally to evaporate their water, or transfer that sensible heat, or radiate it in infrared.

The first analysis is of the low thick clouds, as in Figure 30. The Earth-space system is discussed (both for day and night and for clouded and unclouded areas) through percentages of solar radiation to determine the overall role of low thick clouds, which are the usual opaque types of cloud, distinct of the high altitude thin (transparent) clouds.

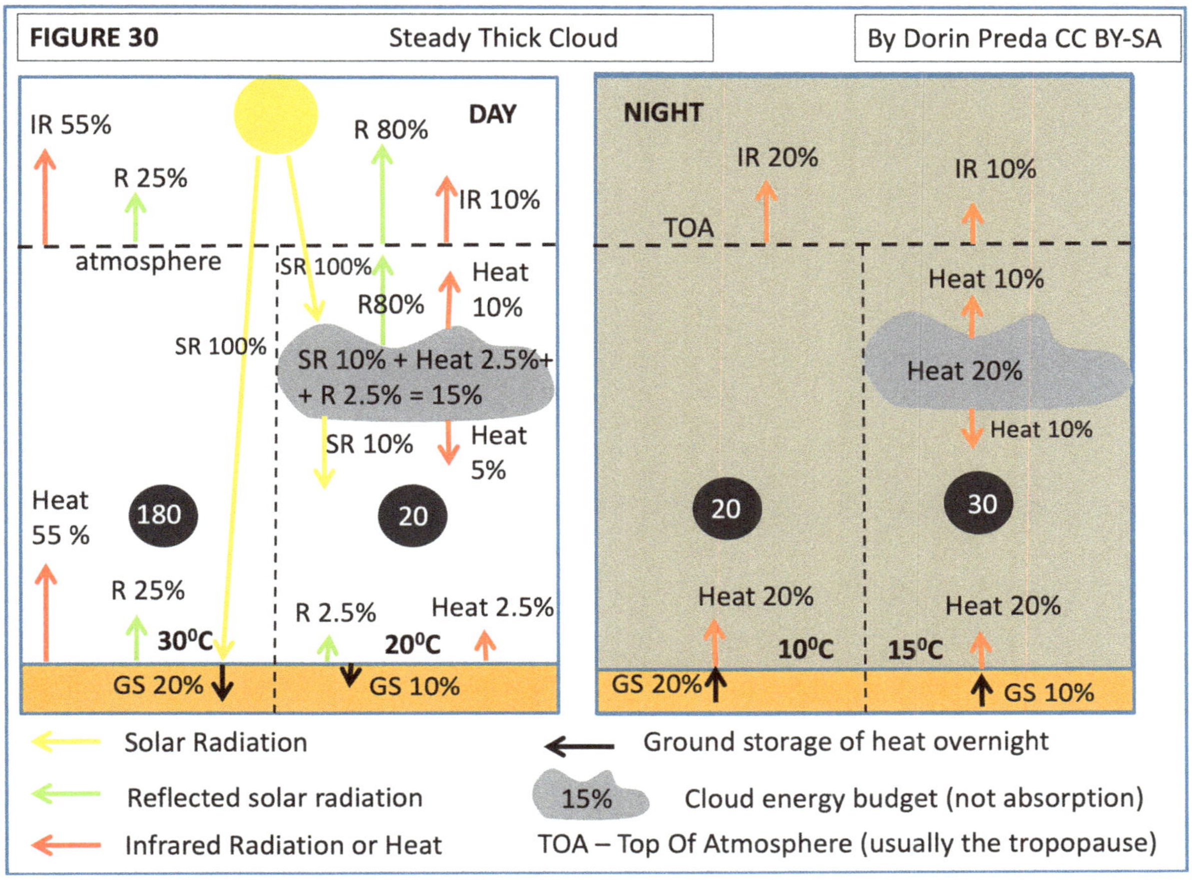

We introduce the concept of "Heat" = infrared radiation + sensible heat. There are three systems that must be balanced: cosmic space, clouds and the ground. I will use reflectance (R) as the proportion of reflected solar radiation (not albedo, or brightness, which are insufficiently precise). In Figure 30, R 80% represents the strong reflective cooling effect (more accurately "shielding from insolation") of low thick clouds.

We begin by setting some realistic values in the clear sky daytime area of Figure 30. Then in the quarter panels on the right we introduce a thick cloud that reflects 80% of solar radiation, while the rest penetrates the cloud, with SR 10% being absorbed and SR 10% exiting the cloud and hitting the ground. Similar to the clear sky area, one quarter of the solar radiation is reflected by the ground, resulting in R 2.5%. The ground will absorb and store GS 20% in the sunny area, and GS 10% in the clouded area, where radiation input is smaller.

Therefore, in the clouded area the ground absorbs SR 10% - R 2.5% = SR 7.5% and Heat 5% from the cloud; it uses this energy for GS 10% and to lose the rest as Heat 2.5%. The cloud has absorbed SR 10%, R 2.5% and Heat 2.5% from the ground, resulting in 15% absorbed heat, which it emits as Heat 10% upwards and only Heat 5% downwards because of the gravity-suppressed convection.

The sky provides SR 100%; in the clouded area it receives R 80%, IR 10% during the day and the balance IR 10% at night. In the clear sky area, more energy is lost to space as infrared than as shortwave reflection because the ground is hotter (and so is the above ground air temperature which is the benchmark of global warming). It results that, within the meaning of global warming, thick clouds cool the Earth.

Computer models could not determine the top-bottom energy budget of clouds. However, based on our conceptual simulations in Figure 30, it logically follows that the heat emitted upwards is greater than heat emitted downwards. Additionally, since the air under the cloud is warmer than the air above it, more energy is transferred upwards.

Another unresolved issue from computer models was the overall impact of clouds on climate (cooling versus greenhouse warming). <u>We find that the greenhouse effect is included in the discussed heat exchanges and it is much smaller than the reflective cooling of the low thick clouds.</u> The cooling effect of thick cloud reflection outweighs any balance of the Heat exchanges.

Therefore, during the day, a small black object situated above the ground under the thick cloud, will receive a total energy flux (from above and below) of 180 in an unclouded area, but only 20 in a clouded area. This demonstrates (confirms) the significant cooling effect under thick clouds, besides the strong above-cloud cooling through reflection. At night, the ground under a thick cloud cannot release more than GS 10% it has received during the day. However, due to the greenhouse effect of the cloud, the total energy flux is 30, thus larger than in the unclouded area, despite the latter storing GS 20%. It is confirmed that thick clouds slightly heat the area underneath at night.

Overall, the total energy flux over 24 hours is 200 in an unclouded area and only 50 in the clouded area, indicating a strong cooling effect of the low thick clouds, larger than their greenhouse effect.

Computer models suggest differences between low thick clouds and high thin clouds, so I also analyse a thin cloud in Figure 31 during both day and night, in cloudy and uncloudy conditions.

In the clouded area, a thin cloud reflects R 20% of solar radiation, while the rest penetrates the cloud where SR 5% is absorbed, while the rest SR 75% exits the cloud and hits the ground. Then 1/4 of that solar radiation is reflected by the ground, resulting in R 20% of which R 5% is absorbed by the cloud, while R 15% will pass through the cloud to be used by a satellite to photograph the ground through the thin cloud. The ground absorbs SR 75% and cloud's Heat 35% and use them to store GS 20% and to eject upwards Heat 70%, which will be absorbed by the thin cloud, together with SR 5% from the sky, and R 5% as part of the reflection from the ground.

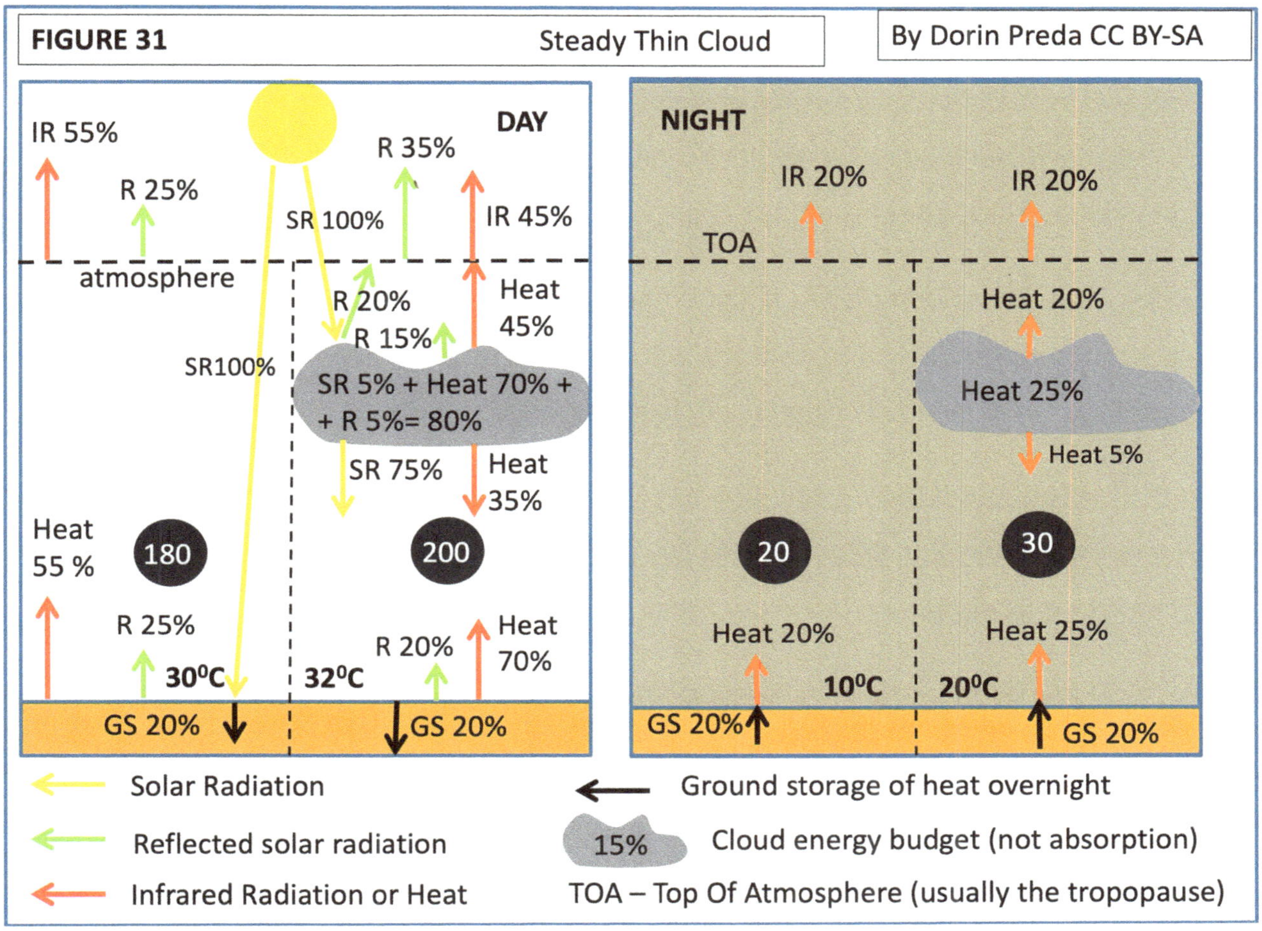

FIGURE 31 Steady Thin Cloud

The cloud uses this energy to emit Heat 35% towards the ground and Heat 45% upwards. Again, the difference between the Heat upwards versus downwards is because of the gravity supressed convection under cloud.

It is evident that thin clouds absorb more energy than thick clouds, so they also send much more energy towards the ground, not only as transmitted SR 75%, but also as Heat 35%.

During the day, the total energy flux under thin clouds is 200, higher than in cloudless areas, and also higher at night (30 versus 20). Over 24 hours, thin clouds have a substantial warming effect of 15% (230 versus 200), confirming previous literature that high thin clouds warm the area underneath more than a cloudless sky.

It should be observed that using very different values of the above ones impedes the correct balancing of the three systems (sky, cloud and ground).

The significant difference in greenhouse effects between thick and thin clouds is due to their different energy absorption and transmission rates.

Therefore, we confirm that thin clouds strongly heat underneath them, while slightly "cooling" above them by reflecting R 35% versus the average Earth (R 30%). However, thick clouds strongly cool underneath at day and slightly heat at night, while they strongly cool above them (R 80% versus Earth's 30%). For the ground system, the cooling total flux of thick clouds (50 – 200 = - 150) is stronger than the (+30) warming of thin clouds.

Taking the above into consideration, if half of the Earth clouds are now thick and half are thin, the daily energy flux in cloudless areas would be 200, but only 140 in cloudy areas, resulting in an overall strong cooling of 30% by clouds. Therefore, clouds now have an overall cooling effect on Earth.

However, if in the future there were 80% thin and 20% thick clouds, the daily energy flux in the cloudy areas would become 194, similar to 200 in cloudless areas, in which case the overall cloud cooling effect would disappear. Obviously, these are just general estimations first presented in [52].

Further, I will add other clouds effects, and investigate different parameters in the energy budget.

The previous discussion focussed on the local energy budget when clouds are present. Let now consider the global perspective. According to climate literature, clouds reflect an average of about 50% of solar radiation, while cities and grassland reflect 25%, tree areas 15% and the ocean only about 10% (though it reflects a lot at low angles). Therefore, clouds "cool" through reflection:

- by approximately 25% (50% - 25%) when they are above cities and grasslands,
- by about 35% (50% - 15%) when they are above tree areas,
- and by 40% (50% - 10%) when they are above ocean areas.

As a result, I calculate that the reduction in solar radiation received by Earth due to cloud reflection is: 25% (cloud effect above cities) x 50% (proportion of cloud cover) x 65% (proportion of cities and grassland on land) x 29% (proportion of land areas on Earth) + 35% (cloud effect on tree areas) x 70% (proportion of cloud cover on tree areas) x 33%(proportion of the tree area on land) x 29% (proportion of land areas on Earth) + 40%(effect of clouds above the ocean) x 50% (proportion of cloud cover) x 71% (proportion of ocean on Earth) = 19%. This is in line with data from literature that, currently, clouds have a global radiative cooling effect of about 20%. All the above refer only to the radiative effects of clouds, while the sensible and latent processes will be discussed later.

2.2.4 The Exclusive Albedo

Albedo indicates the fraction of the incident light that is reflected by a celestial body. It is a dimensionless parameter between 1 (when all light is reflected) and 0 (when all is absorbed) [133].

The equilibrium energy equation for a planet is: solar radiation = reflected solar radiation + infrared emission. Thus, if reflected solar radiation increases, infrared radiation (hence temperature) should decrease. This is why it is claimed that an increase in reflected solar radiation (albedo) through deforestation would mean a decrease in Earth temperature. I argue that this is not correct when:

- other relevant parameters are not considered
- applying this planetary rule of the outer system to the complex Earth inner system
- overstating the importance of this phenomena while downsizing the opposite effects.

An astronomical example when albedo reasoning is incorrect is the planet Venus, which has much higher albedo than Earth, but its ground temperature is much higher than Earth's [45]. The reason is the thick atmosphere of Venus, which produces stronger greenhouse heating than its cloud albedo cooling. Earth also has an atmosphere with clouds and greenhouse gases like Venus, so the albedo reasoning does not either apply to it.

Unfortunately, most recent climate science indicates that tree areas have a low albedo, warming up the Earth more than devegetated areas. This implies that it would be better without the trees. Some papers mention the existence of other relevant factors besides albedo, but fail to identify them and compare their effects to albedo. Because of the modelling errors introduced by exclusive albedo, in 2006, scientists were puzzled that the albedo of Earth was increasing, while the Earth was warming [1].

Radiative forcing takes over the exclusive albedo errors, leading to the mistake that land use (deforestation) results in negative radiative forcing (cooling) of the Earth (see Figure 38A). Therefore, based on the literature from the last 20 years (A4.4 in [16]), forests are shown to contributing to global warming, implicitly making them seem undesirable.

Some studies like [88] compare the benefit of trees capturing CO_2 to their albedo "disadvantage", ignoring many other parameters of the complex reforestation equation.

<u>When models based on albedo exclude other relevant factors from their analysis, they lead to what will be further referred to as the "exclusive albedo" error.</u>

2.2.5 Surface Energy Transfers

Figure 32 displays the daily energy budget of Earth and the main energy transfers.

FIGURE 32 Essential Daily Energy Budget of the Earth

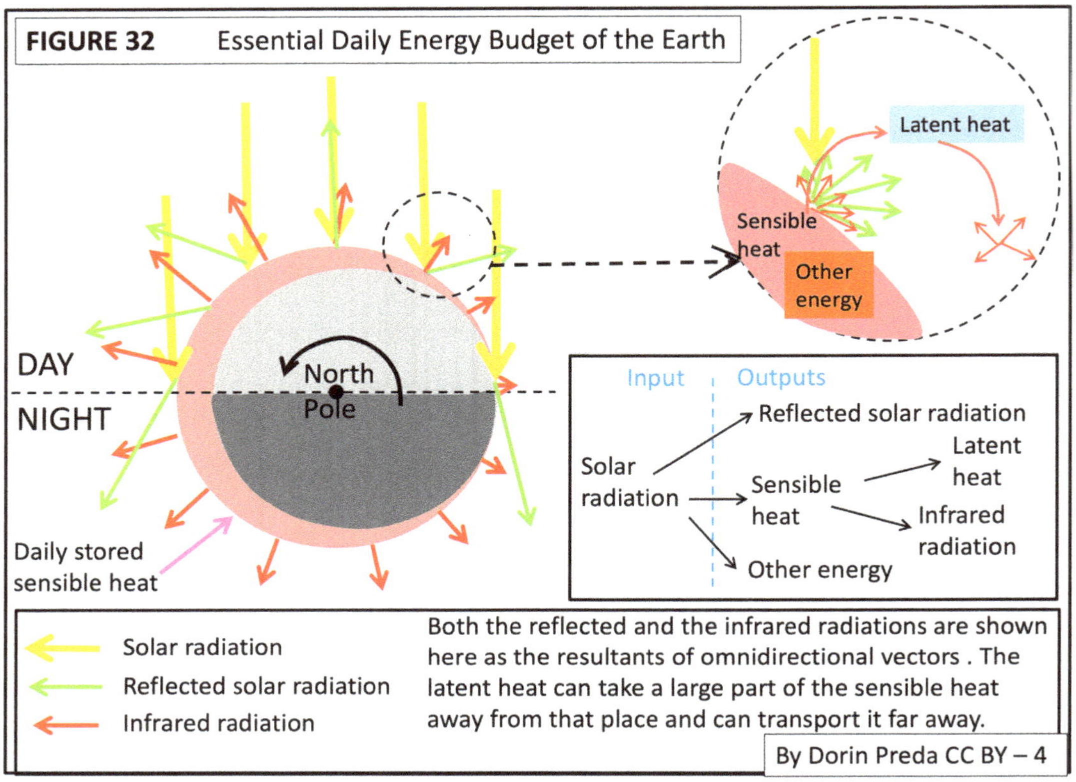

Once solar radiation penetrates the atmosphere, it comes into contact with water, the ground and any objects present. Upon impact, part of the solar radiation is reflected, while the remainder is absorbed and converted into sensible heat (molecular vibration), or other forms of energy such as chemical (photolysis, photosynthesis, oxidation), as illustrated above. Due to the sensible heat gained, molecules emit infrared radiation, depicted by the red arrows in the figure.

When water molecules are heated, those near the surface vibrate intensely and detach as water vapours, carrying a significant amount of latent heat with them. Therefore, deforestation at local level may slightly cool by absorbing less solar radiation than the tree area, but will significantly warm because of less latent cooling, thus net warming, contrary to the albedo interpretation.

The wind transports water vapours over large distances where they condense, heat those regions, and ultimately emit infrared radiation which escapes into space. Therefore, deforestation slightly increases reflectance at a planetary level, but this does not mean that air temperature decreases. This is because greenhouse gases emit the infrared that reaches the space, not the ground or the trees which are the focus of albedo. Figure 32 shows that the energy budget can only be balanced at planetary level.

2.3 The Radiative Heat Transfer

Most climate studies wrongly attempt to establish simplistic relationships through exclusive albedo between the planetary equilibrium temperature T_P and the ground temperature T_G, despite:

a) They belong to different systems (Section 1.11.1), with different varying parameters, so even if an indirect relationship works in one place, will not work later or in another place.
b) Unlike ground temperature, the planetary temperature depends only on solar radiation (SR), albedo (A) and radiative forcing (RF), thus $T_P = (SR - RF – SR(A))/\sigma)$ ^ - 4.
c) Earth should have a high planetary temperature, because it emits a large quantity of energy, when in fact it has a low overall temperature because it emits at low temperatures from multiple overleaped sources.
d) Vegetation influences directly RF calculated at the top of stratosphere [108] through ground albedo and indirectly through cloud albedo, usually in different directions, as seen in 2.5.2, complicating the relationship.
e) RF that always exists on Earth due to the night-day cycle even at average equilibrium, interposes between albedo and planetary temperature and hampers calculation of T_P.
f) The direct connection (better than through albedo) between the planetary and ground temperatures is through the ground infrared emission and the greenhouse effect GHE, which, as shown in Section 2.3.1, is independent of albedo. The relation is: $T_P = ((GHE – IR_G)/\sigma)$ ^ - 4.
g) T_G is different of the above-ground air temperature that is measured meteorologically.
h) Unlike qualitative ground temperature, planetary temperature is quantitative, so incorrect.
i) Satellites cannot properly measure the global infrared that defines planetary temperature.
j) Computers cannot simulate water phenomena essential for ground temperature.
k) We cannot distort real ground temperatures because of imaginary planetary temperature.
l) A few local ground temperatures are not representative for Earth, but all of them are.

m) The cloudy Venus (like Earth) proves that the exclusive albedo reasoning does not work.

n) The T_P variation in time is inverse to that of T_G and the relationship is not linear.

o) At equilibrium, T_P is independent of T_G and of the GHE, thus T_G is independent of albedo.

p) It is incorrect to pick only inner system albedos of tree areas and of devegetation, and use them into the outer system to assume that Earth could be cooled by increasing its albedo through devegetation. In reality, Earth's albedo is dominated by the much larger albedo of clouds that make ground albedo differences insignificant. Additionally, the exclusive albedo reasoning ignores the GHE and the water cycle, which are major factors which influence ground temperatures. Real world observations that of course include the effect of clouds and ground humidity and vegetation show that $IR_F < IR_D$, and $T_F < T_D$ thus, opposite to the assumption of the albedo model. Any relationship between albedo and Earth's temperatures must also consider cloud reflection and the water cycle influence.

q) An unreliable conceptual model should not be preferred to observational evidence.

r) The exclusive albedo is insufficient to correlate planetary and ground temperatures.

There still are significant radiative energy problems to solve in Earth computer simulations [135].

2.3.1 The Albedo – Temperature Disconnect, I find to occur not only on cloudy Venus and Earth, but even on planets with a transparent atmosphere.

As shown in Figure 33, two planets have clear air atmosphere as seen from afar, so their ground is perfectly seen from space. The left one is darker; thus, it has a lower albedo. Through the exclusive albedo reasoning, astronomers would say that the planet on the right is colder because it reflects more solar radiation, which implies a smaller infrared emission to space. In reality, if the planet on the right has CO_2 in its atmosphere, it would have a greenhouse effect that would give it a higher ground temperature above that of the planet on the left. The planetary equilibrium temperature cannot be used as a logical proxy for ground temperature because they can vary differently.

The ground temperatures disconnect from the albedos and the planetary temperatures because of the GHE, despite there are no clouds on these planets like on Venus or Earth. Interestingly, no matter how hot the ground would become if we add more CO_2 to the planet on the right, its infrared emission to space, thus its planetary temperature remains constant (at steady state).

An increase in GHG, thus in GHE, does not produce an increase of the infrared emission to space of its planet, or an increase of the planetary temperature, neither before equilibrium, nor during it. The reason is that infrared emission to space depends only on solar radiation and on albedo.

FIGURE 33 The Albedo - Temperature Disconnect

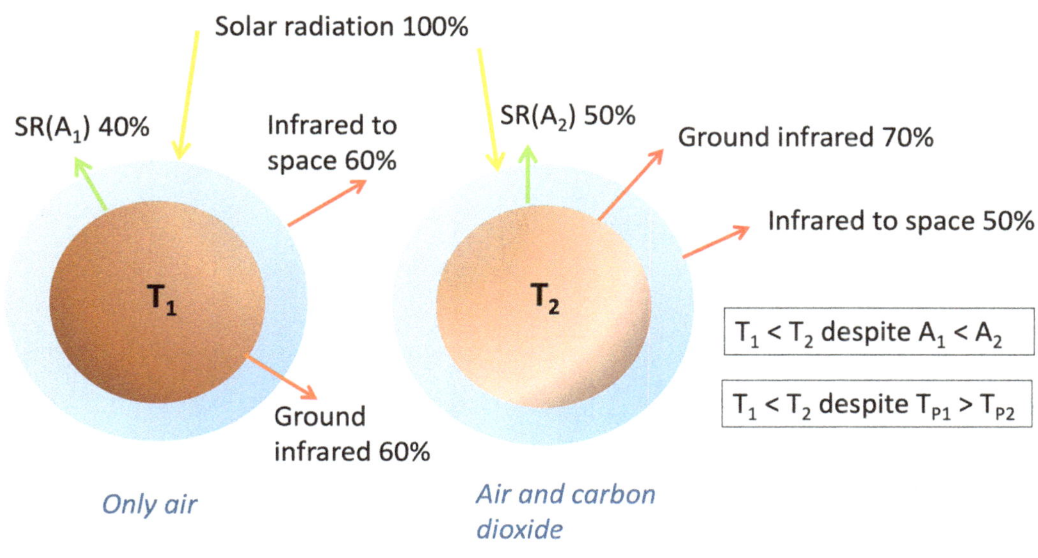

By Dorin Preda CC-BY-SA 4.0

However, the energy retained through GHE can be radiated back to space only when solar radiation diminishes, when the RF becomes negative. The same is valid for planets with other transparent GHG.

It results that we cannot infer a planet's surface temperature from its albedo if we do not know if it has an atmosphere, and what is the chemical composition and density of that atmosphere. The planetary temperature is not a suitable concept for planets with an atmosphere.

Albedo could still be used outside its limitations [134] if all the other factors are also considered.

2.3.2 The Lifetime of Greenhouse Gases

It is known that the greenhouse effect (GHE) of water on Earth is much larger than that of all the other greenhouse gases (GHG) combined, thus water dominates the radiative processes in the atmosphere.

Most climatologists [136] argue that CO_2 would have a more powerful GHE than water, due to water molecules having a shorter residency time in the atmosphere compared to CO_2. This argument is flawed, because power is already the result of the division of energy by time. As a water molecule condenses, another evaporates, maintaining the same quantity of water in the atmosphere, and thus the same GHE of water. We calculate the GHE on the mass of water existing at that time in the atmosphere, not on individual molecules. Even if we could calculate for each molecule, the GHE of a condensing molecule would be instantaneously replaced by that of an evaporating molecule.

The same principle applies to methane: a molecule is released on the ground, while another is dissociated in the ionosphere, yet the overall greenhouse effect of the gas remains constant.

Therefore, the greenhouse effect of a gas depends solely on its concentration in the atmosphere, not on the lifetime of its individual molecules.

2.4 Sensible Heat Transfers

If Earth had no atmosphere, its average surface temperature would be - 19^{0}C (as shown at 1.11.1).

Let us add an air-only atmosphere to Earth (without water or CO_2), quite different from atmosphere model with different emitting layers (https://biocycle.atmos.colostate.edu/shiny/2layer/) which does not perform realistically. However, we can analyse a conceptual model as in Figure 34.

Although air does not absorb solar radiation, it is in contact with the ground and is heated by it, creating convection currents that heat up the atmosphere. The heated air emits infrared radiation, cools down and sinks, then it is reheated by contact with the hot ground.

FIGURE 34: THE COOLING THROUGH DIVIDED HEAT EMISSION | By Dorin Preda CC BY-SA

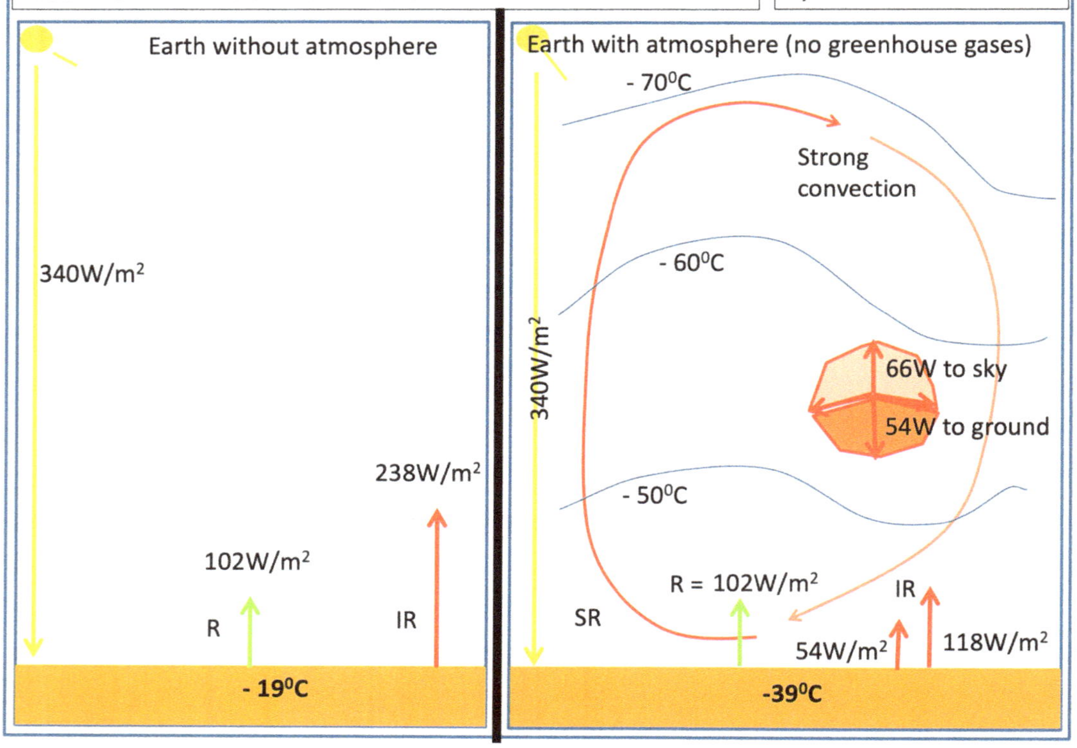

These sensible heat transfers are important on Earth and other planets with an atmosphere. Since albedo and solar radiation did not change after adding the perfectly transparent atmosphere, the planet must emit the same total infrared radiation to space. However, it is now divided between the ground and the air, each emitting separately. Thus, after adding an atmosphere to a planet, its ground irradiates less energy, corresponding to a decrease in its temperature.

Air in the atmosphere emits infrared equally in all directions; about 45% towards the ground and 55% upwards and sideways where it does not intersect the ground, thus it is lost to space because there are no greenhouse gases in this atmosphere and air does not absorb infrared.

We know from solar air heaters practice that about half of the heat is absorbed by air; thus 120J/s (120W) are absorbed by air from one square metre of Earth ground, and 118W remain for ground emission. Of the 120W of heat lifted by convection air, 120 x 0.55 = 66W are emitted and lost to cosmic space, while 120 x 0.45 = 54W are sent towards the ground, re-absorbed and re-emitted to the sky. The 118+54 = 172W/m^2 emitted by the ground result in an Earth ground temperature of 234^{0}K (-39^{0}C). The air will be even colder than the ground, as suggested in Fig 34.

Hence, the air cools the ground by taking part of its heat, and they both radiate their divided (smaller) heat portions at smaller temperatures than the planetary equilibrium temperature.

I argue that this new cooling through "Divided Heat Emission" is important because the initial heat can be divided amongst many agencies: ground, tree canopy, clouds, air, greenhouse gasses, each having its own small radiative flux. According to the Stefan-Boltzmann law, a smaller radiative flux means a lower emission temperature. Thus, dividing a quantity of heat between separate emitters decreases all their temperatures.

This means that the surface temperature of a planet with an atmosphere, but without greenhouse gases, should be inversely proportional to the density of its atmosphere. It also means that if we do not know if a planet has an atmosphere and what its density is, we cannot know its temperature.

Therefore, the exclusive albedo does not apply to planets with an atmosphere.

Since Earth's real average temperature is 15^{0}C as shown in [70], then in light of these calculations, the greenhouse effect on Earth would be larger than previously known (54^{0}C versus 33^{0}C).

The Divided Heat Emission (DHE) process appears to be a significant cooling process that should be applied to all situations when the emitter is more than a solid surface in a vacuum.

2.5 The Latent Heat Transfers

Water plays a crucial role in moderating the Earth's energy balance due to its exceptional properties, such as condensation in clouds, its high specific heat capacity and latent heat. Changes in Sun or Earth conditions lead to large shifts in climate due to water's influence in the global energy budget as discussed in [135].

More heat is transported by water through latent heat transfers than by air through sensible heat.

Carbon gases, including carbon dioxide and methane produced through burning fossil fuels and deforestation have led to a 1.5^{0}C global warming and climate change effects, as IPCC reported.

However, water appears to have a larger role in the climate effects (droughts, storms, floods, forest fires) that appear to be directly produced by global water cycle disturbances.

The important infrared emission of clouds (Figure 35A) appears not to be fully understood, confirming that the insufficient astronomical view again claims dominance upon ground certitude.

FIGURE 35A The "missing" infrared energy

The CERES satellite images of a cyclone in visible light (left) shows its reflective "cooling" by its white clouds. The image in infrared on the right was taken at the same place and time, and shows that the cyclone would emit much less infrared radiation than the surrounding clear sky areas.

Image by Giorgiogp2 - Own work, CC BY-SA 3.0, https://commons.wikimedia.org/w/index.php?curid=15581032

The previous literature describes storm clouds (see Figure 7) and even cyclone clouds as producing global warming. For example NASA explanation of the infrared image on the right in Figure 35A https://ceres.larc.nasa.gov/instruments/satellite-missions/ reads: "Without these tropical cloud systems, the Earth would lose a large amount of heat to space as seen by the surrounding clear-sky regions (red and yellow areas in the right image)."

In fact, clouds reflect 50% - 80% of solar radiation, even if they block emission of 20% solar radiation as infrared emission to space, thus they produce more reflective cooling than greenhouse warming.

If clouds would warm underneath, it would become warmer and warmer in a cloudy period, while in fact it becomes cooler as cloudiness persists for more days.

A large cyclone has a heat transfer power comparable to all the world's power plants combined; hence claims that clouds would not emit significant energy are false because at that scale the conduction rate is minimal, thus most of that large energy must be dissipated through radiation.

There are two possible explanations for this "missing" power: either we are unable to measure all the cloud infrared, or it is distributed differently. Let us discuss the first one:

The infrared images of clouds depict them as blocking the escaping infrared, but these images are typically captured in the 5 – 10 microns band, while top cyclone temperatures are around 200K, thus peaking at 14 microns. This could explain the missing infrared energy shown in Figure 35A.

A usual climatology explanation for the supposed low emission in infrared of the top of the clouds is that they would have lower temperature than the ground, so they would emit less. This is incorrect, as quantity of energy (radiative flux) is different of quality (temperature). A cool cloud can emit more than the hot ground because they are not a single emission surface, but a superimposing of multiple emission surfaces which all emit (DHE) through the same atmospheric window a very large quantity of energy in far infrared, as shown in section 2.4.

2.5.1 General Convective Heat Transfer as illustrated in Figure 35B in my interpretation refers to the second hypothesis of the above radiation distribution issue.

The heat (1) originates from the solar-warmed surface of the Earth, increasing air temperature and evaporating water, thus expanding their volumes. This expanding parcel of air and vapours near the ground physically lifts the atmosphere above it, providing the energy for convection kinetics. This atmospheric column has become unstable, with less dense hot air near the ground, and cool dense air above it, which gains gravitational energy from the hot air and water expansion through evaporation. Any disturbance initiates the convection process, causing the heavy cold air on top to descend and push the hot air upwards.

As the warm air and vapours rise, they expand and cool almost adiabatically (if that parcel of air is large) while emitting little net infrared radiation to the sky because of the strong greenhouse effect.

The air takes heat from the ground during its horizontal lower path, expands and ascends adiabatically while its temperature drops. Therefore, it does not emit much net infrared on the ascending part of the loop. Further up in the ascending part of the loop there is cooling due to adiabatic expansion that promotes water vapour condensation that heats the air and promotes further ascending.

When all vapours have condensed, the air heating and therefore the ascension stop. The condensed droplets of the formed cloud reflect most solar radiation, cool through conduction and by emitting infrared, and merge to create raindrops which fall to the ground.

During its horizontal trip at altitude, the parcel of warm air moves in the clear sky area and produces a stronger net infrared flux because there are fewer greenhouse gases above it.

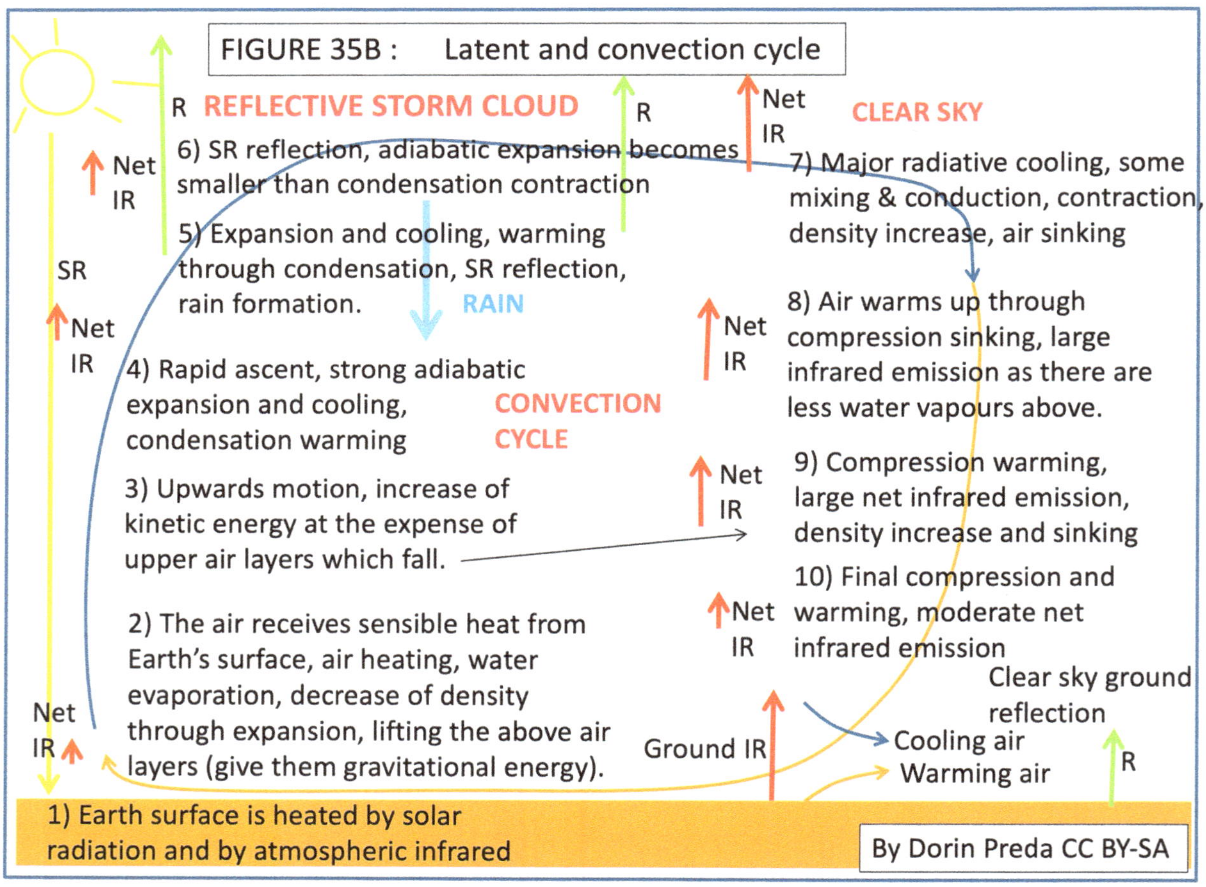

FIGURE 35B : Latent and convection cycle

This way, the air can become even colder and denser. Hence, the parcel of air falls gravitationally and warms up through compression, emitting more net infrared because the air is drier and warmer after precipitation.

I think (and Figure 35 A confirms) that the cloudless areas adjacent to storm clouds emit more in infrared because they have received the condensation warm air of the nearby rain clouds. As seen in figure 35B, rain condenses from the air and the dry and warm air moves away from the cloud and emits in infrared the heat taken from the ground by the evaporating water.

Radiometers and indirect methods for cloud measurements are in continuous development [137].

The improved radiometers will confirm the clouds' large emissions in far infrared.

2.5.2 The Temperature – Albedo Disconnect

Figure 36 illustrates my first proof of global warming through deforestation, contrary to the exclusive albedo theory. At the top left of the figure, tree areas reflect R_1 (less than R_2 of the deforested planet), resulting in a larger infrared IR_1 compared to IR_2, as per exclusive albedo theory.

Water cools the forest by absorbing most of its heat to create water vapours, which rise and disperse into the atmosphere, condense into clouds and emit water infrared radiation (WIR).

FIGURE 36 Warming through Deforestation

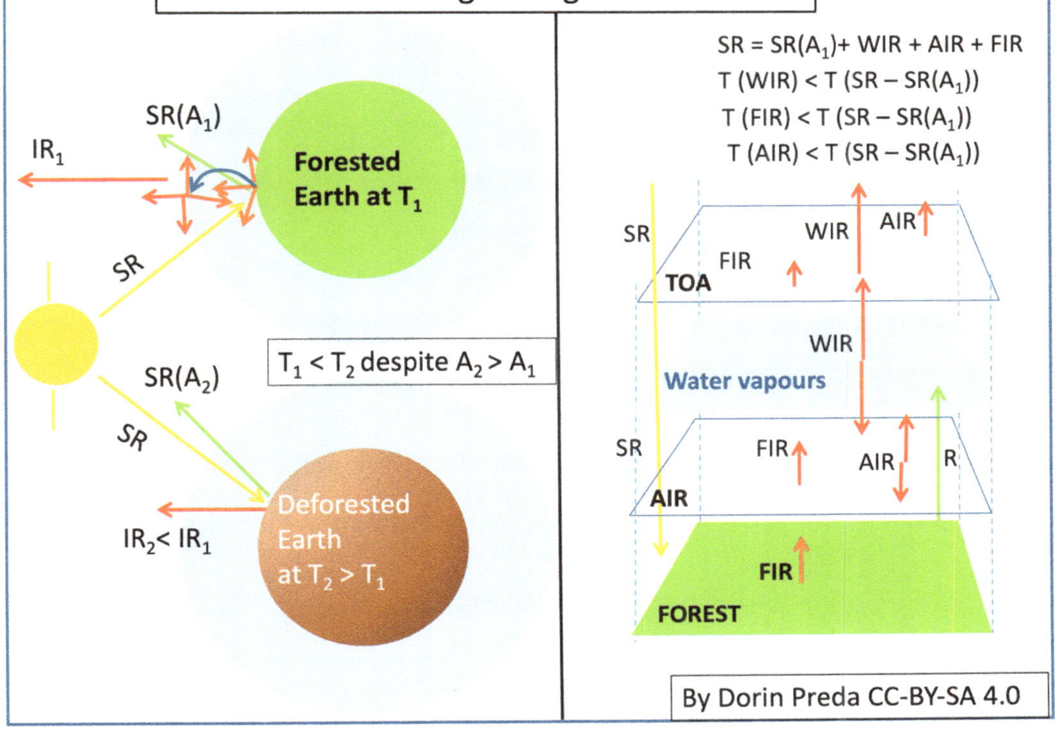

Only half of the heat taken by water is re-radiated towards the forest, and the same happens with the heat taken by the air (AIR), while the rest is lost to space. The remaining heat is emitted as forest infrared (FIR). As discussed in Section 2.4 all these sources emit their radiation through the same atmospheric window at the top of the atmosphere (see the right of Figure 36). Despite the total infrared flux IR_1 being greater than IR_2, the actual temperature T_1 of the forest has decreased to less than half due to the divided heat emission (DHE) between forest, air and water. This way, the temperature of the forested planet is smaller than the temperature of the deforested planet, proving that temperature does not depend only on albedo for planets with an atmosphere. The deforested planet has only air to decrease its temperature, while the forested planet has air and water, where water takes much more heat than air. This is obvious from direct observations: a forest is cooler than the adjacent hot bare ground. Both areas emit in infrared the same total energy from the Sun, but the forest emits it at a lower temperature. Even at medium and high altitude, the air temperature profiles have lower values above the forest than above the baren ground. Due to DHE, the energy fluxes can be inversely proportional to their temperatures.

Additional to the "Divided Heat Emission" effect, I also describe here a "Dispersed Emission" effect, produced by water absorbing heat as it evaporates from an area, vapours then rising and dispersing over a much larger emission area, this way reducing even more their radiative flux and emission temperature. We can combine the two into a "Divided and Dispersed Emission" effect (DDE) that indicates a larger cooling effect of water than its greenhouse effect.

An important concept derives from the DDE: overleaping layers of water vapours and water droplets at the top of a cloud behave in the DDE way, emitting more as a group than a solid black surface of the same area according to the Stefan-Boltzmann law. This reinforces the distinction between the temperature of a body and its radiative flux.

Therefore, the exclusive albedo theory cannot be applied to planets with an atmosphere, especially with water, as they produce opposite and larger effects than those produced by albedo variations.

2.5.3 Comparison Forest - Desert (the end result of deforestation) is presented in Figure 37 to check again if deforestation would cool the Earth (as believed through the exclusive albedo theory).

Some authors have mentioned that parameters other than albedo are also important for a planet's temperature, but there has been no analysis comparing these parameters to the albedo effect.

FIGURE 37 : Forest and Desert Heat Transfers Comparison

By Dorin Preda CC BY-SA

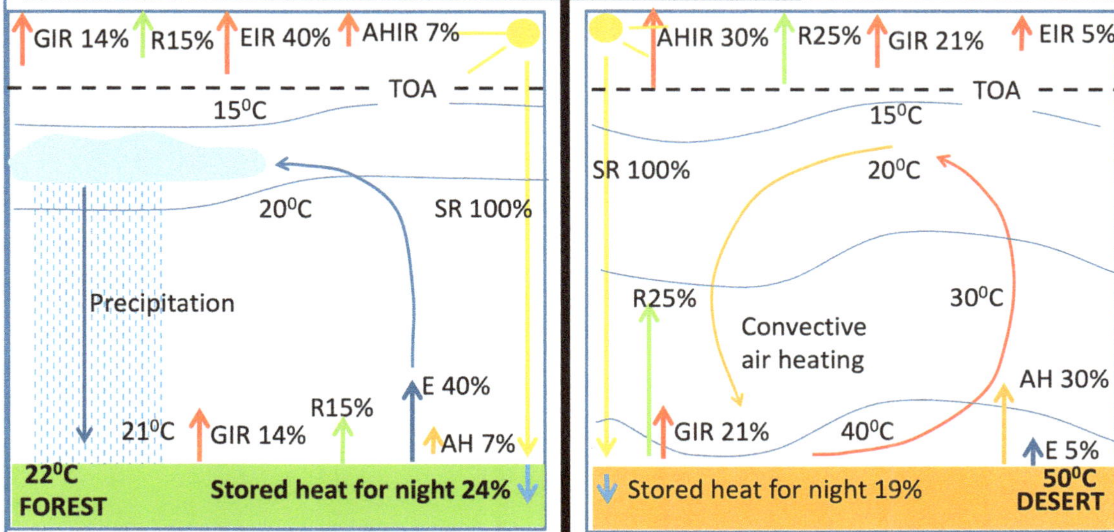

Solar radiation 100% = Reflection (R15%)+ Stored heat for night 24% + Ground Infrared (GIR14%)+ Air Heating Infrared (AHIR7%)+ Evaporation Infrared (EIR40%).
Despite the much larger heat transported through evaporation, only 230 kg/s of water (400m³/s of vapours) are needed for 1km².

Solar radiation 100% = Reflection 25% + Stored heat for night 19% + Ground Infrared (GIR21%) + Air Heating (AHIR30%) + Evaporation (EIR5%)
For 1km² of desert and 10⁰C difference of air temperature, 26,000m³/s of air remove heat from land, resulting in strong wind. Without air convection temperature would be about 100⁰.

Therefore, all literature discusses the albedo cooling effect of deforestation as in a vacuum, a conclusion reflected in the blue rectangle labelled "Land reflectivity change" in Figure 39.

However, the above ground air temperatures in a forest (about 25⁰C at day, 20⁰C at night) are more desirable than the above ground air temperatures in a desert (about 40⁰C at day and 10⁰C at night).

The excusive albedo theory assumes that variations in the above ground air temperature would correspond to variations of the planetary equilibrium temperature in a vacuum. Then ground temperature variations are discussed (while air and water effects are ignored) relative to changes in albedo when a forest is replaced by bare ground in a vacuum. Obviously, the planetary temperature refers to a different system than the ground one, and it is stated in climatology that local changes do not significantly influence the global system.

In Figure 37, we discuss the thermal effects of deforestation, including changes in albedo, by comparing forests to desert in terms of percentages of solar radiation. There is not enough room in the figure to represent the greenhouse effect, so the ground infrared is decreased proportionally.

"E" stands for evaporation, while EIR is the evaporation infrared energy which dissipates at high altitude the heat of condensation transported there by water vapours. Similarly, AH represents air heating and AHIR is the air heat dissipated as infrared radiation at high altitudes.

By using the values in Table 1, we can perform basic calculations for areas of 1km^2 of desert and forest. The forest receives 18MJ/m^2*day of solar radiation, has abundant water and a large plant exchange surface. Evapotranspiration is 1.5m/year, keeping the forest cool due to the constant low temperature (about 25^{0}C) during this latent heat transfer. Evaporation utilizes about half of the solar energy received, while air heating in tree areas is minimal (7%) due to the low temperature.

More heat can be stored for night time in the massive body of the forest than in the few top centimetres of desert soil which are heated during the day.

Deserts lack water, and can only use 5% of solar radiation for evaporation. This leads to a rise in temperature, with the remaining energy (30%) used for air heating due to the lack of other outlets for heat. Unlike latent heat transfer, which occurs at a constant low temperature, sensible heat convection begins only when the temperature differential exceeds 10^{0}C, initiating buoyancy. Inefficient sensible heat exchange increases temperatures, resulting in hot ground and powerful vertical air currents that heat the troposphere above deserts, preventing rainfall.

The small volume (400m^3/s) of vapours on the left in Figure 37 carry upwards 40% of solar energy flowing at less than 0.1m/s. This is much slower than the many metres per second of the 26,000 m^3/s air convection in the desert that can transport only 30% of solar energy. Obviously, it is not desirable to replace forests by hot and stormy deserts.

<u>Unlike hot deforested areas, water cools forests by absorbing heat at low and constant temperature and transfers it gently to high altitudes, where it is dissipated at a lower temperature than deserts.</u>

This serves as a second proof of the warming effect of deforestation, contrary to the exclusive albedo theory.

Earth climate studies should always properly consider the water cycle, like: [40], [41], [42], [43].

2.5.4 Benefits of Latent Heat Transfer

As shown at point 1.2, the troposphere has inverse density stratification (light hot air layers under heavy cold air as on the right in Figure 37) due to air being heated by the ground, not directly by solar radiation. This makes the troposphere potentially unstable and violent through large convective currents, as seen in Sahara's sand storms. <u>In tree areas there is no density inversion or instability because plants provide evaporative air cooling near the ground from where heat is taken and moved by the water vapours to the top of the troposphere where it heats the air and</u>

establishes a stable air column (see left in Figure 37). For millions of years forests have diminished the gravitational and thermal energy gradients of the troposphere through evapotranspiration, creating a calm climate and favourable living conditions under the clouds.

Humans reduced evapotranspiration through deforestation, creating desert conditions and the increased air heating near the ground and the great convective currents seen on the right in Figure 37. These violent air currents can be an increased a threat to airplanes due to increased devegetation. This also means more frequent and powerful sand storms and tornadoes. Therefore, we have changed through deforestation much of the calm latent heat transfers into violent sensible heat transfers.

The TOA energy budget is about the same above tree areas and deserts (if we ignore the beneficial reflection of tree area clouds), but tree areas are cooler and calmer than deserts. Other advantages:

- Evaporative cooling is more effective than convective cooling, as shown in Figure 37 comparing a 22^{0}C tree area to a 40^{0}C desert, creating more favourable conditions for life.
- Ground heat is extracted from under greenhouse humid layers, bypasses them and delivers heat above them, for easy emission in dry air.
- It can transport a large amount of low latitude heat to high latitude areas where it can be dissipated more easily, while also warming those regions (Figure 8).
- Evapotranspiration removes heat from the ground during the day, storing it into the clouds, that can radiate it at night, potentially resulting in the beneficial night rain.
- Temperature remains consistently low and favourable for life through latent heat energy transfer, unlike sensible heat transfer, which only can transfer significant quantities of energy over large temperature differentials.
- Humid air parcels move slowly, beneficial for living beings, unlike the desert storms.
- Evaporation from salty marshes or from ocean purifies water and delivers fresh rainwater from otherwise unusable sources.
- Rainfall spreads water over vast areas of the land, which is essential for life.

In conclusion, the latent heat transfer of vegetated areas is more preferable than the sensible heat transfer of deforested regions, highlighting the importance of halting deforestation.

Earth is a complex system, with biotic and abiotic factors interacting sometimes paradoxically, therefore comparisons are difficult. Albedo is an entropic concept, while plants actively resist entropy by regulating their energy and mass balances despite natural inputs. For instance, desert plants, despite receiving less water contain more water than rainforest plants. Additionally, plants, despite absorbing more solar radiation than nearby stones, maintain cooler temperatures.

Therefore, exclusive albedo is not suitable for vegetated areas, and should not be applied to them.

2.5.5 Latent Heat and the Greenhouse Effect (GHE)

The greenhouse effect (GHE) is a passive warming process: infrared radiation from the hot ground and lower air layers is absorbed by greenhouse gases (GHG) in the atmosphere which heat up the air around them and re-emit that heat including downwards, further heating the ground, and so on. A conceptual GHE equation would be: **GHE = (IR + AH) x GHG,** where AH is similar to IR, as is sensible heat delivered by opaque objects to nearby air which also radiates in infrared, while GHG depends on the nature and specific density of greenhouse gases. Thus, if (IR + AH) is small in a place, GHE is also small, even if GHG is large.

Accordingly, if the ground has a lot of water (as in a tree area), it will evaporate it, keeping ground and air temperature AH low, and therefore reducing infrared emission IR and GHE. For example, cooling an emitter from 40^0C to 20^0C (like cooling a desert with water) results in around 23% less infrared emission, and proportionally diminishing the GHE.

It is important to understand all the decreasing effects of tree areas onto the GHE:

a) Unlike urban areas, which only have high thin clouds that intensify the GHE, tree areas are most of the time covered with low thick clouds that reflect in average 66% of solar radiation.
b) Moist tree areas transform all solar heat into latent heat of evapotranspiration, instead of infrared radiation or air heating, thus tree areas produce almost no greenhouse effect, despite the many water vapours present there. Cool forest temperatures confirm low GHE.
c) The ground heat removed through evapotranspiration is delivered in the clouds as warm air at low temperatures and infrared radiation and re-radiated in far infrared between the low thick cloud and the top of troposphere, where average water content is 10 times less than under the clouds, thus the GHE of water is around 10 times smaller than near the ground.

d) The (IR + AH) above tree areas is smaller than at the same altitude of the thermals above deforestations, <u>thus the GHE of CO_2 is smaller above tree areas than above urban area</u>.

In Figure 38 is shown a comparison of GHE in tree area (forest or plantation or orchard) on the left versus deforested area (grassland/crops and urban) on the right. As known, in the intertropical area, tree areas are covered 70% of time by low thick clouds, while high thin clouds cover 40% of time the deforested areas, resulting in further division of four areas in the figure. Two of them shall be further divided in above and below clouds areas, hence we shall examine the 6 areas which describe the main situations influenced by clouds (the real weather-makers). The two above TOA areas over deforested and forested areas complete the picture. There is strong positive feedback between a type of land cover and the type of clouds that form about it, increasing each other.

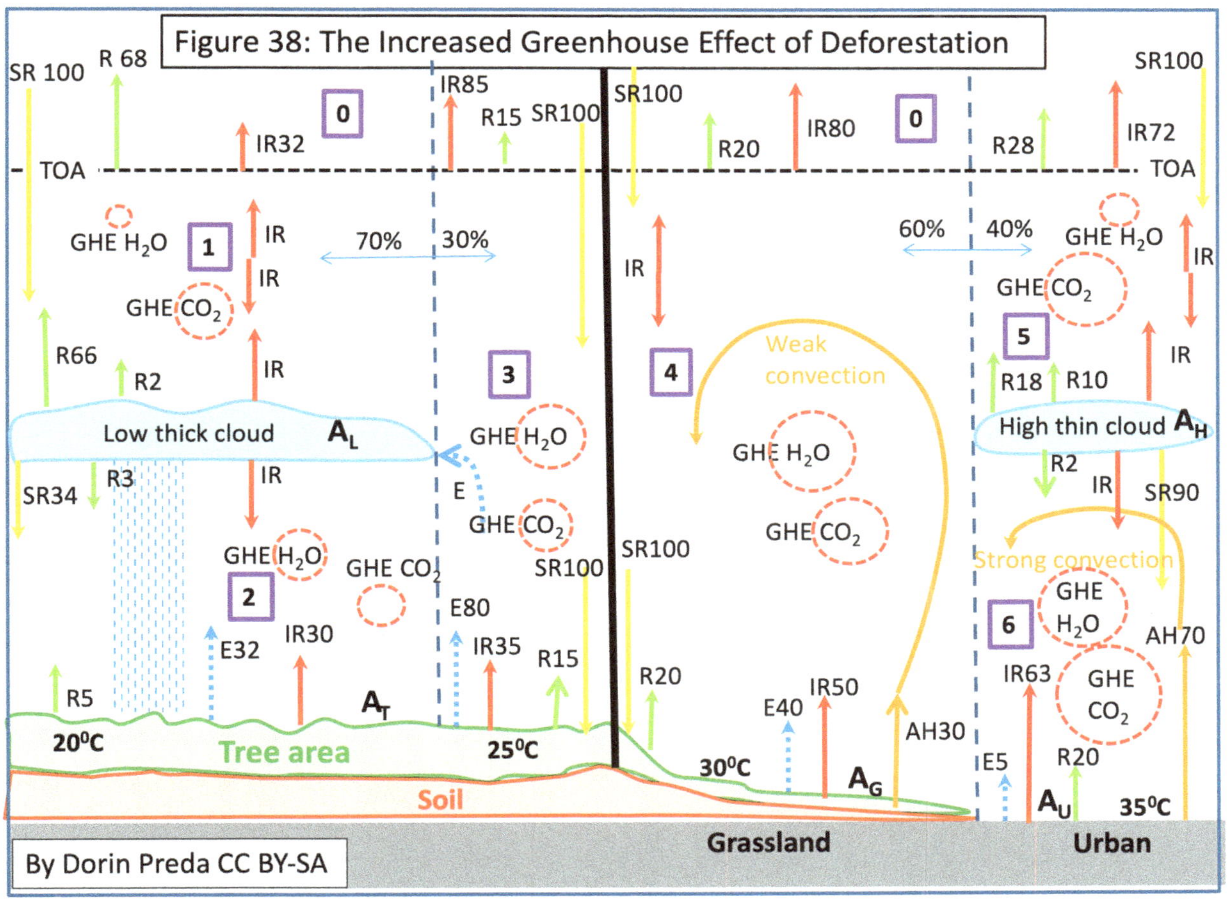

0) The average infrared emission at TOA above tree areas is (32% x 70%) + (85% x 30%) = 48%, while that above deforestation is (80% x 60%) + (72% x 40%) = 77%. Planetary temperatures are directly proportional to the infrared emission fluxes, therefore <u>planetary temperature increases with deforestation,</u> it does not decrease as stated by mainstream climatology.

However, as we said, we must also check the GHEs below TOA to make sure that the trend is similar.

1) Above the tree area clouds infrared radiation is not high, because only IR32 exits that area, thus GHE CO_2 is moderate, while the GHE H_2O is much smaller because there is 10 times less water at 7,000m than at 2,000m.
2) Evaporation of tree area water and their attraction of ocean water through humidity pumps create conditions of cloud condensation. Precipitation from the cloud delivers plenty of water to the tree area for evapotranspiration that uses all available heat. This heat pump extracts heat from the ground and delivers it without a significant GHE to the clouds. There is only 37% of solar radiation under the clouds of the forest, because of high cloud reflection, and evapotranspiration has also transformed almost all that energy in latent heat, thus there is no excess infrared emission or solar heating, resulting in very low GHE indicated by the low temperature in the tree area.
3) In the unclouded area of the tree area, most solar energy is transformed into latent heat of water and only a smaller part in infrared, thus the GHE is moderate both for CO_2 and H_2O. It has to be observed that, despite there is more than twice the solar energy here than under the cloud, the GHE is not proportionally large because most of the sensible heat is transformed into latent heat, which does not produce GHE if there is no condensation. It is to mention that, during the iterative calculations in the forested area, a quantity of vapours from the unclouded area was necessary to balance the cloud infrared emission.
4) In the unclouded area of the deforestation, the albedo is higher than that of the tree area, while the grassland evaporates less water than the tree area, thus, besides the infrared requirement, there is energy available for heating the air. The AH30 results in tall but weak convection currents (thermals), as observed in savannah and grasslands. This hot air emits additional infrared to the IR50 of the ground, and increases the GHE to large values, while near the ground maintaining the normal ratio 60% GHE H_2O and 40% GHE CO_2.
5) At the high altitude above the high thin clouds of the deforested area, there is little water, therefore GHE H_2O is rather small, despite the infrared radiation is high; however, it would be larger than above the tree area. The GHE CO_2 is quite large because of the large infrared.
6) Under the high thin cloud, both GHE CO_2 and GHE H_2O are large due to the intense infrared emissions of the high thin cloud and of the overheated ground (because lacks evaporation water). However, there is water as absolute humidity for the GHE, despite the relative

humidity is low due to the high temperature, as explained in Section 2.2.1. The strong thermals propel hot air with some humidity at height, where it forms high thin clouds.

The above is a conceptual simulation of GHE at low latitude areas. There is variability due to specific conditions in different places, but the GHE circles have proportional relative sizes, and show that air temperature and greenhouse effect are larger in deforested areas than in areas with large trees.

Despite having about 6 times more CO_2 than at present, Earth during the time of the dinosaurs did not overheat because it had many more forests which evaporated more water, keeping ground and air temperatures low, emitting less infrared radiation and producing a proportionally smaller greenhouse effect than without water (planet Venus is hot because it has CO_2 and very little water).

Currently, due to deforestations, there are fewer forests evaporating less water, resulting in more heat available to heat the ground, leading to increased infrared radiation and larger passive global warming through the greenhouse effect. Additionally, deforested and hotter ground heats a lot of air, generating large, hot convective currents that heat the troposphere, preventing formation of low thick clouds that could cool the Earth through the reflection of solar radiation. Instead, the powerful hot air convection in deserts shoots vapours too high, creating high thin clouds that do not condense into rain, but warm the planet through their strong greenhouse effect.

Hot deserts like the Sahara have relative humidity of 25%, which at 40^{0}C means a high content of around 10g/m^3 of water, producing a strong greenhouse effect due to high ground and air temperatures, while the air temperature is too high for those vapours to condense out of the air.

Trees regulate the local climate through the latent cooling processes of the water cycle, reducing the greenhouse effect. Therefore, it is crucial to stop deforestations and plant new trees.

2.6 Calculation of Deforestation Effects

Now we can calculate to check albedo interpretation that average forests would warm the Earth because they have an albedo of 0.15, while deforested land being a combination of dried grass (0.21) and bare soil (0.17) [https://en.wikipedia.org/wiki/Albedo], thus having an albedo of 0.19.

Following is an estimation of albedo and temperature effects in case of total Earth deforestation.

If the energy budget is done in an exclusive albedo style, cutting down all forests would increase Earth's reflectance by: 0.04 (difference in albedo between forests and deforested land) x 0.3 (proportion of forests on land) x 0.29 (proportion of land on Earth) = 0.0035. This means that about 0.35% less absorbed solar radiation would result in planetary cooling.

In reality, it is known that the hot deforested land would be covered by clouds (as seen in Figure 6) only for 10% of the time, while forests are known to be covered about 70% of the time by clouds that have an average albedo of 0.65 [66]. The 0.35 of solar radiation that passes through the clouds is reflected by the 0,15 albedo of forests, resulting in about 0.05 of solar radiation that passes again through clouds and adds to their 0.65 initial reflectivity, to result in a total reflectivity of clouds above forests of 0.7. Similarly, the cloud cover of the deforested land would have an albedo of 0.72.

Therefore, the average albedo of forests at present is: 0.7 (proportion of clouds on forests) x 0.7(albedo of clouds above forest) + 0.3 (forests without clouds) x 0.15 (albedo of forests) = 0.535

The average albedo of deforested land would be: 0.1 (proportion of clouds on deforested land) x 0.72 (albedo of clouds above deforested land) + 0.9 (deforested land without clouds) x 0.19 (albedo of deforested land) = 0.243. Therefore, deforestation would decrease, not increase albedo by 0.292.

The global warming of deforestation would be: 0.29 (decrease in albedo upon deforestation) x 0.3 (proportion of forests on land) x 0.29 (proportion of land on Earth) = 0.025 of solar radiation.

This 2.5% increase in solar radiation that is reaching the ground means strong warming, or an increase of the radiative flux of 2.5% x 1361W/m^2 / 4 (the planetary intercept rule) = 8.5W/m^2.

To the above we should add the real warming effect of deforestation because water is a second source of radiation (Figure 36) and helps emit more infrared at low temperature through the "Divided and Dispersed Emission" effect (DDE) of point 2.5.2. Its cooling effect similar to irrigation in Figure 1 of https://en.wikipedia.org/wiki/Radiative_forcing would be lost through deforestation.

For two adjacent large areas, one covered with forest and the other deforested, we can measure simultaneous above-ground air temperatures of 25^{0}C in the forest and 35^{0}C in the deforested area. Through the Stefan-Boltzmann equation they result in a differential radiative flux of deforestation: (514-450) W/m^2 = 64W/m^2 on land areas. We do not apply the ¼ Earth intercept rule because this is not solar radiation, but infrared radiation emitted night and day. However, we have to account for the proportion of forests on land (0.3) and land on Earth (0.29), resulting in a radiative flux of total deforestation due to the DDE effect of water of 5.6W/m²; this adds to the 8.5W/m² cloud effect of total deforestation, to result in a total warming of around 14W/m².

This is a very rough estimation and does not account for the more complex interplay of oceanic clouds especially due to their redistribution. The much larger quantity of water evaporated from the ocean will have to be discarded somewhere, and land features like mountain ranges are preferred. Therefore, there will be precipitation as floods on land even if all forests are cut down. Global temperature would indeed increase, and the temperature increase would be higher on land.

I conclude that deforestation does not cool the Earth, but results in strong planetary and above-ground air warming. Cutting down all remaining forests would heat the Earth around six times more than the carbon dioxide emissions since 1750 as stated by IPCC. Since the exclusive albedo interpretations of deforestation are proved wrong, deforestations and forest fires must be stopped, and new trees must be planted in their place.

2.6.1 Deforestation Since 1750, as shown in Figure 39 under "Land use", was reported by IPCC as producing a radiative cooling of – 0.16W/m² (the blue rectangle at Land reflectivity change).

A similar reasoning and calculation as in above Section 2.6 of the albedo effect of deforestation results in a decrease in albedo of 0.29.

FIGURE 39 : RADIATIVE FORCING

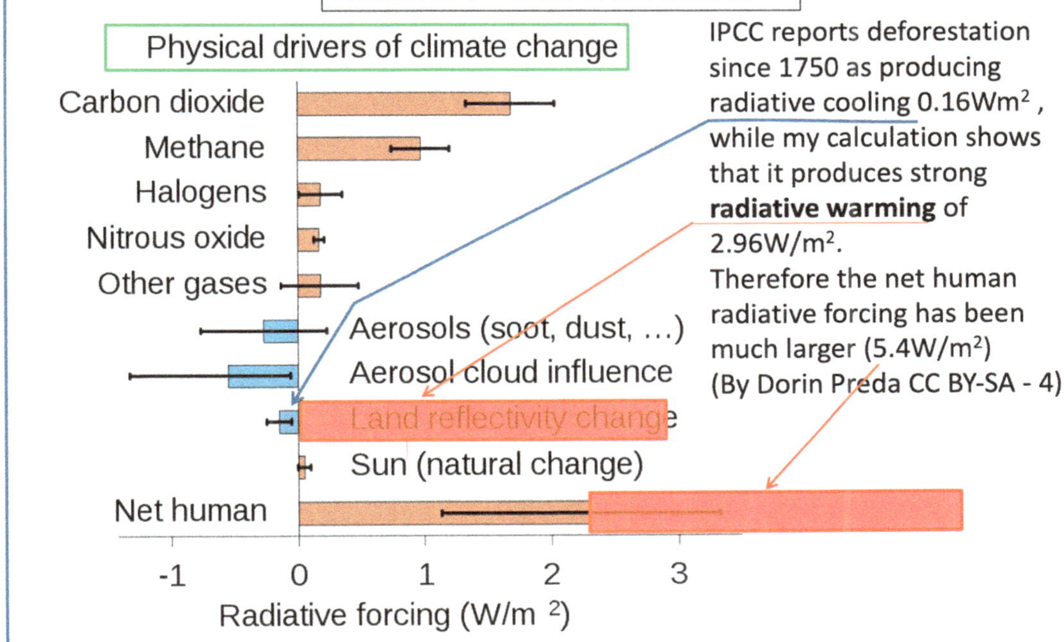

Addapted by Dorin Preda from Femke Nijsse and Eric Fisk, Original: IPCC 2013: Radiative Forcing 1750 – 2011 https://commons.wikimedia.org/w/index.php?curid=81034563

An area equal to 0.29 x 0.71 x 0.14 = 0.03 of Earth was forests transformed into grasslands since 1750 (https://ourworldindata.org/deforestation). I then calculate the radiative forcing of deforestation due to clouds effect since 1750 as: 0.03 x 0.29 x 1361W/m^2/4 = 2.96W/m^2.

Therefore, deforestation since 1750 did not cool, but has warmed the Earth as much as all other human causes combined (see Figure 39).

This was my third proof of warming due to deforestation contrary to the exclusive albedo theory.

2.6.2 Deforestation, Water and Radiative Forcing

There were many attempts, partially successful to quantify the role of land use change in climate like [138]. At present. climatology does not consider water (the most powerful energy carrying fluid of Earth) a driver of climate chance, but only a follower to CO_2, thus water is not included in the radiative forcing shown in Figure 39. However, I consider that anthropogenic water changes on Earth are a significant driver of climate change due to water dominance, and to the large-scale human action:

- hydrotechnical works (dam reservoirs, canals) significantly increase global evaporation
- all burning of fossil fuel and vegetation generates more water vapours than CO_2
- evaporative power plant cooling releases more water vapours than CO_2 and water through burning, while conductive cooling warms that water body and increases evaporation
- irrigation is practiced more and more as dryness increases at low latitudes.
- the future evaporative cooling (EC) as described in Chapter 7 below.

Additionally, human actions on vegetation indirectly produced a very large climate change through altering the water cycle, leading to the present climate emergency that required this book:

- extensive past deforestations reduced global evapotranspiration, altering the climate
- present reforestation at medium-high latitudes further shifts the climate

It is impossible to correctly characterise the climate through other parameters if water effect is not correctly assessed, because its large variations overshadow those of the smaller parameters. Computer simulation problems might be caused by not treating water (the main player) adequately.

Carbon dioxide is generated and used both by Nature and people. Similarly, water is generated and used both through natural and human activity. It would thus be logical to also include anthropogenic-induced water variations in the radiative forcing (RF).

An additional reason to include water amongst the main drivers of climate change is that it can directly influence the planetary temperature through clouds, while CO_2 cannot do more than to increase the above ground air temperature through the GHE (water also can do that).

Usually, not only does one parameter vary, but others are also interconnected. For example, deforestation increases albedo, but also reduces local water availability. Normally, water absorbs heat from the surface (more than air can), cools the air above ground including the benchmark of global warming (meteorological stations), and releases that heat at cloud altitude, above the densest greenhouse layers. Most of that heat energy penetrates more easily upwards to space because there are fewer greenhouse gases (especially water) at altitude, while the downward emission is absorbed by the denser greenhouse layers. Thus, through evaporation, and divided and dispersed emission (DDE), water has a real cooling effect on the global warming benchmark, equivalent to a negative radiative forcing. Conversely, by reducing water availability, deforestation has a measurable global warming effect, as seen in its energy budget shown in [135].

The global radiative forcing balance is calculated at the top of the troposphere or stratosphere. As expressed in 2.1.1, climatology is concerned that water enhances the greenhouse effect in the troposphere. However, water also has cooling effects, extracting heat from the area where the above-ground air temperature is measured and delivering that heat at high altitude, thus it is substantially changing the distribution of heat in the atmosphere, just like the carbon gases. Therefore, since the tropospheric greenhouse warming is considered a forcing, the tropospheric water-cooling effects of divided and dispersed emission (DDE) should also be included in the global radiative forcing.

At point 2.6 it was assumed that the ground radiative flux deforestation due to DDE would be around 64W/m^2 on land areas. From (https://ourworldindata.org/deforestation) we know that 0.03 of Earth has been deforested since 1750. This results in the radiative flux of deforestation due to DDE since 1750 being 1.9W/m^2. Calculating the radiative forcing is beyond this study.

The total radiative flux of deforestation since 1750 is calculated as 3 W/m^2 (cloud effect of point 2.6.1) + 1.9W/m^2 (DDE water effect) = 4.9 W/m^2. Thus, the warming effect of deforestation since 1750 has been about 5 W/m^2. The radiative forcing would be larger than the radiative flux.

Therefore, considering its DDE effect (section 2.5.2), water has a much larger cooling effect than its greenhouse effect.

This is represented in Figure 39 as the reddish highlighting above "Land reflectivity change", resulting in doubling the total radiative forcing from human activity since 1750. This significant warming justifies that we should now be in a glaciation, as other authors assumed previously. [107]

2.6.3 The Radiative Forcing of Global Cloud Redistribution

Above it was estimated the effect of overall deforestation, by decreasing the cloud cover as global percentage, resulting in more solar radiation reaching the ground all over the Earth and overheating the atmosphere.

An added change in climate, is the latitudinal redistribution of clouds, which is due to the fact that humans first lived at low latitudes and changed those lands which receive the most solar radiation.

As shown in Figure 40A, by transforming Sahara in a desert through devegetation (as explained in the next sections), clouds cannot form above Sahara anymore, but those ocean vapours have to condense at higher latitudes, where they meet colder air. In a basic representation, we see that solar radiation of same spatial cross section (a) is more intensely distributed on the tropical ground (c) than on the temperate area (d), thus if we move the same dotted cloud cover from low latitudes (a) to high latitudes (b), we block less solar radiation, and create a heating of (a - b). Obviously, the real processes are more complicated, but they obey this basic principle and have similar results.

FIGURE 40 A: The Cloud Redistribution Effect

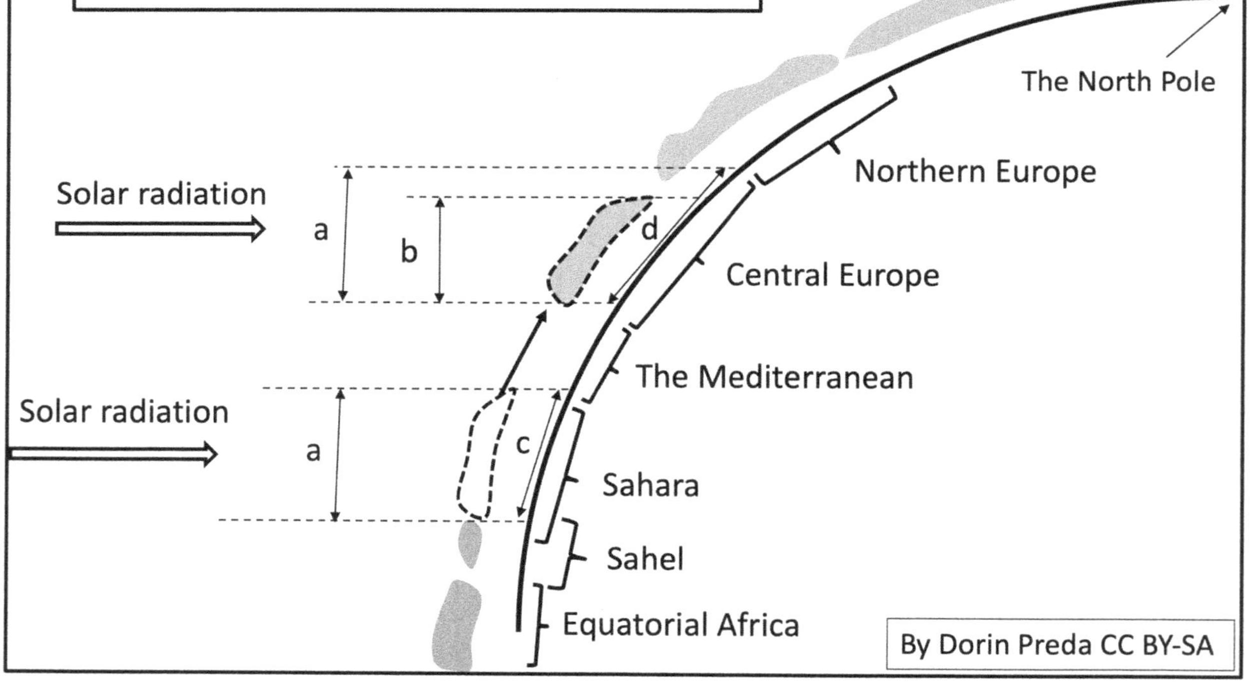

This process happened in the small latitudes all over the world and at present deforestation intensified in the equatorial South America and Central Africa, threatening to reduce cloud cover in the entire low latitudinal belt of the world. The Earth receives this way more and more net solar radiation, rising temperatures and sending the clouds and excessive precipitation to medium latitudes, while also overheating the polar regions.

We can estimate the global heating contribution of this effect by the average shift of clouds to higher latitude. A rough formula would be: Difference = Cloud area x (1/cos λ - 1), where λ is the difference in latitude with the old place. Therefore, if Sahara (9 x10^6km^2) was forested, and had a 70% cloud cover, and those clouds were moved 20 degrees North over Central Europe, then the reduction in cloud shade area due to latitudinal shift would have an area of 404,300km^2. Multiplication by the solar radiation of 340W/m^2 and division by ¼ of the area of the Earth results in a radiative forcing of around 1W/m^2, which is a significant warming and confirms that the cloud latitudinal shift effect must be considered in the assessment of radiative forcing of deforestation.

There is abundant proof of changes in clouds distribution, for example [139].

2.7 Ground – Space and the Water Solution

A. The climate and the benchmark of global warming (the meteorological stations) are in the lowest half of the troposphere, where human life and interests are. The most accurate meteorological and climate measurements continue to be made directly, but since the advent of space instruments, the largest data source is remote sensing, increasing a lot data density and area.

Together with the weight of measurements, modelling also changed to an outside Earth frame (like the assessment of Earth temperature through the planetary temperature, RF and albedo). Unfortunately, as shown in this chapter, the relationship between the space modelling frame and the ground measured one is still unreliable, despite computer simulations are improving.

Greenhouse gases disconnect albedo from temperature by increasing temperature (section 2.3.2), while water disconnects temperature from albedo by decreasing temperature (section 2.5.3).

As seen in Figure 40B, the planet equilibrium temperature T_P only depends on albedo, while the ground temperature T_G is linked to internal radiative fluxes; thus, the two are varying differently:

$f(SR - SR(A_G)) = T_P \neq T_G = f(IR_0 > (SR - SR(A_G))$

Additionally, when ground temperature increases (like now, due to increasing CO_2), the planetary temperature decreases, confirming that the planetary temperature model is not consistent.

Ground measurements and conceptual simulations are more reliable than space models.

B. On the left in Figure 40B is depicted the simplest atmosphere that illustrates our present problem, comprising only air and CO_2, that is the main driver of global warming. We investigate at equilibrium (steady state) first the atmosphere without water, because it is stated by mainstream climatology (Section 2.1.1) that water vapours and clouds would enhance global warming.

Present average land albedo is AG = 0.25. Therefore, R1 = SR(AG) = 0.25SR, while the part of shortwave absorbed by the ground is 0.75SR. That power is radiated in infrared to the atmosphere, where it is emitted by CO2 half upwards and half downwards followed by more absorption and re-emission until the ground emits IR0 = 0.75SR + GHE1, where GHE1 is the greenhouse effect.

On Earth: GHE1 = 0.4 IR0 = 0.4 x 0.75SR = 0.3SR, and IR0 = 0.75SR + GHE1 = 1.05SR.

At equilibrium, IR3 = SR – SR(AG) = 0.75SR, and TP1 = (IR3/σ) ^ - 4 = (0.75SR/σ) ^ - 4.

FIGURE 40 B: The Space – Ground Relation and the Water Solution

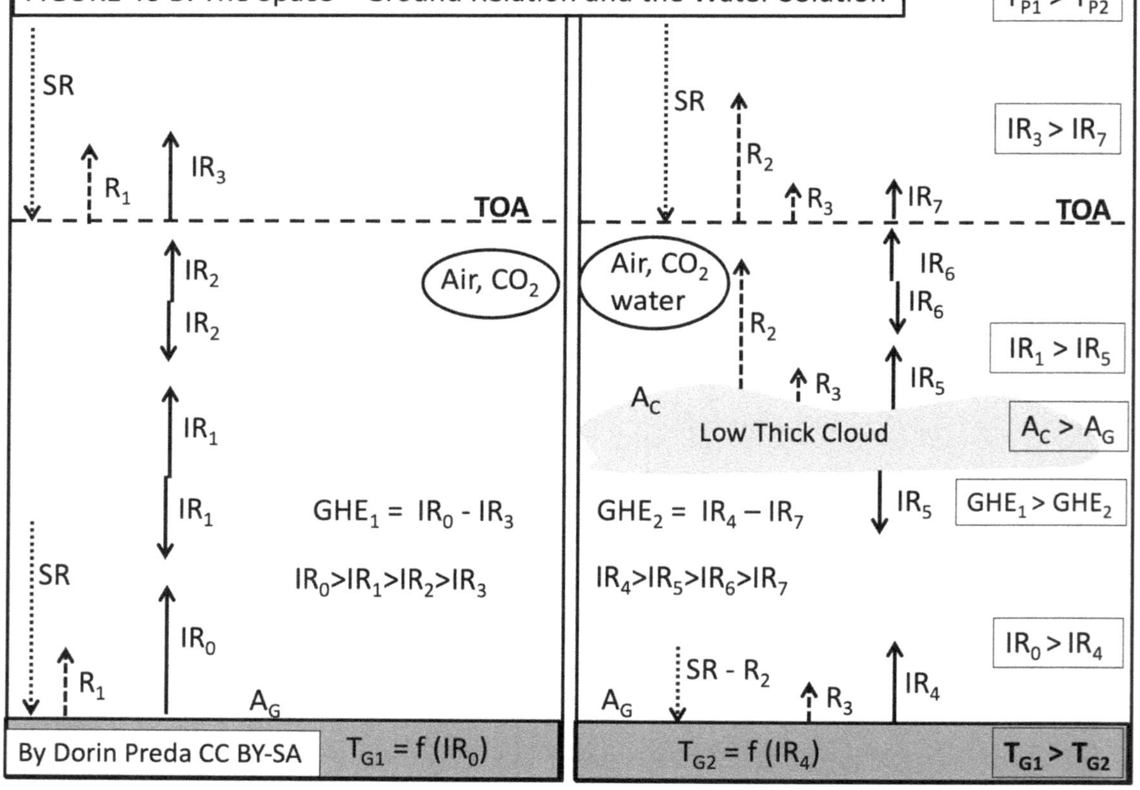

Let us test what happens if we add water to the Earth atmosphere. Clouds with an albedo of A_C = 0.65 will form, reflecting R_2 = 0.65SR and allowing to pass through 0.35SR of which 25% is reflected back up by the ground as R_3 = SR (1 – A_C) (A_G) = 0.09SR and produces a total loss of shortwave through the TOA of 0.74SR. As result, at equilibrium the longwave escape to cosmic space becomes: IR_7 = SR – 0.74SR = 0.26SR that causes planetary equilibrium temperature: T_{P2} = (IR_{G7}/σ) ^ - 4 = (0.26SR/σ) ^ - 4.

If SR = (1360W/m²)/4 = 340W/m², then T_{P1} = -14⁰C, while T_{P2} = - 75⁰C, indicating a strong global cooling due to cloud reflection. As discussed, planetary temperatures are irreal and cannot be fully trusted, so we have to check and make sure that there would also be real cooling in the atmosphere, and especially just above the ground, where our interests are.

The greenhouse effect under the cloud becomes:

GHE_2 = 0.4(IR4) = 0.4(SR - R2 - R3) = 0.4 x 0.26SR = 0.104SR, thus GHE is smaller under the clouds.

$IR_4 = IR_7 + GHE_2$ = 0.26SR + 0.104SR = 0.364SR = 124W

IR_O = 1.05SR = 357W > IR_4

T_{G1} = (IR_O/σ) ^ - 4 = 8⁰C, while T_{G2} = (IR_4/σ) ^ - 4 = -57⁰C, thus clouds decrease ground temperature.

To these radiative effects, the latent ones add up, together with the DDE effect, as shown in the previous Sections, to further increase the beneficial role of water in Earth's climate.

It results that water also cools in the interval ground-clouds, where our interests are.

Despite climatology models consider that water and trees would increase global warming (Section 2.1), in this Chapter it was demonstrated that trees actively use water to substantially cool at local and at global scale, and that they regulate Earth's climate, as they did for eons.

3 CHAPTER THREE: Deforestation, Desertification and Global Warming

Studies from the 1970s such as [57] and [58] suggest that prehistoric deforestation increased albedo, cooling the Earth with significant implications for climate change. In contrast, recent climatology studies over the past 20 years have indicated warming due to the greenhouse gases effect. More recent studies, like [28] have shown that land clearing for agriculture dating back 8,000 years has led to carbon dioxide emissions and global warming. However, the ocean absorbs about 2GT of CO_2 per year, thus those gradual CO_2 emissions alone could not have caused global warming, while cloud and evapotranspiration change could, as modelled in [134], but disputed subsequently.

As will be shown and justified further, human-induced deforestation has indeed produced global warming since pre-historic times, through combined land cover changes at global scale.

3.1 The Most Powerful Tool

Around 300 million years ago, during the Carboniferous and Permian periods, just before the Mesozoic (the time of dinosaurs), the climate was warm and humid, with forests covering almost all the land due to intense plant growth allowed by the much higher concentration of CO_2 in the air than at present (see Figure 20). It is logical to assume that those primitive plants (gymnosperms) gradually absorbed the carbon dioxide in the atmosphere up to the level they could not function properly anymore. There are theories [115] of nearly reaching a glaciation due to the reduction in CO_2, but that seems exaggerated. However, forests decreased in size and density, as observed in geological records from Jurassic to the end of Cretaceous, inducing also a decrease in the size of herbivores. The decrease in forest size and density probably reduced the quantity of water on the continents, which is proven by the much-reduced number of coal deposits formed at the end of the Mesozoic, and by the records of climate dryness.

Approximately 125 million years ago, long before humans existed, that climate dryness coupled with the reduced CO_2 levels induced the disappearance of some plant species, while new species that required less water and CO_2 emerged and spread worldwide. This coincided with a great animal extinction, possibly due to the known asteroid impact. However, I assume that the drop in CO_2 produced a significant reduction in early plants, that contributed to the large animal extinctions. Then the modern plants that thrived in less CO_2 started their Earth conquest. Much later, about 2.5 million years ago humans evolved, and after 1 more million years they invented artificial fire.

Homo Sapiens left Africa in a few successive waves since around 70,000 years ago. Initially, they stayed in the intertropical area likely due to the cold weather during the glaciation period. They first settled in the Middle East, India, Southeast Asia and Australia. By 12,000 years ago, humans have populated the entire Earth (see Figure 41A).

Analysing the map, it can be calculated that their rate of expansion was generally less than 1km/year, suggesting that war and competition were not the driving forces behind their movement. This slow rate of spread of hunter-gatherers means that they just migrated due to population growth. This gradual spread aligns with the idea of conquering a rich, untouched environment that provided resources easily without competition. It is possible that their population was much larger than currently believed, as reasons will be given for the lack of prehistoric human traces.

Fire, initially obtained from natural sources was one of the earliest tools used by humans and undoubtedly the most powerful. While they may have used fire earlier, there is concrete evidence of humans use of fire approximately 1.5 million years ago. Numerous studies, like [140], show the role of fire in shaping ecosystems and its historical and ongoing impact on climate.

FIGURE 41A Spread of Human Populations (thousands of years before present)

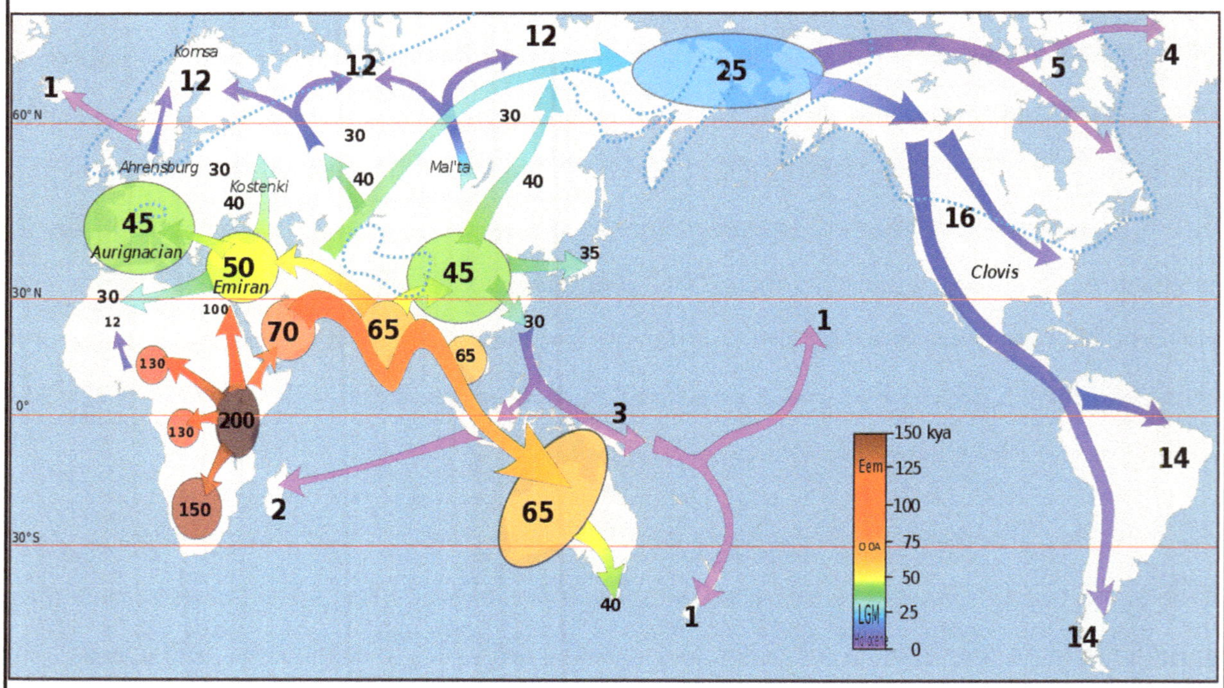

By User:Dbachmann - Own work;Map: File:World map blank shorelines.svg (the shoreline map is {{PD-USGov-USGS}}, derived from a mercator svg map posted in 2005, extracted from an original file released by pubs.usgs.gov), CC BY-SA 4.0, https://commons.wikimedia.org/w/index.php?curid=68921860

Pristine forests had dense undergrowth, making it challenging for human hunters to run through. Additionally, humans observed that herbivores gathered to feed on the grasses growing in forest clearances. Therefore, setting fire to forests appeared logical to create open spaces in the forests.

Humans and their precursors have been using stone axes for around three million years, likely utilizing them not only for hunting, but probably for creating pathways in the forest, making shelters, fences, traps and gathering firewood.

Humans are extremely adaptable and resilient beings, occupying diverse environments, including the Arctic regions, where fire was crucial for survival.

3.2 The Hunter-Gatherers

"Palaeolithic humans preferred sparsely wooded areas and avoided dense forest cover" [14]. They observed small tree clearings produced by natural fires, and imitated them. Although they attempted to start fires on greenwood, they lacked the power of lightning to burn a tree sharply, so they resorted to dead and dried wood which burn easily. They probably just peeled off the trees' bark with their stone axes, allowing the trees dry off before starting a fire. Another tree clearing method I imagine could have been to pile up a lot of dry wood around the trunk of a living large tree to set it alight this way. We must not underestimate the so-called "primitive people" who were in fact as intelligent and knew how to make the most with the means they had.

In those tree clearings, richer grass attracted herbivores, making hunting easier. Human populations that maintained the same hunting territory also managed it, by preventing with periodic fires new tree growth in the clearings they created, as the aboriginal peoples now do.

Grazing animals and especially browsers like goats, giraffes and camels also instinctively know that tree shoots and saplings are not only comestible, but eating them will enlarge their grazing territory by pushing the trees back. Such animals cannot put down a tree or set it alight, but time will; after a tree dies off, without saplings no other tree takes its place, so the forest slowly retreats. This way, grazing animals can make up savannas from forests, while if grazers numbers are diminished for example by carnivores, forests tend to expand. There is this balance which allowed both forests and grasslands survive, because they are at competition over land and sunshine. Natural balance allows all forms of life coexist: trees, grasses, grazers and carnivorous animals.

Humans cannot digest cellulose; hence trees and grasses are not directly useful to them. However, they can pray on grazing animals, which depend on grasslands. This way, forests gained a powerful enemy, and the area of land occupied by forests has continuously diminished since primitive people appeared. Humans are more effective at forest destruction than grazing animals, because by repeated fire lighting, they not only prevent saplings from taking over the open land, but they can attack the forest itself. Most trees were useless to primitive people, while grasslands were feeding their grazing pray.

Obviously, humans observed these facts and acted accordingly, especially after they occupied all land and stopped their territorial expansion. They had to increase the food production of the territory they could maintain against the other groups. Due to inter-groups competition and possibly wars, the survival chance of each group was directly proportional to their number of individuals, which in turn depended on their harvest of grazers, that in turn depended on grazing area. Clearing out forests was therefore of vital importance for the individuals of each human group, and probably an assiduous and programmatic work.

Evidence shows that the repeated use of fire by pre-historic and proto-historic people led to a significant reduction in forest cover [48],[49]. Some studies suggest this practice dates back 20,000 years ago in Europe [50], and 60,000 years for Australia.

While direct evidence in areas out of Africa supports these claims, we can attribute the lack of direct proof in Southern Africa to the erosion of soil and vegetation. Despite this, it is believed that humans mastered fire 1.5 million years ago, and likely invented early the slash and burn method, reshaping Africa in the process. Total soil erosion of these very old deforestation areas has removed all proof that humas produced the South African deserts, while proof survived of the more recent deforestation and desertification of Northern Africa.

The human cause of deforestation is proven by the natural expansion of forests into savannas when human activity ceases in that area.

Diminishing forests and expansion of grasslands by human fires led to decreased evapotranspiration, resulting in less rainfall which in turn further decreased tree areas. Additionally, the clearings allowed more wind around trees, drying them out and making them more susceptible to fires. This gradual transformation of dense forest into sparse forest, then shrub country and savannah altered the climate and created more drought through feedback loops. The connection between forests and rainfall will be further explored.

Continuous use of fire has changed the landscapes all over the world, from Northern America where indigenous people created prairies for buffalos, to the ritualistic tool called "the fire stick" practiced by aboriginals in Australia since around 60,000 years ago and continued to present [140].

The "fire stick" culture demonstrates the long-standing practice of the hunter-gatherers of using fire to transform forests into savannahs and subsequently into deserts. This traditional, repeated burning of the land has had a large impact on the environment, leading to the desertification of once rainforests.

3.3 The agriculturalists

"Mesolithic foragers cleared the forests to make room for game, while subsequent Neolithic populations used fire extensively to make room for agriculture" [13]. About 10,000 years ago humans invented agriculture, which involved the domestication and sustained propagation of the main plant and animal species. They already knew that the easiest way to clear a patch of forest was through fire, noticing that the burnt terrain became more productive for pastures or agriculture. This led to the repeated practice of slash-and-burn which continues today and has increased in frequency and area [141].

The cooking of food likely originated from consuming animals and plants burnt by forest fires.

In the Middle East, especially in the Fertile Crescent, permanent settlements like Catalhoyuk in Turkey emerged around 9,000BC, with agriculturalists growing plants and domesticating animals before 10,000BC. Despite the current desert landscape, the Middle East from Libya to Turkey was once covered in lush vegetation that sustained a large variety of wildlife, including antelopes, water buffalos, elephants, lions and birds as depicted in artifacts.

The presence of elephants in the region suggests high vegetation productivity. Elephants lived in Syria for 3million years until Roman times, when they were used as war elephants. Hannibal even attacked Rome with war elephants from Carthage in present-day Tunisia. Historical records indicate that over 2,000 elephants were killed in Roman circus shows during the reign of Augustus around 20AD. These historical facts demonstrate the past abundance of vegetation in the Middle East and North Africa, which never had a cold climate, neither suffered intense desertification like at present. Humans produced total loss of plants, animals and soils, resulting in irreversible desertification.

Agriculture has a destructive impact on soil, primarily through tillage, leading to erosion rates that are about one hundred times faster than natural soil formation rates. As a result, soils become exhausted relatively quickly, forcing population to migrate elsewhere, as seen in various historical migrations. One exception is the extremely thick loess in China, which was eroded over thousands of years, but its remains are still productive at present.

Although Catalhoyuk in Anatolia was highly productive around 7,500BC, it was abandoned around 4,000BC due to soil productivity loss, marking the beginning of desertification in the Middle East. This indicates the resilience of Catalhoyuk's soil, likely due to its thickness, which lasted for 5,000 years, probably with good care. Soils in modern day agriculture areas cannot last 500 years.

3.3.1 Desertification

Therefore around 4,000BC, Middle East desertification started due to excessive agriculture around the Fertile Crescent [142]. This desert continued to expand and became permanent about 1000 years ago due to the environmental pressure of the great ancient civilisations that thrived there.

Some proto-Egyptians depleted their former lands to the East of the Red Sea and retreated to the Nile Valley [48, 49]. Other well-known civilisations emerged one after the other in the Middle East, flourished, depleted their lands, and eventually disappeared, leaving deserts in their wake. In Mesopotamia, there were the Sumerians, Akkadians, Babylonians, Assyrians, Mitanni, and Hittites, followed by the Persians. To the south were the Semitic tribes and on the Mediterranean coast were the Phoenicians and the Greeks.

Only the Egyptian civilisation was spared from soil loss and desertification due to the regular flooding of the Nile which brought water and fresh sediments from Central Africa, enriching the soil and ensuring beautiful crops. The annual flooding of the Nile also allowed people to engage in activities like monument building.

Northern Africa also had abundant vegetation a few thousand years ago, with patches of savannah and occasional deserts. The Sahara was home to a variety of wildlife, including giraffes, antelopes, lions, and elephants. Evidence suggests that there was a chain of large freshwater lakes in the region less than 7,000 years ago, extending from the Mediterranean Sea to the Congo Basin [33], Lake Mega Chad being as large as the present Caspian Sea. A major river called Tamanrasset flowed east to west in the Sahara up to 6,000years ago, sustaining rich vegetation and fauna.

The elephants of North Africa were a distinct sub-species than had lived in the region for much longer time than the last Ice Age (100,000 years). Interestingly, their name was "the North African forest elephant", certifying that Sahara might have had some dryland patches, but it mostly was forests and savannah. Cave paintings depict scenes with gazelles, giraffes, crocodiles, and people swimming in deep lakes, indicating a humid period in central Sahara that ended about 6,000 years ago [46]. Despite this, the Northern edge of the Sahara remained fertile and rich in vegetation until Roman times [150].

Further evidence that the northern Sahara was not always a desert comes from the fact that after the Roman conquest around 200BC, Carthage was known as "the breadbasket of the Roman Empire" for its abundant grain crops sent to Rome. However, after centuries of intensive agriculture, that area also lost its soils, its vegetation and rainfall. It will be shown in the next chapter that the Romans initiated desertification of Sahara by demanding high grain tribute from Carthage and from Egypt which at that time was also cultivating Eastern Libya.

<u>The decline of agriculture due to land over-exploitation, and subsequent desertification in the Middle East and Northern Africa may have contributed to the fall of the Western Roman Empire.</u>

3.4 The pastoralists

Other human populations have transitioned from hunter-gatherers to more pastoral lifestyle while still maintaining the nomadic way of life due to their reliance on their animals. The lower vegetation density in pastures compared to forests requires larger areas of land. <u>Some of the earliest pastoralist societies occupied the Sahara region, where they transformed the vegetation cover through extreme pastoralism, leading to a shift in the local climate towards permanent dryness about 6,000 years ago [46].</u>

Other ancient pastoral regions include the former agricultural lands of the Fertile Crescent, which were repurposed after land depletion through intensive farming.

In the Gobi area, the land consisted of medium-high altitude (1000-2500m) pastures with thin soil, more trees than at present, and a climate characterised by cold nights and mornings alternating with hot sunny days. The fragile soil was composed of altered bedrock and cosmic dust held together by grass and small tree roots on a rocky foundation. Vegetation relied more on dew than scarce precipitation for survival. Herbivores naturally migrated to the area for grazing, followed by human hunters who subsequently domesticated them, leading to the emergence of the Proto-Mongols and the Xiongnu pastoralists.

In a natural environment, herbivores cannot deplete all vegetation, as it would force them to move, allowing pastures to recover. However, humans supplemented their herds during times of low vegetation, and exploited the pastures intensively, causing irreversible damage. Herders cut down locally adapted trees for fire and to collect water from the dew in their bark, resulting in soil erosion, similar to agricultural degradation and desertification [151].

It is suggested [51] that humans altered the grasslands of the Eurasian steppe by replacing wild herbivores with domesticated animals not adapted to wild plants. Human management of herds, fodder and water resources resulted in soils depletion, excessive nitration and erosion. The extraction of firewood to support growing human populations and to increase pastures, leading to large migrations, was a primary cause of desertification.

The Gobi Desert and the Eurasian steppe are human-made deserts, started about 2,000 years ago through excessive pastoralism, resulting in three major waves of human migration with significant historical implications.

Concurrently, as Gobi pastures lost productivity, the Proto-Mongols and Xiongnu (probably the Proto-Huns) frequently raided the Chinese Empire and pillaged its resources for centuries, leading to the construction of the Great Wall of China.

There is a vast East-West mountainous ridge stretching thousands of kilometres from the Quilian Mountains in Western China to central Turkey. To the North of this ridge, encompassing most of Eastern Europe and Asia, there is a vast northward sloping region, which before humans was mostly forest with rich grassland valleys like the Taklamakan, Margo, Lut, Kavir, Qizilqum, Betpaqdala, Karaqum, etc. I assert that hunter-gatherers initially hunted game in these valleys, then they used these valleys for pastoralism. They exploited the surrounding forests initially for timber and firewood, but gradually cleared them out to make room for more pastures. The entire region of grassland valleys surrounded by forests was changed into the Eurasian Steppe by 3,000 BC. The pastoralists had by now domesticated pigs, cattle, sheep and horses. Proto-Mongols resided high on the Gobi Plateau, while the Proto-Huns lived northwards from them. Indo-European peoples inhabited areas north of the Black and Caspian Seas, and the Proto-Germanic peoples resided north and east of the Eurasian Steppe. While tribes engaged in trade, significant inter-mixing of populations was limited due to their differences in lifestyle and culture.

Starting around 2,000 BC a historical cycle began on the Eurasian steppe, characterised by tribes overgrazing pastures until they could no longer support them, hence they migrated westwards towards Europe because there are the mountains in the south, other warring peoples in the east and frozen Siberian forests in the north.

Between 2,000 – 1,000 BC the Indo-Europeans expanded both towards the West and the East. In the East, they occupied the large Taklamakan mountainous depression (1,000km x 350km) where the much thicker soil had accumulated, supporting lush vegetation of tremendous and constant productivity. The name of the desert comes from the Turki language "taqlar makan", meaning "the place of ruins"[26]. The pastures of Taklamakan were so rich that the Indo-Europeans did not have to move their herds and follow them, so they built cities, and created a proto-culture since 1,800 BC, from which artifacts and mummies remained. However, overgrazing of Taklamakan and steppe pastures forced the Indo-Europeans to migrate between 1,500 - 700 BC westward into Europe, where they mixed with the Proto-Europeans.

The following 1,000 years the vegetation of the Euro Asiatic steppe recovered and attracted the Proto-Huns who advanced westward. Unlike the Indo-Europeans, the new inhabitants were fierce warriors who relied on horses for power and sustenance. The Taklamakan cities transitioned to an Asiatic population, while the Taklamakan pastures declined and desertification became permanent.

As grasslands recovered between the Caspian and Black Seas after the departure of Indo-Europeans, the Huns and Germanic populations grew. However, excessive pressure on the pastures led to their collapse once again. The Germanic peoples first migrated into Europe after the split of the Eastern and Western Roman Empires. As the Hun population grew and increased its pressure, between 300 AD and 600 AD a second wave of Germanic peoples entered and conquered Europe alongside the Huns, resulting in the fall of the Western Roman Empire.

In the meantime, from the last centuries BC to the first centuries AD, the Chinese resistance remained strong, leading the Mongols expanding westwards from the Gobi area and occupying the Taklamakan. After the second wave of migrators departed, the Eurasian steppe slowly began to recover. However, the process was hindered by the depleted soils and the rocky outcrops which were the first signs of desertification. As some pastures began to recover, different Mongol tribes expanded westwards, eventually occupying the entire Eurasian steppe. This vast territory allowed the Mongols to increase their power and refine their warrior culture.

Around 1,200AD the Mongol hordes were united by Genghis Khan and, with significant population and military pressure began a universal expansion. They firstly attempted to conquer Europe, but finding too large resistance in Eastern Europeans, they turned and found success in other directions. In the South they conquered the Middle East, Egypt, and established a Muslim state in northwest India, setting the conditions for the later separation of Pakistan from India. In the north, the Mongols conquered Russia, while in the South-East they reached Indonesia. In the East they made attempts to conquer Japan. China managed to resist them until 1279, when a grandson of Genghis Khan established the Yuan Dynasty for less than 100 years.

Unfortunately, just 200 years after the Mongols left the Eurasian steppe, the Turkish people followed suit. Merging with the remnants of the Mongols, they attacked Europe and successfully conquered and destroyed the Eastern Roman Empire (Byzantium). This led to hundreds of years of wars, hardship and loss of progress for the Eastern European peoples, who nevertheless managed to prevent the Turks from pillaging Western Europe.

Over time, the once forested and rich pasture land of the vast Eurasian steppe was depleted through excessive pastoralism, resulting in bare rock patched with sandy deserts instead of the former fertile valleys: The Taklamakan Desert, Margo Desert, Lut Desert, Kavir Desert, Qizilqum Desert, Betpaqdala Desert, Garaqum Desert.

The forests and rich pastures of 10,000 years ago decayed to patches of forests and grassland 2,000 years ago, then turned into patches of rocky desert and poor grassland 1,000 years ago, ultimately becoming sandy desert valleys within a large rocky flat 500 years ago. Thousands of years of mis-management of this vast area (the largest land mass of Earth), has resulted in deforestation and colossal loss of fertile soils, easily observable from satellite as a continuous rocky plateau with sandy valleys.

Peoples who centuries ago raised cattle, sheep and horses, now mainly raise goats, donkeys and camels which can feed on the much scarcer vegetation patches amongst the vast expanses of rocky or sandy desert of the Eurasian steppe. The greatest loss was the eroded soil, because without it plants cannot easily re-occupy those areas. The remaining tribes from their great migrations now make up present-day populations of Kazakhs, and Turkmen who occupy the steppe. The Mongols primarily inhabit the Gobi, where they continue their pastoral practices to this day, exploiting the steppe to its maximum capacity. As pastures recover slightly, herders increase animal numbers to elevate their social status. However, this leads to the land collapsing again after a few years, causing a cycle of depleted grasslands. There are historical records of Genghis Khan hiding armies of a few thousand warriors and their horses in the 1.5m tall grass and bushes of the Gobi. To this day, old herders remember that 50 years ago the grass was growing to 1m high when pastoralism was forbidden. In contrast, grass is now overgrazed, because many Mongols returned to pastoralism and endanger the rare desert plants [25].

Besides the Gobi, the large Taklamakan desert generates massive sand-storms which reach North America, and influences the climate through the particular minerals of its dust, favouring the creation of ice crystals of the undesirable high thin clouds.

Without grass, but trampled by herds in that strong wind environment, the soil turns to dust carried away as dust storms. Under the natural high alteration conditions, the rocks transform to sand which is pushed eastwards by the dominant winds. By the time, more sand has accumulated in the Gobi as dunes which slowly but implacably moved East over the Chinese fields. Now the dunes are about 75km West from Beijing, and the desert threatens the city, many times engulfing it in dust. Unfortunately, the sand follows the dust.

There is uncertainty on the original extent of world forests, because after forests were cut down or burned, soils eroded and the remaining scrub-land or savannah attracted less precipitation, which in turn conduced to erosion and disappearance of relicts or proof. The resulting desertification has generated the extreme dry climate, which makes some scientists believe that there was never a forest in that place, and it could not have existed in those rocky and dry climatic conditions.

In parallel with other authors [117][143][145], I think that the Eurasian steppe was once mostly covered with trees of which only the Siberian forests now remain. The geographical setting has the East-West ridge which hampers air circulation and collects rainwater which then flows northwards to evenly feed a huge flat area of potential vegetation, including forests. Trees did cover the land because they could, while people only remember the steppe of the past 500 years.

The lack of trees on the Eurasian steppe is simply explainable: most of the trees were cut for firewood and huts and the remaining bushes were slashed and burnt to make room for pastures. No significant forests remained, but it would have been impossible for the past populations of millions to live without firewood for millennia in that cold area. Subsequently, the herds (especially the goats) grazed away any new tree shoots, preventing forest comeback.

It is easier to keep grazing animals than do agriculture, therefore many national traditions went the pastoralism way and continued to practice it even when they destroyed the land completely, without thinking that meat eating is a luxury that requires much larger area than cropping.

It is a sorry paradox that the most inefficient type of agriculture is practiced in the poorest lands.

3.4.1 Goats and Desertification

The issue of goats is critical, was described in [147], but cannot be stressed enough, as goats can feed on anything and thus are highly valued by herders. Throughout human history, unobservant and careless individuals have benefited from the extreme efficiency of goats:

- Goats will eat any vegetable matter when they are hungry, including the bark of trees, this way killing them
- They climb trees and consume leaves and new branches, hindering tree growth.
- They eat new tree shoots emerging from the ground, preventing forest regeneration.
- Their pointed hooves damage grass and its roots
- They not only graze on grass, but also eat its roots, ruining pastures
- Due to their effectiveness as grazers, versatility and lack of pretentions, they are utilized by the poorest herders in areas most vulnerable to desertification, which never recover
- The goat population can be directly correlated to desertification.

In areas more than 50% on the path to desertification, goats are the single way to benefit from the poor land. In the Sahel (just south of the Sahara), Middle East and Australia, herders and their goats extract the last remnants of vegetation before retreating, leaving behind a growing desert [153]. This is not a sustainable strategy for a confined farm. If a landowner keeps goats, the land will become entirely barren and lose its agricultural value within a few years.

Goats should not be allowed in vulnerable and semi-desert areas, and ideally, they should not be used at all due to their strong feedback effect with desertification.

A significant issue with desertification is its slow and progressive nature, causing people to adapt, and even consider desertification normal, making their poor lifestyle traditional, thus they fiercely oppose improvement. Only in the final stage, when the land becomes totally unproductive, and people must retreat, they realise that something is wrong. People tend to blame small natural factors accounting for few percentages of climate change, while ignoring their own responsibility for drastic changes of tens of percentages.

Interestingly, in parallel with being attracted to the goats' capacity of bringing benefit from a depleted land, people also observed the total destruction of land by goats, therefore ancient wisdom associated goats with evil gods, or in Middle-eastern religions with the Devil himself.

The only reasonable solution to the goat problem is to prevent them from feeding freely in areas with scarce vegetation. Instead, goats should only be kept in paddocks, with food brought to them.

Goat herding produces desertification; thus, goats should not be allowed outside of their paddocks.

3.5 Population, Deforestation and Desertification

Historical records show that areas like Mesopotamia, though smaller than the Eurasian steppe, sustained large armies and populations. This land was historically rich in vegetation, but is now a sandy desert due to overexploitation [149]. Mesopotamians left archaeological evidence of their existence by building monuments and stone walls, but their timber structures have disappeared. Archaeological artifacts are preserved by burial, which is impossible in a rocky desert.

We cannot assume that the past human population of Earth was small just because we lack archaeological artifacts, or because a region is now a desert that cannot sustain many people.

The Eurasian steppe, now mostly a desert, once supported large populations like the Turkish army, which had over a million soldiers, besides their families and the rest of their people. These nomadic populations of millions lived in tents and did not leave stone buildings as proof of their existence. They needed for hundreds of years pastures for their herds and firewood due to the cold climate, leading to deforestation and desertification of the Eurasian Steppe, the largest land mass of Earth [145]. Unfortunately, their timber and leather artifacts disappeared through decay and total soil erosion, but that does not mean that they did not exist. It would be more reasonable to make population estimations based on the former capacity of that land to sustain them.

This reasoning may also be valid for other regions and populations, thus Figure 41B shows that Earth population by 1500AD might have been twice as large than previously thought.

The arrows indicate the beginning of desertification in a region, which continued over time. Pastoralist populations in North Africa initially deforested the region [46], then Roman intensive agriculture has initiated desertification that now expands southwards due to the use of goats.

FIGURE 41B Estimations of World Population, Deforestation and Desertification

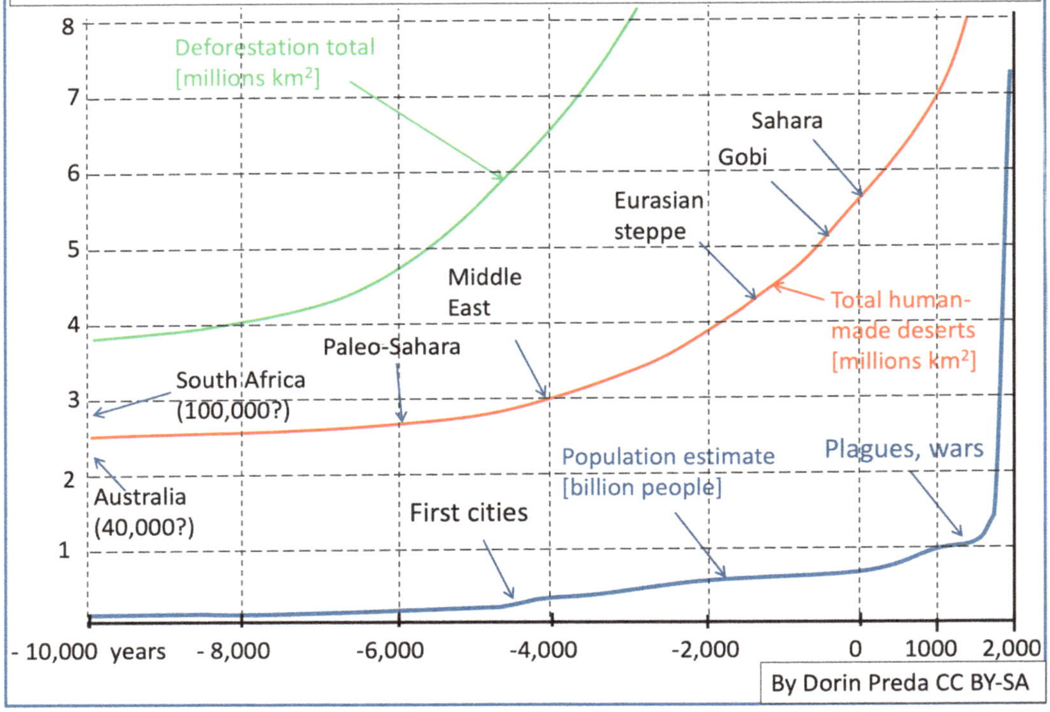

In [28] it is assumed that those deforestations have increased CO_2 emissions, which raised temperature, delaying the next glaciation; this is less likely, because the oceans had the capacity to absorb that excess of deforestation CO_2.

In studies like [118] it is stated that the warming of the climate led to the retreat of forests, followed by setup of pastures, but I assume and will prove logically below that the generalised local warming and drying induced by multiple deforestations was the cause of warming of the climate. This was not a local effect, because human population around 2,000 BC was substantial, they had metal tools and extensive experience in using fire for deforestation. By this time, humans had already created great deserts in Africa, Middle East, India, Australia and probably cleared by that time most of the Eurasian forests, amounting to around half of the Earth's land area.

<u>Therefore, I assume that the warming of the climate at the end of Holocene was caused by the deforestations and desertification induced by human activity.</u> As shown in [52] and detailed further, the true cause of global warming and dryness in the past few thousand years was likely caused by the excess solar heating caused by deforestation; the reductions in rainfall changed those climates.

<u>Concluding observations:</u>

- Deforestation for hunting-gathering purposes have likely created the South African, Namibian, the Thar Desert and Australian deserts, and many present-day savannas.

- Deforestation for pastoralism resulted in most of the world's grasslands, the Eurasian steppe and its deserts, and created dryness conditions in the Sahara region.
- Excessive agriculture has started the Middle East and subsequently Sahara deserts.
- Excessive pastoralism led to desertification of Gobi, Eurasian steppe, extension of present-day Sahara and Eastern Australia.
- Goat herding and repeated, traditional fire completes desertification.
- Desertification and forest fires raise the above-ground air temperature that is further amplified by carbon dioxide emissions from burning of fossil fuels and deforestations.
- Deforestation stops the humidity pumps, reduces precipitation and leads to desertification.
- Humans deforested most of Earth, leading to desertification and climate change.

3.6 Global Warming

Unlike the speculative variations of the planetary temperature inferred from out space, the real global warming is the long-term variation in temperature of the above-ground air temperature, as measured by the meteorological stations in the lowest part of the troposphere.

3.6.1 Carbon Gasses

Carbon gases (mainly CO_2 and methane) emissions result from all our activities. Besides the well-known burning of fossil fuels, deforestation and fires consume enormous quantities of vegetation which partially regrew (taking CO_2 from the atmosphere) only to be re-burned later. However, forests were not allowed to recover and gradually diminished with the loss of soils until they became a permanent loss. Over historical times, the size and density of the remaining trees have significantly decreased. Forests have turned into shrubs, transformed into grasslands, and due to human desperate use of goats to extract the last piece of life from them, became deserts.

The land area of Earth is about 150 x 10^6 km^2, with possibly 75% of it initially covered by forests that now only occupy 30%. Since measurements resulted in 20,000 tonnes of carbon stored per square kilometre of forest, 150x10^6 km^2 x 0.5 (areal and thinning losses and difference with grasslands) x 20,000t/km^2 = 150x10^{10}t.

In conclusion, Earth might have lost through deforestation 1,500 Gt of carbon, which was gradually, permanently locked in the oceans at a rate smaller than 2Gt per year (the absorption capacity).

Interestingly, the present rate of deforestation is estimated at 100,000km^2 per annum. If forest's capacity of carbon storage would be 20,000t/km^2, and if all that carbon would be instantly fed into the atmosphere, it would mean 2Gt/year, equalling the ocean's capacity of absorption. This would mean that all the 9Gt of carbon emitted from fossil fuels burning would be accumulating in the atmosphere, which now holds 800Gt as shown in Figure 22, and unlike estimations of [72].

The above-land global temperature rise of 1.4^{0}C over the past 200 years is attributed to carbon gases emissions since the industrial age [4]. Carbon and other greenhouse gases warming is only a delay of heat escaping the atmosphere; it is not active heating.

3.6.2 Solar air heating

This is a central notion of this book, and very important for what happens to the world climate, but it is not addressed properly by the climatology simulations. While it is important everywhere on Earth, its importance in the intertropical zone is paramount.

As seen in Figure 37, solar radiation (SR), which is the energy source for all climate processes (except the rare volcanoes), passes through the top of the atmosphere, and if it reaches a humid part of the globe (not the deforested areas or the deserts), it has good chances to meet and be largely reflected by low thick clouds. Only about 1/3 of SR passes through the clouds and barely keeps the ground at a reasonable temperature, it never blankets and overheats the ground, as per the climatic statements, because 2/3 of SR was already reflected to space.

When it does not meet a cloud, solar radiation is reflected and absorbed by the ground, and transformed into sensible heat (see Figure 32). If the ground objects have enough water, all available energy is transformed into latent heat that is taken away by water vapours, keeping the ground at low ambient temperature. Like under the clouds, there is no ground or air overheating. Vapours then condense in the atmosphere above, from where their heat is evacuated to space, without overheating the ground.

However, if the ground does not have water (which is almost always connected to vegetation), solar radiation will start overheating it, increasing the ground infrared emission (GIR in the desert on right in Figure 37). Additionally, the air around the hot ground objects is heated and rises in convection currents, which heat the entire troposphere and emit in infrared, increasing the greenhouse effect with the much larger ground infrared radiation than when there was water.

Solar heating is active heating (unlike the greenhouse gases heating) and it has replaced most of vegetation evapotranspiration cooling because we deforested Earth. There is more than about 30%/7% = 4 times more solar air heating in a desert than in a forest. When solar radiation is 1000W/m^2, this means an additional heating power of 300W/m^2 – 70W/m^2 = 230W/m^2 in a desert than in a forest. It is important that we realise the power of solar heating. For example, cutting down a medium-size tree is equivalent to heating at noon with a large cooking hotplate (2,500W) the land it was shadowing and to which it was providing surficial water from the depth.

Unlike the passive carbon gases global warming, the active solar heating is much larger, variable (with solar radiation), and dependant on the local land use. There are two heating effects: firstly, the hot ground emits a lot of infrared radiation upwards, where it meets the GHGs and is absorbed as greenhouse effect (GHE). Secondly, the air that is in contact with the hot ground is heated, raises and heats and dries the entire troposphere while emitting in infrared its energy and contributing again to the GHE. The hot convective currents generate the high thin clouds and (like Sahara) they also heat the adjacent ocean and increases evaporation, promoting the hurricanes.

We know from statistics that at present: 30% of land is forest, 33% are deserts, 3% are urban areas, 29% is grassland/crops, and 5% are rocky areas.

From common experience, I estimate the average temperatures of air above close-up areas with different covers as being: 30^{0}C desert, 28^{0}C rocky areas, 25^{0}C city, 20^{0}C grassland/crops, and 15^{0}C forest. The actual values are not very important, but the differences between them are.

Let us assume that before humans there were no deserts or urban areas, but only 5% rocky areas, 20% grasslands and 75% forests. The Eurasian steppe, Australia and the American plains most likely were forested.

The average temperature before humans would have been: 5% x 28^{0}C + 20% x 20^{0}C + 75% x 15^{0}C = 16.6^{0}C. The calculation for the average present temperature would be: 5% x 28^{0}C + 29% x 20^{0}C + 3% x 25^{0}C + 33% x 30^{0}C + 30% x 15^{0}C = 22.3^{0}C.

Therefore, the increase in Earth air temperature due to deforestation and desertification might have been around 5.7^{0}C. However, the ocean while adjusting to the new air temperature, would have absorbed a lot of heat. We could consider that the ocean would cool the air by say 2.7^{0}C.

Therefore, deforestation might have produced a global warming visible now of around 3^{0}C, presented in Figure 42 as the orange area immediately above the blue line that is the assumed temperature if humans did not exist and Earth would have cooled due to a new glaciation.

Deforestation decreases the reflective cloud cover and increases ground infrared emissions and air temperature in the under-the-clouds system.

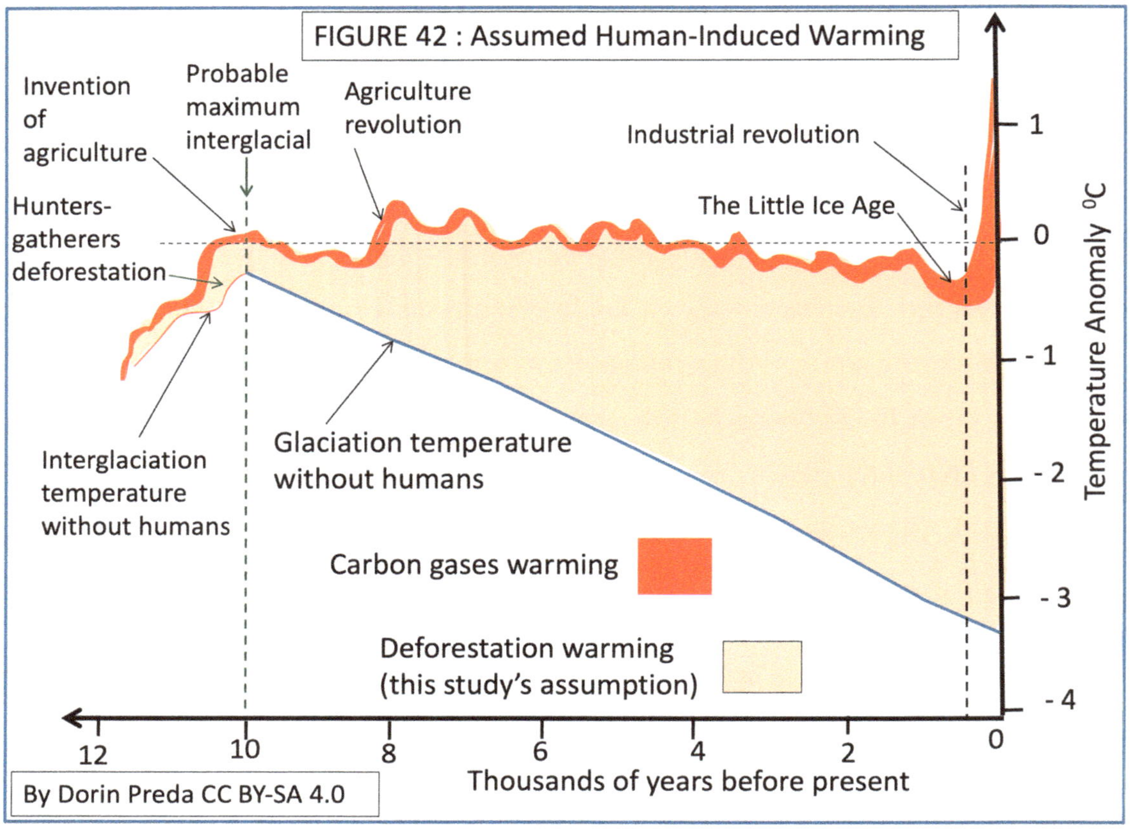

FIGURE 42 : Assumed Human-Induced Warming

The thick red area on the right of the "industrial revolution" line is the 1.4^{0}C carbon gases warming since 1750 as stated by IPCC, while the red area on the left represents carbon gases warming through agriculture according to [28]. The thin red line above all red areas is from https://commons.wikimedia.org/w/index.php?curid=466269.

After projecting my 3^{0}C deforestation warming onto the graph of Figure 16, and extending it with the same slope of 1^{0}C for 3,200 years of the previous interglaciation peak in Figure 17, there was a pleasant surprise to find coincidence with the oldest temperature peak, while leaving reasonable deforestation warming for the pre-agricultural times. This confirms my 3^{0}C deforestation warming.

We cannot have the interglaciation maximum at the 7800BP peak, because it would result that the rate of deforestation would have been much larger before agriculture start than after it. It results that the known invention of agriculture at around 10,000BP probably coincided with the interglaciation maximum, but it took about 2500 years of agriculture dissemination and implementation delay until the 7,800BP temperature peak marked the agriculture implementation. Despite the global warming induced by humans, Earth was still cooling and reached critical coldness during the Little Ice Age. The high rate of carbon gasses emission [28] was assumed to delay the glaciation, but it was too small compared to ocean sequestration.

Rather, I assumed in [52] that the avoidance of the glaciation was due to the 3^{0}C solar heating of deforestation and its associated cloud cover decrease. Therefore:

- If humans did not invent fire, a glaciation would have started 10,000 years ago.
- If humans did invent artificial fire, but not agriculture, a glaciation would probably have started about 7,000 years ago.
- If humans did invent artificial fire and agriculture, but did not progress to industrialisation, a glaciation would have started about 300 years ago.

3.7 Vulnerability and Feedbacks

Unfortunately, present official assessments of desertification do not consider the direct actions of people, leading to inaccurate forecasts.

FIGURE 43 Desertification Vulnerability Error

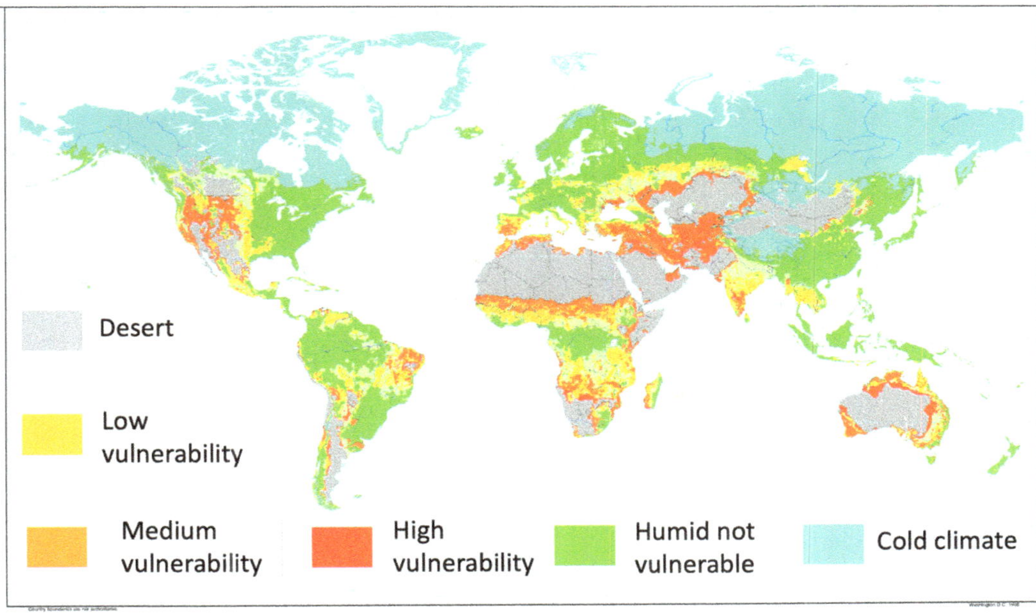

By USDA employee - https://www.nrcs.usda.gov/wps/portal/nrcs/detail/national/nedc/training/soil/?cid=nrcs142p2_054003, Public Domain, https://commons.wikimedia.org/w/index.php?curid=1349739

According to Figure 43 from the U.S. Department of Agriculture, the green areas were deemed not vulnerable to desertification, yet those areas are now experiencing wildfires in Europe, South America, North America, Southeast Asia and Australia.

Desertification is no longer just a slow progression due to mismanagement, but it is now rapidly advancing due to programmatic human actions. It is crucial to understand the chain of actions and reactions and predict where we will end up as a result of critical feedback. Feedback mechanisms intensify desertification by incorporating its past outcomes into its present input, leading to cascading effects, as seen with the red curve at the right of the industrial revolution line in Figure 42.

3.8 Small Settlements in Forest Clearings

A large, dense, natural forest, with bushes and a thick layer of twigs and fallen leaves between trees permanently retains humidity, thus a lightning strike cannot burn more than the tree it had struck, otherwise humans would not have found large and old trees. There were no mega-fires to burn entire forests before humans, who created the conditions and put fire to forests.

Human settlements typically start with individual houses or a small number of dwellings encroaching on a pristine natural environment. When this environment is a forest, people clear a patch of forest for the settlement and construct a few roads leading to it.

In a pristine forest, a tree is shielded by neighbouring trees from solar radiation and wind, which only affect the top of the tree. However, a tree on the edge of a clearing receives more sunlight and wind, leading to dryness. The fewer tree neighbours a tree has, the drier it becomes due to increased exposure to solar radiation and wind (see Figure 44). Isolated trees are significantly more vulnerable to wind damage and can be broken by storms.

The larger the forest clearance, the more area is exposed to solar radiation, heating the ground and subsequently the air. The wind then carries this heat to surrounding trees, drying them out.

Trees are highly susceptible to groundwater extraction, with water drawn for irrigation lowering the groundwater level under the tree roots, leading to dryness and increased flammability during fires. As water wells are located near houses, nearby trees (the yellow trees in Figure 45) are starved of water and become fire hazards.

FIGURE 44 Tree Vulnerability to Solar Radiation, Groundwater, Wind, thus Fire

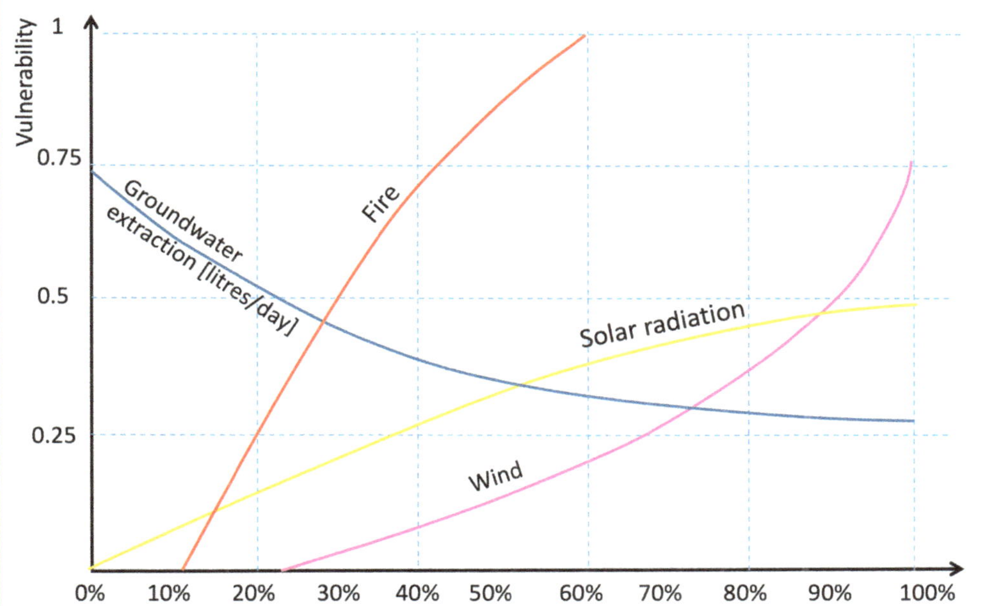

By Dorin Preda CC BY-SA

<u>As more trees are removed to make way for settlements and roads, the forest shrinks and thins causing increased dryness due to wind and solar radiation, the driest trees being those near the clearings.</u> Forests become this way fuel depots that burn in firestorms, their hot thermals generating the strong wind which feeds the fire with oxygen, further increasing its intensity.

FIGURE 45: TREE DRYNESS AND FIRE DANGER

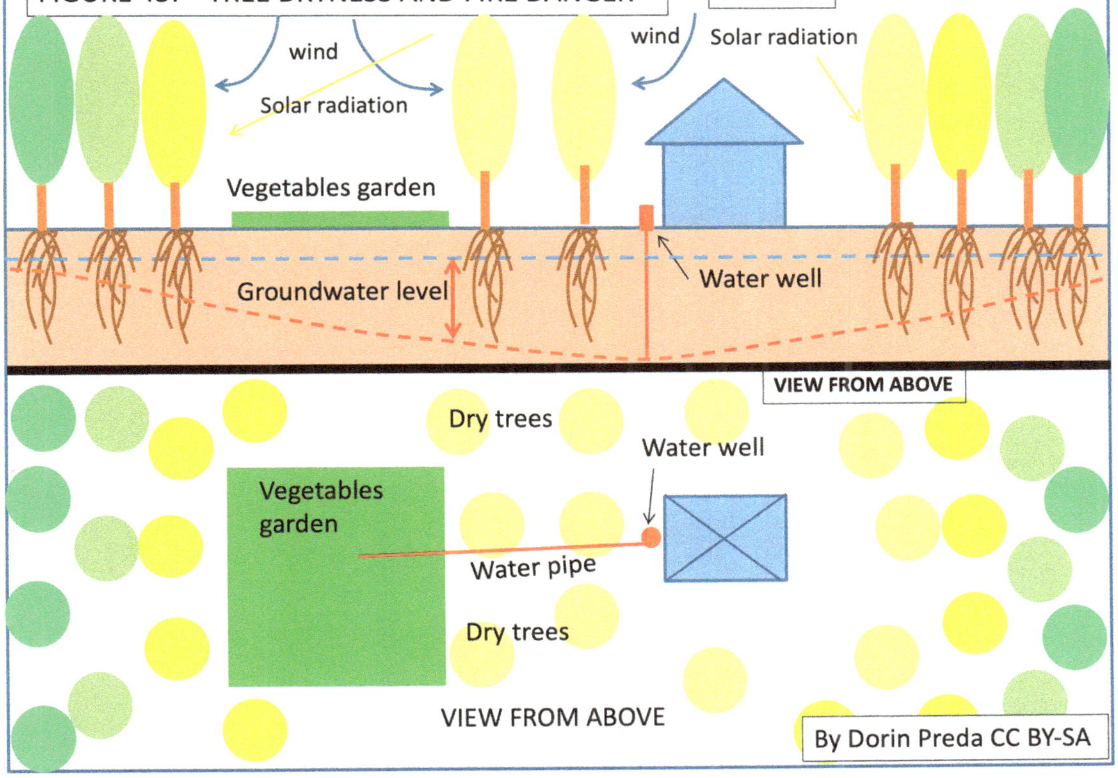

By Dorin Preda CC BY-SA

After a fire, people usually rebuild houses again using combustible materials and continue extracting groundwater, without addressing the root causes of excessive fires, leading to a vicious cycle.

Eventually, forests disappear as all trees are cut to protect houses from fires. Without the trees and their roots and twigs which protect the ground from sunlight and erosion, the soil is gradually lost, making any forest recovery impossible. Also, because of forest disappearance, the local climate becomes drier, and the desertification feedback is established.

As will be shown in chapter 5, it is not easy "to leave in nature", but to live in a mature, untouched forest without risk of fire and without cutting down any tree, a suggestion would be to:

1) Build a house among trees, spreading rooms between the tree trunks.
2) Prefer a single-storey house, to avoid cutting tree branches.
3) Use non-combustible materials for the house exterior.
4) Install internal sprinklers to extinguish house fires.
5) Create a narrow driveway between trees by laying lose stone to protect tree roots.
6) Collect rainwater for use in a large tank and avoid extracting groundwater.
7) Use composting toilets to conserve water.
8) Use only biodegradable soap and detergents, to enable discarding water nearby.
9) Leave twigs, shrubs and tall grass intact on the ground.
10) Avoid clearing trees for gardens.

This method ensures that the forest remains moist and fire-resistant.

Another solution is to build a sturdy house in a small natural clearing and cover it with soil and plant bushes or at least with a green canopy to diminish the solar heating.

A more comfortable lifestyle surrounded by nature is suggested In Chapter 6.

4 Chapter Four: Deforestation, Desertification and Climate Change

Human society has become very powerful in the past thousands of years and especially in the past 100 years, quickly making considerable changes to natural conditions [155]. We now see how the climate reacts according to the fundamental law of action and reaction. Some people still doubt that the human-induced changes are significant, but unfortunately, they are very wrong. We show below that the changes we have initiated long ago propagate and may amplify themselves like avalanches through what climatology calls "feedbacks" [161].

UNESCO estimates that one-third of the land area is threatened by desertification because overgrazing depletes the land of its last vegetation, urbanisation replaces the ground with asphalt, concrete or tiles, overdrawing groundwater, climate change (heat, droughts, floods that erode the soil), and unsustainable agriculture practices like tillage.

Therefore, the investigation into the effects of deforestation is well justified, as it may provide more insight into a larger climate change effect than previously known [154]. Containing this change through improved specific measures could not only avoid climatic collapse, but bring economic and social benefits, especially to countries in tropical and subtropical regions.

Deforestation destroys ecosystems, and leads to desertification through imperfect agriculture practices or urbanisation. Of interest to us are the deforestation effects on climate change that are unfortunately still debated among climatologists because of:

- the open nature of the atmosphere (the causes in an area disperse their impact over a larger area, or affect a remote location),
- inadequate anemometers
- inadequate use of exclusive albedo theory
- exclusive reliance on insufficient computer models and ignoring water cycle studies.

As a result, most climatologists believe that deforestation would cool the Earth, would increase precipitation and it would prevent forests from becoming carbon sources in case of climate dry out. This is a dangerous point of view, because it presents the solution to climate change as its cause.

Unfortunately, the IPCC has mistakenly reported positive effects of deforestation and urbanisation:

- Figure SPM 5 (I.P.C.C. 2013 LAND) shows negative 0.15W radiative forcing (therefore global cooling) because of deforestation [17]
- "Increased urbanisation can also intensify extreme rainfall events over the city or downwind of urban areas (medium confidence)." [16]
- "The urban heat island warms cities 0.6 to 5.6 °C (1.1 to 10.1 °F) above surrounding suburbs and rural areas. This extra heat leads to greater upward motion, which can induce additional shower and thunderstorm activity. Rainfall rates downwind of cities are increased between 48% and 116%. Partly as a result of this warming, monthly rainfall is about 28% greater between 32 to 64 kilometres (20 to 40 mi) downwind of cities, compared with upwind. Some cities induce a total precipitation increase of 51%." [16]

Such statements may conduce to the present deforestation policies in intertropical regions which increase desertification and amplify the causes of climate change. In fact:

- those extreme events are rare and local,
- in average and in the long term, cities decrease regional precipitations,
- that rainwater does not originate in the city, it has a different cause, as explained further
- in fact, cities dry out and heat up their downwind areas,
- the urban heat island effect is detrimental.

Cities are unnatural and they change their regional climate, as explained further.

4.1 Energy Budgets

We have to describe the basic physics of the processes at the above-ground level, where we live and where the meteorology stations measure air temperature. We make reasonable estimations, representative of observations, as depicted in Figure 46, rounded to the nearest 5%.

The energy input is solar radiation which hits opaque objects and transforms into other forms of energy, including infrared radiation, while the remaining energy heats the air that rises in the atmosphere and also emits in infrared.

The main energy components are:

- **SR** – input solar radiation
- **R** – reflected solar radiation
- **EH** – Evapotranspiration (latent) heat
- **IR** – Infrared radiation (as the difference between outgoing and GHE incoming radiation)
- **P** – Photosynthesis energy (without the sensible and latent heat used by the plant)
- **AH** – Air heating is the energy that remains from SR after using the other components above.

AH = SR – R – EH – SH – IR – P

We are interested in air heating (AH) and infrared emission (IR), as active causes of global warming; greenhouse gases are the passive cause because they only delay the escape of IR to space.

Deserts are old devegetated areas, but people do not create deserts directly; they only remove vegetation mechanically, with fire and with goats (that results in the loss of precipitations), then by the time the soils are lost through erosion and the desert ensues in a dried-up climate. Initially, people only create bare ground and sometimes cover it with the asphalt and roof tiles of a settlement we can call "city" if it does not have much vegetation. Tropical cities behave very much like deserts, and we shall use their name, although it should be understood that in time, cities are abandoned (especially when their surrounding area becomes a desert), and they also become deserts buried by sands. From the point of view of this study, hot cities and deserts are similar.

Let's study the four main types of land use: desert (city), bare ground (without vegetation), grassland (savannah or cropland) and forest (natural forest, or tree plantation, or dense orchard) in an average position at 40^0 latitude, 5,000MJ/m^2*year solar radiation, and 1,600mm/year precipitation. The formulas from Section 1.11.3 and the values in Table 1 allow for simple calculations.

The small percentage of P is the actual photosynthesis, while most solar radiation absorbed by plants is used for evapotranspiration.

City or desert:

- R - reflected solar radiation is 25%
- EH to evaporate 110kg/m^2 including for parks (the rest of rainfall is runoff) is 5%

- IR is 20%
- P is 0, as it is less than 1%
- Thus AH = 100% – 25% – 5% – 20% = 50%

Bare ground:

- R - reflected solar radiation is 20%
- EH estimated to evaporate 340 kg water/m^2x year (the rest is runoff) is 15%
- IR is 20%
- P is 0, as it is less than 1%
- AH results in 100% – 20% – 20% – 15% = 45%

Grassland or crop land:

- R Reflected solar radiation is 25%
- EH estimated to evaporate 910 kg water/year (the rest is runoff) is 40%
- IR is 15%
- P is 5% (for a grazed grassland or two crops per year)
- AH results in 100% – 25% – 40% – 15% – 5% = 15%

Forest:

- R - reflected solar radiation is 15% (deciduous forest)
- EH for forest evapotranspiration of 1350 kg water/year (the rest is runoff) is 60%
- IR is 15%
- P is 5%
- AH results in 100% – 15% – 65% – 10% – 5% = 5%

The infrared radiation (IR) is absorbed by the GHG, which in turn heat the air around them, so IR has a similar role to AH, only that the heating happens above the ground. Not only that water concentration increases downwards, but also does gases density, which means that by far most of the GHG is near the ground, and only little above the clouds.

The IR and the AH are the active energies of the GHE (deriving from the sensible heat gained by the ground) in the conceptual equation: **GHE = (IR + AH) x GHG**, where the content in greenhouse gases GHG vary in quality (nature) and quantity. I argue that the variation of the first term of the equation is as important for the overall greenhouse effect (GHE) as the GHG.

<u>Evapotranspiration reduces infrared radiation in the lower troposphere, thus diminishes the GHE.</u>

Figure 46 illustrates the opposing climatic effects of forests versus deserts or tropical cities.

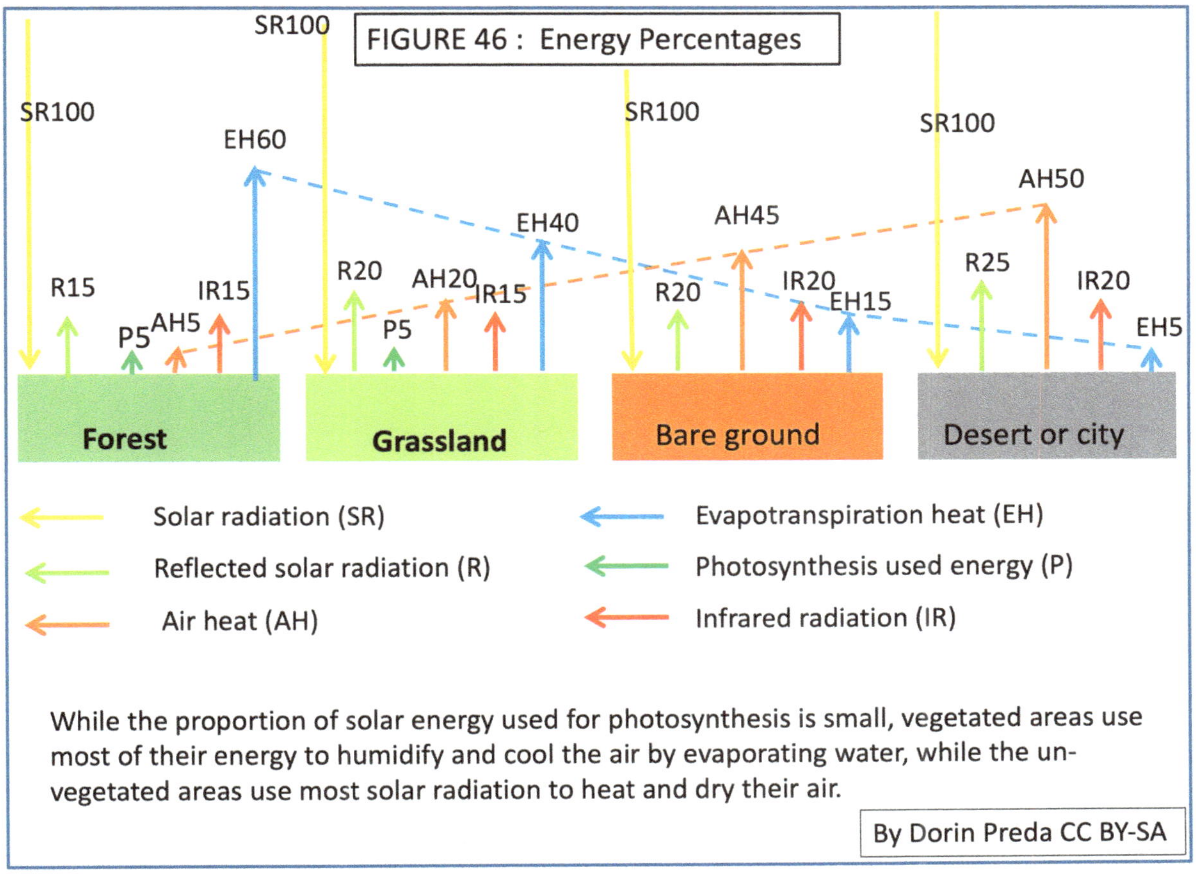

FIGURE 46 : Energy Percentages

While the proportion of solar energy used for photosynthesis is small, vegetated areas use most of their energy to humidify and cool the air by evaporating water, while the un-vegetated areas use most solar radiation to heat and dry their air.

By Dorin Preda CC BY-SA

<u>Therefore, though water, plants strongly influence the greenhouse effect and the radiative fluxes.</u>

IR and AH are part of the GHE, and their effects in different scenarios are calculated as below. Deforested areas use about 70% of solar energy to heat and dry their air, while forests use about 70% of solar energy to cool and humidify their air through evapotranspiration. Unfortunately, present climate studies obtain the opposite result by considering only albedo in their analysis.

Table 2 helps to approximate how much solar heating varies when we change land use from one of the four categories to another. For example, if we replace the grassland of an area with bare ground, we should enter Table 2 at Grassland on the left and read the value of 20% at the intersection with the exit at Bare ground. Thus, it results that through deforestation for a city we increase the solar heating by 50% x 5,000MJ/m^2*year x 10^6m^2 = 2.5x10^9MJ/year*km^2.

TABLE 2	⇧ Forest	⇧ Grassland	⇧ Bare ground	⇧ Desert/City
⇨ Forest	**X**	10%	45%	50%
⇨ Grassland	– 10%	**X**	35%	40%
⇨ Bare ground	– 45%	– 35%	**X**	5%
⇨ Desert/City	– 50%	– 40%	– 5%	**X**

Conversely, if we replace the bare ground of an area with grassland, we should enter Table 2 at Bare ground on the left and read the value of – 35% at the intersection with the exit at Grassland. In this fortunate case it results – 35% x 5,000MJ/m^2*year x 10^6m^2 = – 1.75x10^9MJ/year*km^2. A negative result means a cooling effect.

The following estimation highlights the magnitude of the change introduced by people through land use (deforestation and desertification). The assumptions are as follows:

- The energy components are given in Figure 46.
- Before humans, the Earth was: 0.71 oceans and 0.29 land area, of which 0.75 might have been forests, 0.2 grasslands, and 0.05 rocky areas.
- The real statistical figures for present Earth are: 0.71 oceans and 0.29 land area, of which 0.3 forests, 0.29 grassland, 0.33 deserts, 0.03 urban, and 0.05 rocky areas.
- The maximum solar radiation at ground level on the horizontal is 1,000W/m^2.
- To allow comparison with the computer models, we have to use their assumption of planet intercept, thus solar radiation becomes a continuous 1000/4=250W/m^2
- The oceans just evaporate water, do not heat the air, but will have a differential IR = 10%.
- The rocky areas and the desert would heat as much as the city: AH + IR = 70%

The average solar heating of the lower troposphere before humans would have been: 0.71x10% + 0.29(0.75x20% + 0.2x30% + 0.05x70%) = 14% of solar radiation.

The average solar heating of the lower troposphere at present, after deforestation, would be: 0.71x10% + 0.29(0.3x20% + 0.29x30% + 0.33x70% + 0.03x70% + 0.05x70%) = 20% of solar radiation.

The difference produced through deforestation might have been 20% - 14% = 6% of solar radiation x 250W/m^2 = 15W/m^2 of warming by increasing the GHE in the lowest part of the troposphere. Deforestation decreased evapotranspiration cooling, and the resulting infrared of the ground and of the hot air now act at the low part of the troposphere onto the dense GHG layers near the ground and produce a much stronger GHE there than before deforestation.

Reducing evapotranspiration, deforestation also reduced the clouds, therefore the albedo decreased depending on different land covers albedos versus the 0.19 albedo of baren soil (0.17) with old grass (0.21). The albedos as in Figure 46 are: 0.15 forest, 0.22 grassland, and 0.3 desert/tropical city. The clouds' albedo is 0.65, the ocean's is 0.05, and we could assume based on experience that clouds make a 40% cover above the ocean, 70% cover above forests, 50% cover above grasslands, 30% above bare ground and 10% above deserts/cities in the tropics.

Therefore, the average albedo of the ocean is: 0.4 x 0.65 + 0.6 x 0.05 = 0.29. Similarly, the average albedo of forests is 0.7 x 0.65 + 0.3 x 0.15 = 0.5. Albedo of grasslands: 0.5 x 0.65 + 0.5 x 0.22 = 0.43. Albedo of deserts is: 0.3 x 0.9 + 0.65 x 0.1 = 0.33. Therefore, Earth's albedo before deforestation was: 0.71 x 0.29 + 0.29 (0.75 x 0.5 + 0.2 x 0.43 + 0.05 x 0.33) = 0.34.

Similarly, Earth's albedo after deforestation is: 0.71 x 0.29 + 0.29 (0.3 x 0.5 + 0.29 x 0.43 + 0.33 x 0.33 + 0.03 x 0.33 + 0.05 x 0.33) = 0.32. The decrease in albedo after deforestation is 0.34 - 0.32 = 0.02. Thus, the cloud albedo effect of deforestation might have increased the radiative flux by: 0.02 x 250W/m^2 = 5W/m^2.

Totalling, the radiative flux results to be around 20W/m^2, that could confirm the 3^0C assumed increase in global temperature due to deforestation from Section 3.6.2. under the assumption 0.7W = 0.1^0C.

On the entire area of Earth, the additional solar heating power resulting from anthropogenic land use (vegetation clearing) would mean about 20W/m^2 x 10^6x $5x10^8$km^2 = 10^{16}W, which is about 500 times larger than the power of all forms of energy used by human society ($2x10^{13}$W).

4.2 Deforestations and the Urban Heat Island (UHI) Effect

The dense, tall forests that once occupied most of Earth were cleared and replaced with crop fields or urban areas. Despite proper earlier analysis of the UHI, like [162], the recent ones are only partial as they mostly rely on albedo. As seen in Figure 47A, a deforestation for crops in the middle of a forest (as happens now in the Amazon basin, replaces about half of evapotranspiration of the forest E_F with infrared of the ground (IR) and air heating (AH) as explained in the previous Section. AH forms only weak convection thermals because there is still significant evapotranspiration E_C of the crops. The crop field thermals are significantly warmer than the air of the forest, thus they cannot condense their water vapours during their adiabatic ascension. However, sometimes they push up from underneath the cold and humid wind coming from the forest to condensation height, promoting precipitation of its water vapours. This effect was not properly explained until [52], and wrongly presented as useful increase in rainfall through deforestation. Previous to deforestation, this humidity was providing rain to the rest of the forest, while after deforestation it disrupts the even distribution of rainfall and creates areas of excess dryness and areas of excess rainfall. This is exactly the climate change effect we see now all over the world due to extensive deforestations. The deforested areas themselves experience less and less rainfall and gradually lose their crops due to aridity, while after decades or centuries they lose their soils though desertification.

FIGURE 47 : Deforestations and The Urban Heat Island Effect By Dorin Preda CC BY-SA

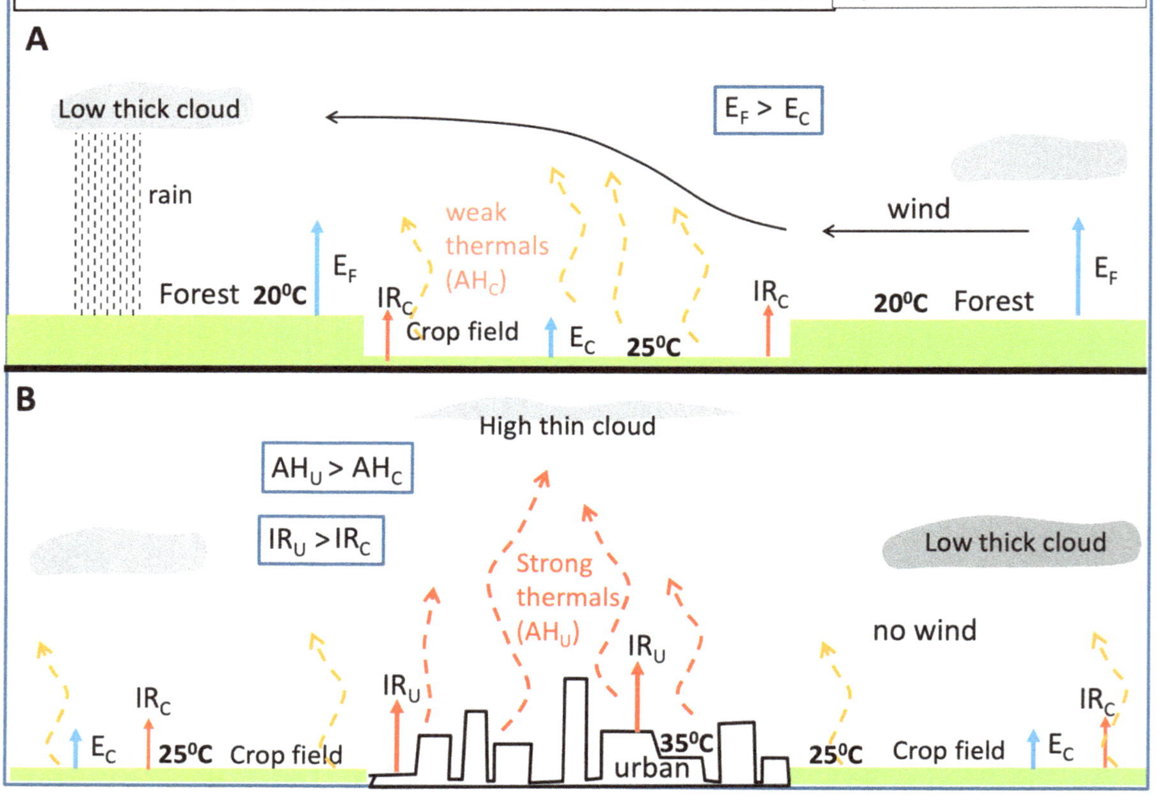

In Figure 47B it is shown a usual urban area surrounded by crop fields. Most modern cities and urban areas like the sprawling low-income suburbs have lost all their trees mainly due to the transportation system which eliminated the side of the road trees and the tree parks in favour of parking spaces. Since all urban surfaces are waterproofed, they cannot retain significant quantities of water for evaporation (E), thus basically all solar energy is transformed in sensible heat (IR and AH). This change had significant repercussions, that unfortunately became the usual "normality":

- Increased air temperature both as averages and maximums; decreased relative humidity
- Decreased solar radiation attenuation (without temperature correlation)
- Increased heat retention for night (without temperature correlation)
- Significantly decreased water retention and evapotranspiration
- Diminished annual rainfall, but increased rainfall intensity
- Apparition of tornadoes and increased likelihood of hail
- Decreased annual runoff but increased rate of runoff
- Increased local wind and altered wind patterns
- Diminished and significantly changed biota
- Soil erosion in uncovered areas

Due to almost nil evapotranspiration, solar radiation overheats the city area, creating an urban heat island effect (UHI) where air temperature is a few degrees higher than its vegetated surroundings.

The temperature records of the meteorological stations close to cities are adjusted by many degrees to eliminate the UHI effects to allow the conclusion: "Therefore, the well-known global temperature time series from in situ stations is not significantly impacted by urban warming". [74].

Most climatologists explain the UHI effect quite differently of the above, as being caused by the lower albedo of city surfaces [89] (opposite to climatology's albedo theory) and by the reduced emissivity of city canyons. Reduced city ventilation is considered a cause instead of the real strong air heating and convection of cities compared to forests. As result, ground sensible and latent heat are ignored, while the radiative (space) exclusive albedo is ill-used and overused in UHI studies.

The classical explanation of the urban heat island through city albedo or city canyons is not sufficient or correct when we compare Memphis (USA) and Baghdad (Iraq): they have similar albedos, are both land-locked, are situated at the same latitude, and both have a river flowing through the city. There is around 10^0C air temperature difference between those cities because a desert surrounds Baghdad. Therefore, it is not in Memphis's interest to lose all its surrounding vegetation. This is an additional reason for the increase in urban temperatures in the past 200 years, as settlements gradually lost their surrounding vegetation (thus their evapotranspiration cooling).

Differences in albedo are an order of magnitude smaller than differences produced by devegetation in evapotranspiration versus solar heating, which are the real cause of heating, including the UHI.

In low wind or no wind situations as shown in Figure 47B, the strong thermals (AH_U) generated by urban heat propel hot air at high altitude to generate the high thin clouds responsible for global warming (as calculated in Section 2.2.3). This results in deforestation replacing the cooling low thick clouds with warming high thin clouds.

In normal wind situations, urban areas behave similar to the above deforestations, its thermals pushing the cool humid wind up to its condensation height, increasing local rainfall and decreasing it elsewhere. Literature abounds of papers showing that due to the urban heat island effect there is an increase in downwind precipitation, but they fail to show the humidity imbalance they create.

Cities (heaters) are pleasant in winter or cold climates, while forests (coolers) are pleasant in summer or warm climates. However, cities are unpleasant in a warm climate or summer, while forests are not unpleasant in a cold climate or winter. Score 3:1 for the forests. It will be shown further that this antagonising behaviour has climatic consequences, not only air conditioning effect.

4.3 A Historical Glimpse

For millennia, people did not give any consideration to their influence on the natural environment [146], as the changes in the local climate were slow, and long-distance and generational transfer of information in society was minimal before the invention of printing.

The explanation for the hydrological cycle was not understood until 300 years ago, long after some ancient civilisations destroyed their lands, and disappeared. As seen here, even now we are re-discovering essential aspects of the Earth's water cycle and the role of vegetation in climate.

My observations below present the stages of development and decay of lands and civilisations [52].

The first known city Çatalhöyük, was established around 7,500 BCE in the northwest of the Fertile Crescent by hunter-gatherers and pastoralists who did not intensively work the land.

The first civilisation was created by Sumerians who appeared in The Fertile Crescent (Mesopotamia) around 5,000 BCE benefiting from the fertile lands of the deltas of Tigris and Euphrates (Figure 48, stage 0). As agricultural techniques improved, former hunter-gatherers dried up marshes for the new practice of agriculture and domesticated the water buffaloes.

By the time, they expanded away from the river, having to clear forests and savannas that sustained large animals including lions. This deforestation significantly reduced the rate of evapotranspiration and consequently the rainfall (Figure 48, stage 1). Therefore, those lands did not last long and dried up, decreasing crop production. People had to bring water to the crops, thus inventing irrigation.

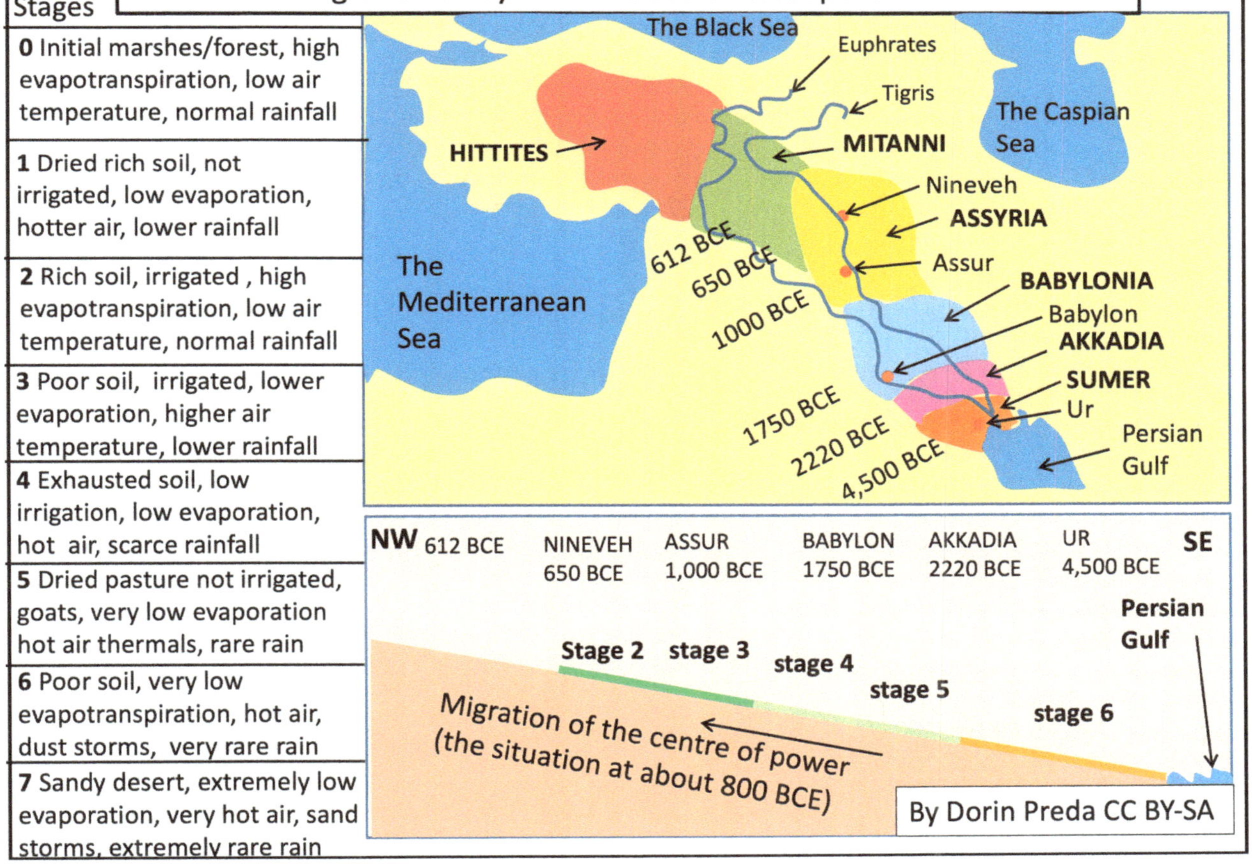

FIGURE 48: Migration away from the sea of Mesopotamian civilisations

The first Sumer cities (Eridu, Uruk, Ur) appeared in the delta, near the sea, around 4,500 BCE, benefiting from greater agricultural surplus due to improved irrigation, (Figure 48, Stage 2).

FIGURE 49

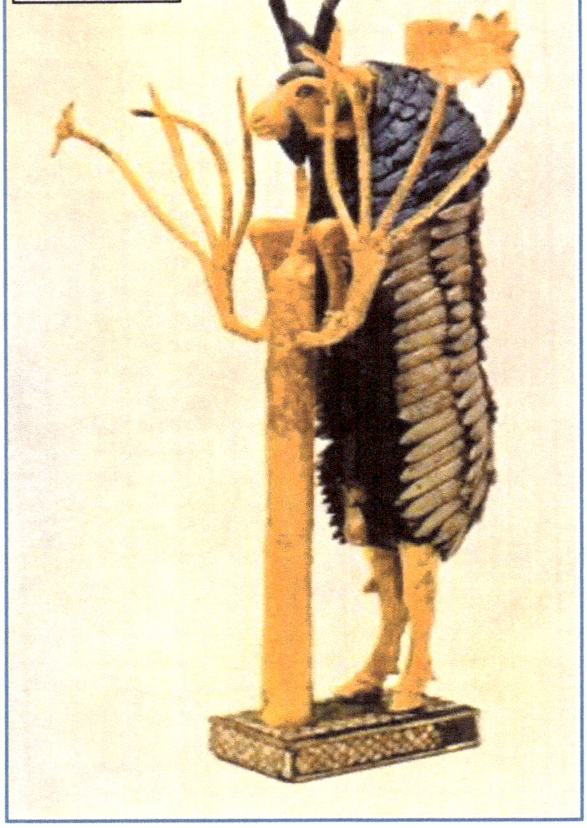

Irrigation increased water for evapotranspiration, leading to increased precipitation conditions for clouds coming in summer from the Persian Gulf and in winter from the northern seas, resulting in consistent rainfall and abundant crops for over 2,000 years [149].

However, no matter how good land management is, soil quality declines over time (Figure 48, Stage 3), resulting in smaller crop plants that evaporate less water; therefore, evapotranspiration decreases. This lowers air relative humidity, which decreases rainfall that reflects into even smaller crops, in an avalanche effect. Lower crops make irrigation work unjustifiable (Figure 48, Stage 4), which further reduces relative humidity and rainfall. Additionally, following an invasion by the Gutian tribes, irrigation systems were damaged and neglected, thus Sumerian lands rapidly degraded.

The exhausted land gradually decayed further despite attempts to repair the irrigation system during the Ur III period, leading to inconsistent crops, abandoned fields and desertification. As a result, irrigation is discontinued, and lands are used as pastures, similar to nearby fields that did not receive irrigation. Due to the increasingly dry climate, only shrubs could grow in such lands, leading people to keep goats which could feed on dried shrubs and provide milk, meat and wool, but in turn totally degrade the land (Figure 48, Stage 5).

As in all animistic cultures, useful animals are associated with gods. The Sumerians had hundreds of deities with one of the two gold-laden statuettes of worshipped goats found in a tomb at Ur from 2250 BCE, now in the British Museum (Figure 49). The goats, by consuming shrubs and small trees, destroy the land irreversibly. This leads to decreased evapotranspiration, hotter and drier air, and further degradation of inland vegetation as the wind blows during the day from the deserted lands near the sea to higher lands. Land degradation progressed from the sea towards inland areas.

Sumer prospered for about 2,500 years, but its oldest irrigated lands near the sea were almost exhausted. Newly irrigated lands further from the sea provided better crops leading to more wealth and power for society, allowing expansion away from the sea. Upstream lands facilitated the growth of the Akkadian empire, taking Sumer under their control around 2,200 BCE.

The Akkadian empire thrived for some time, but then soil productivity decreased, less water became available for evapotranspiration resulting in more solar energy used for air heating, hotter air rising in the troposphere, decreasing rainfall and leading to desertification. The Akkadian civilisation ended with some settlers remaining as goat pastoralists (Figure 48, Stage 5). The soil was blown away, the land became useless (Figure 48, Stage 6), and after a few centuries that lower area of Tigris and Euphrates transformed into a sandy desert (Figure 48, Stage 7).

Meanwhile, most of the population moved upstream to more productive fields in Babylonia, where the cycle repeated and after land degradation population moved further upstream to start the Assyrian civilisation. Land degradation to desert ended up all great civilisations between the Tigris and Euphrates, leaving behind nomad tribes of goat herders. The centre of civilisation (the capital) progressively moved upstream away from the sea (see the lowest profile in Figure 48) leaving behind lands in gradual stages of decay.

Historical sources state that the first Sumerian cities were not only abandoned by 500BC, but nobody knew at that time where they were because sand buried everything. Such a large quantity of sand at 500BC on lands that were lush with vegetation at 5,000BC, confirms my avalanche land desertification hypothesis.

The proto-Egyptian civilisation developed to the West of the Nile Valley, and in present-day Libya. Vegetation clearing for agriculture gradually reduced rainfall, prompting the population to move eastwards into the Nile valley, leading to the start of the ancient Egyptian civilisation [150].

Similar patterns occurred in the West North Africa, near the Atlantic Ocean, with irrigation water from the Atlas Mountains making Carthage fields the grain basket of Ancient Rome for a few hundred years. Soil productivity in North Africa lasted only around 500 years, although we know that soils can last thousands of years; thus, it must have been a different cause than soil quality. As more land was cleared on the slopes of the mountains for pastures, the decreased evapotranspiration might have reduced rainfall in the Atlas Mountains, which were supplying most of the water for irrigation.

4.4 The Avalanche Desertification

We can detail the desertification due to deforestation of [144], by looking into the processes.

Healthy soil is a living biotope of invertebrates and microorganisms in a micro-swamp of organic matter sustained by a mineral matrix that also provides nutrients through its alteration. As the soil ages, the alteration of bedrock deepens, incorporating a larger mass of now friable material into a thick, habitable soil. The microorganisms and organic matter in the soil are vital components, but they cannot thrive at temperatures above 30°C and cannot survive at 50°C.

Most of the time, the air is not saturated with water, creating a powerful "evaporative cooling" effect on moist soil due to the large latent heat of water. The ground is cooler because it continuously "breathes" through its pores under micro-pressure variations from the wind, and loses water through evaporation.

Plants, invertebrates, and microbes that grow in the soil also lose water through transpiration, resulting in the overall process known as "evapotranspiration." In Figure 50, I explain the interaction of solar radiation, clouds, and vegetation cover through the water and energy (latent heat) budgets of the soil and plants.

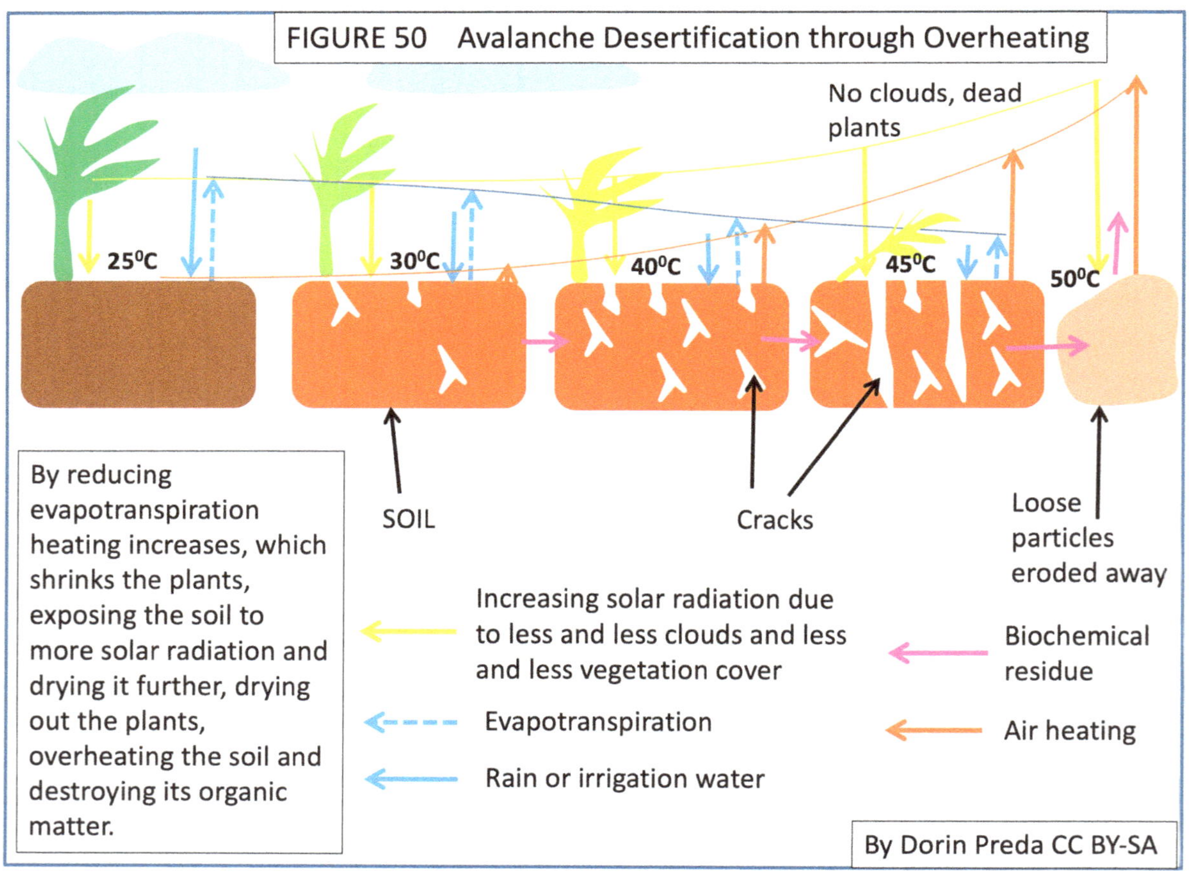

FIGURE 50 Avalanche Desertification through Overheating

In a healthy soil in a large field, only about 20% of solar radiation reaches the ground due to the shadow of the vegetation it supports. Additionally, since the field is humid, it is covered by clouds about 30% of the time, allowing only 20% of solar radiation to pass through. The damp soil loses a significant amount of water through evapotranspiration and cools, resulting in a temperature of only about 25°C on a sunny day, creating an environment where life thrives.

If water input decreases, the soil will still lose water through evapotranspiration, depleting its water reserve. If the water deficit continues, there will not be enough water for plants or for evapotranspiration. This causes the soil and plants to heat up, becoming solar air heaters. The difference between solar energy and evapotranspiration heat losses results in air heating.

Hot air produced in a field of dried plants can be carried by the wind to nearby green plants, initiating an avalanche drying process. As the soil dries out, the plants shrink and die, exposing the ground to more solar radiation. This reduction in evapotranspiration leads to less humid air above the field, decreasing cloud formation and allowing more solar radiation to reach the soil, causing overheating.

As the soil loses all water, it shrinks, cracks, and exposes its organic matter and microorganisms to extreme heat and dryness. Organic matter is destroyed by overheating, and the soil crumbles into smaller particles that are eroded by the wind, leading to desertification. Once drought begins, it rapidly amplifies through an avalanche (positive feedback) process.

4.5 Coastal Effects of Deforestation

People were drawn to settle on coastlines due to the opportunities for navigation and fishing, as well as cooler climate in low latitudes. Therefore, they replaced the coastal forests with waterproofed buildings and streets. Any rainwater quickly drains away and leaves dryness behind.

During the forest fires of 2019, I attempted to understand the causes for their unusual intensity. Eventually, I developed new explanations including HAC and ASH published in [52], by observing:

- Rivers extract water from continents, which are above ocean level, thus only precipitation on land of water evaporated from oceans can balance the river flow.
- Low precipitation on tropical lands produces droughts, reduced river flows and forest fires.

- Excessive heating and flooding of the higher latitudes shows that now tropical ocean humidity is increasingly discarding there.
- Oceanic winds are diverted in an alongshore direction when approaching most continents.

4.5.1 The Hot Air Curtain (HAC)

Diagram A of Figure 51 shows the original forested coastline with high evapotranspiration rates, resulting in cold forest air and slow vapour rise. Temperature examples are from the subtropical area, which was less studied previously. As known, forests are frequently covered by low thick clouds that reflect most solar radiation. The rest of it is gradually absorbed by plant leaves, little used for photosynthesis, but most of it is used to evaporate water for the biological and hydrological cycles. If there is enough water for evaporation, almost no heat is used to heat the air, which remains cool and humid.

Marine air is attracted above the forest by a tree humidity pump, increases in humidity by absorbing forest vapours and condenses into rain, aided by marine aerosols containing salt micro-crystals and tree organic carbon aerosols. Since marine humidity easily travels onto the continent, rainfall is frequent, abundant and sustains a perennial river.

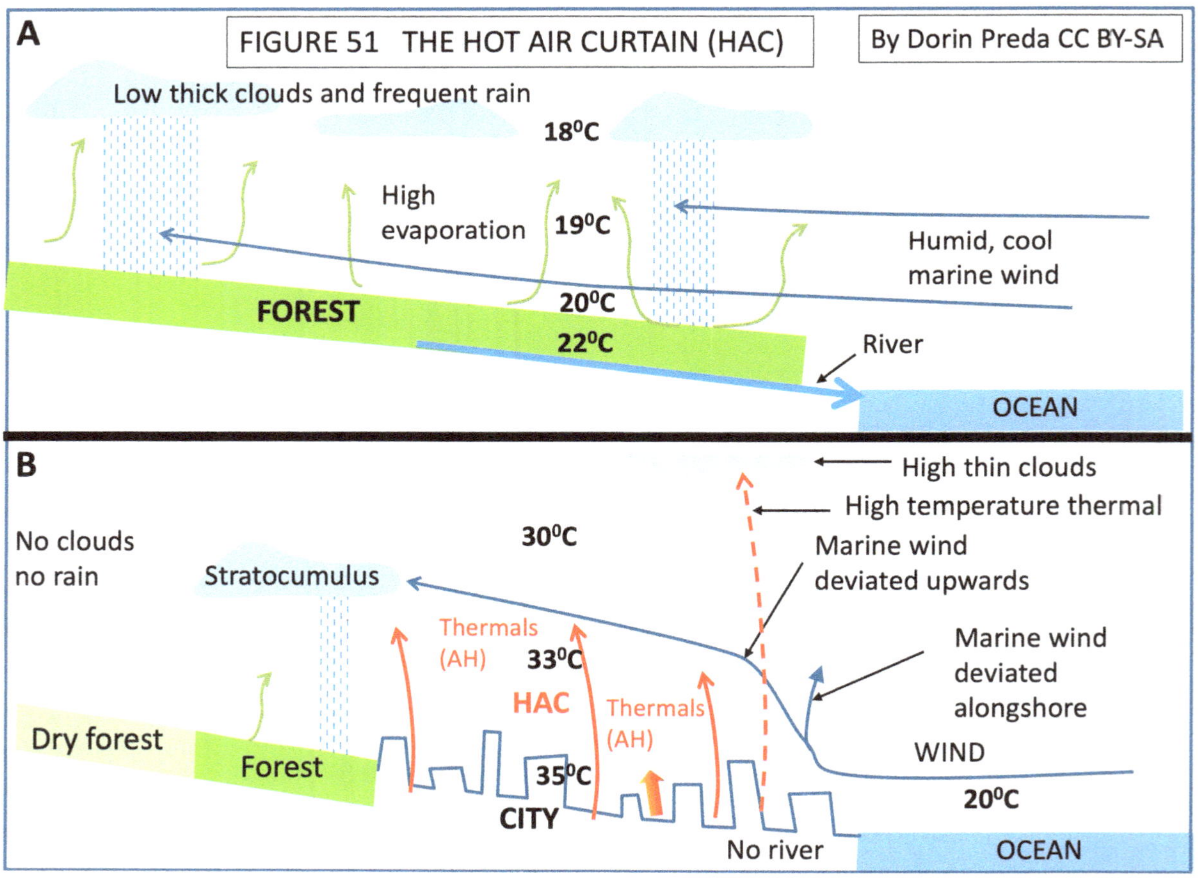

When a large settlement, like a modern city, replaces a forest (see diagram B in Figure 51), its waterproofed streets and roofs cannot retain water. Thus, the absorbed solar radiation heats the opaque objects in the city, which reach a high temperature, emitting highly in infrared. Those objects also heat the air around them which becomes more buoyant and rises in great convection currents (thermals), which discard that heat to higher altitudes through sensible heat transfer and through radiation. Let us first discuss the low-speed case of oceanic wind.

Some of those thermals that developed above very hot ground rise at great height and their humidity freezes forming high thin clouds which have strong greenhouse effect. Most thermals rise and create a hot air curtain (HAC) in front of the incoming marine wind, which is diverted in an alongshore direction, precluding its humidity from reaching the interior lands that this way dry out.

However, when the marine wind is strictly perpendicular to the coastline, city thermals cannot divert sideways the marine air, but lift it from underneath to a height where it may condense into clouds and produce rain from its oceanic humidity. It is the cool humid ocean breeze that produces the rain, not the hot air of the city. The same effect happens inside of forested continents when the thermals of the islands of deforestation or of the urban heat islands propel at height the humid wind coming from the surrounding vegetated lands to produce downwind precipitation. This effect is wrongly presented as air heated through an urban heat island effect which rises and condenses to generate rain: "...Partly as a result of this warming, monthly rainfall is about 28% greater between 32 to 64 kilometres (20 to 40 mi) downwind of cities, compared with upwind. Some cities induce a total precipitation increase of 51%." [16].

A similar mistake was made by the climatologists involved in the most extensive and expensive experimental study [101] in the Amazon, when they observed that increased local rainfall correlates with deforestation, and failed to understand its causes and overall effect, resulting in political leaders believing that they would increase general rainfall with deforestation. The real effect was exactly opposite, and it now leads to the demise of Amazonia.

The air column above cities (and deserts) is hotter than above forests, and although it cannot condense, it still contains water vapours (absolute humidity), that create strong greenhouse effect.

Air curtains exist in air conditioning and refrigeration technology, but they are smaller, blow air downwards, and use cold air. The principle and effect are similar: they separate two regions of air, reducing mixing without a substantial barrier. The HAC functions oppositely (with hot air going upwards) but achieves the same separation of air regions, thus it stops the tree humidity pumps.

This natural effect identified in [52] was not studied previously because only horizontal wind was consistently measured. However, thermals can have vertical speeds of more than 3m/s and transport a significant amount of heat which was not adequately measured or understood.

The HAC has a long-term climate change effect of drying up inland areas by blocking moist marine air. However, it is statistical, occurring most of the time, but not always. Exceptions include strong wind breaching the HAC, night-time weakening, and occasional rain reducing the HAC next day.

Coastal deforestation and development have created HACs globally, preventing marine humidity from reaching inside the continents, leading to increased water deficits. As seen in Figure 26, the water deficit increases on continents, which are already draining their glaciers and groundwater.

The continental effect of HAC is confirmed in Figure 52 that shows how the general East-West air circulation in the intertropical region is deviated sideways when it approaches a continent, as if there would be a barrier or a curtain there. The marine wind turns along the coastline except in South America, where the wind does not change direction, but follows through into Amazonia because that coast still has forests. Brazilian coastal forests are now replaced by wide coastal developments, a HAC started forming there also, drying up the entire Amazonia.

FIGURE 52 World Wind

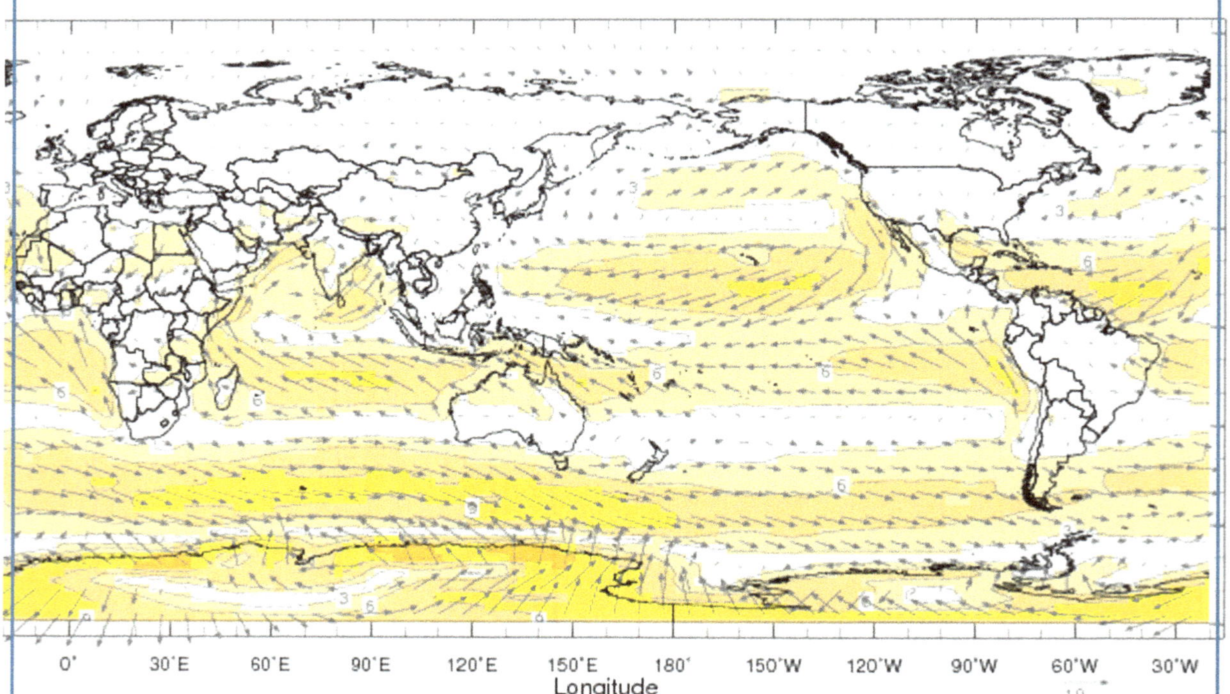

By KVDP - Own work, Public Domain, https://commons.wikimedia.org/w/index.php?curid=8158719

The Madagascar, the Horn of Africa and Australia have been deforested and have strong HACs.

In meteorology, these curling winds are explained as travelling around highs of air pressure, but we should ask "what creates the pressure highs"? Since meteorologists do not measure vertical wind, they refer only to barometers which measure the high air pressure produced through excessive solar heating and expansion over overheating dry lands.

The global winds in Figure 52 are not low altitude winds which could be influenced by the relief, but they are major tropospheric air movements. The North-Eastern Australian coast has very low altitude, the same as Madagascar and the Horn of Africa; thus, it is not the coastline which bends the wind, but the HAC.

The hottest land, which probably has the most intense HAC is in the Horn of Africa and Asia Minor; in that area, the wind certainly hits a barrier, makes 180^0 turn and reaches India. Not even the monsoon winds cannot usually penetrate the Horn of Africa HAC, which results in one of the lowest rainfalls in the world despite being on the Equator and receiving the humid air of the Indian Ocean. That is the same moist air which subsequently produces the abundant Indian monsoon rains.

<u>HACs produce climate change by disrupting the global water cycle, resulting in continental desertification through dryness feedbacks: HAC = less rainfall = less vegetation = even less rainfall.</u>

<u>Clearing coastal vegetation and replacing it with settlements or other land uses which produces HACs decreases the marine humidity input to continents, totally drying them out over time.</u>

4.5.2 The Power of HAC

In Europe forests remained after the Middle Ages mostly in the North, but they gradually were replanted, mostly in Central Europe. However, the coastlines were cleared of vegetation and replaced by sprawling cities under a continuously growing demographic and development pressure.

In North America, the forests never increased in size, but continuously decreased, and similarly in Asia, with the notable exception of China, which is now the champion of reforestation.

In Australia, the continuous use of fire by the first inhabitants resulted in total desertification of most of the continent and the change of most vegetation towards the gumtree (Eucalyptus).

Substantial forests remained on the Eastern and Northern coasts until the arrival of Europeans, who deforested the coast for timber, settlements, agriculture, pastures and mining.

Africa lost all its vegetation in the North, most of the vegetation in the South and on all its coasts, therefore its central areas are highly endangered.

South America is deforested in the South and East and Amazonia is endangered by a line of deforestation near the Atlantic Ocean.

On all continents, coastal deforestation produced strong HACs; thus, rain seldom penetrates inland, resulting in chronic drought that people now consider the norm. When rain does penetrate and stops the HAC, so much more humidity comes in as a deluge, pushed by the excessive pressure of humidity build-up on the ocean. However, after a few rainy days, the HAC restarts.

To show the power of HAC, and what happens when a type of land cover changes to another, I calculate in the style of 4.1 Energy Budget with values for 30 degrees latitude:

- solar radiation 7,000MJ/m^2*year
- evapotranspiration of grassland/crops is 500mm/year
- evapotranspiration of primary forests before HAC was 1,100mm/year.

City:

- R is the reflected solar radiation 15%
- EH estimated to 5% (due to the city vegetation)
- IR is 20%
- SH is 15%
- P is 0, as it is less than 1%
- AH results as 100 – 15 – 5 – 20 – 15 = 45%

Bare ground:

- R Reflected solar radiation is 20%
- EH estimated to evaporate 200 kg water/year is 5%

- IR is 20%
- SH is 15%
- P is 0, as it is less than 1%
- AH results as 100 – 20 – 5 – 20 – 15 = 40%

Grassland:

- R Reflected solar radiation is 25%
- EH estimated to evaporate 500 kg water/year is 15%
- IR is 15%
- SH is 15%
- P is 5% (for a grazed grassland or two crops per year)
- AH results as 100 – 25 – 15 – 15 – 15 – 5 = 25%

Forest:

- R Reflected solar radiation is 15% (deciduous forest)
- EH for forest evapotranspiration of 1100 kg water/year is 35%
- IR is 10%
- SH is 20%
- P is 5%
- AH results as 100 – 15 – 35 – 10 – 20 – 5 = 15%

The calculation of EH for Bare ground: 200kg x 2.2MJ/kg = 440MJ/year; 440 x 100/7000 = 6% = 5%

The calculation of EH for Forest: 1100kg x 2.2MJ/kg = 2420MJ/year; 2420 x 100/7000 = 34.5% = 35%

For a forest with solar radiation input of 7,000MJ/m^2year, and 1 km^2 of forest, the annual solar heating because of changing it to city is: 1,000m x 1,000m x 7,000MJ/m^2 x (50% - 15%) = $2.45x10^9$MJ/ km^2year.

The most powerful heaters we have are the fossil fuel power plants. An average power plant has an efficiency of 33% and delivers 600MW of electric power; thus, its burners have a capacity of 600MW x 3 = 1,800MW. During the daylight, such a power plant produces 1,800MJ x 3,600 seconds x 8 hours x 365 days = $19x10^9$MJ/year of heat.

We make a power analogy between the above fossil fuel power plant and the additional heating power of the land due to deforestation (which is the difference in air heating between the bare land and the former forest). Thus, the solar heating area equivalent to a "power plant" would be 19/2.45 = 8 km^2. Therefore, each 8 km^2 of deforestation adds as much solar heat to the air as an average fossil fuel power plant (see Figure 53). In contrast, water vapours cool the forest by the same quantity of heat at much lower temperature. This calculation did not take into account the additional cooling through reflection of the low thick clouds above the initial forest.

The largest temperature difference is near the ground, where global warming is measured.

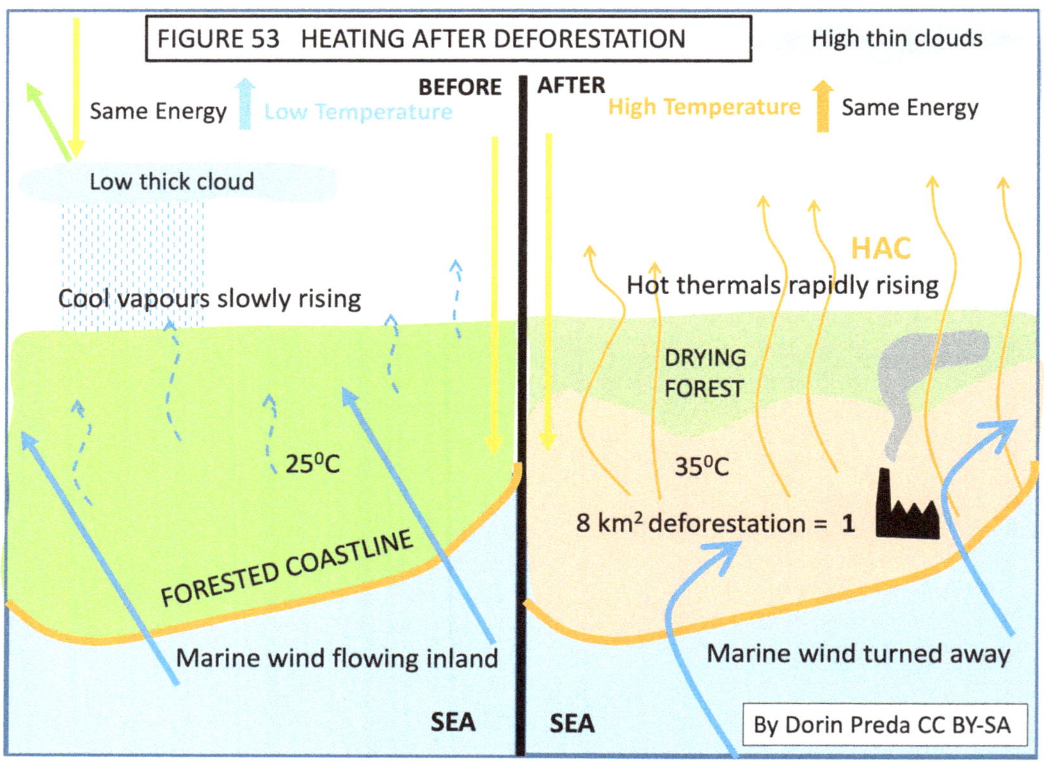

There are land photographs from the 19th century and aerial photographs from the beginning of the 20th century showing coastlines with dense forests on all continents except Europe and Antarctica; those tall and dense forests disappeared, being replaced by settlements, crop fields, pastures, infrastructure works, quarries, or just by shrubs.

A large subtropical coastal city may be about 400km^2 in area (including suburbs and industrial areas), which by replacing the initial forest will generate additional solar heat equivalent to 50 large "power plants". Hence deforestation for a single large city adds as much solar heat to the lower troposphere as all the power plants of that country, unleashing an enormous solar heating power, which raises powerful thermal curtains on that coast and dries up those lands.

If the coastline is deforested 4km inland, it means that one power plant would be placed at each 2km along the coastline. Each morning when the sun rises and heats the ground, those "power plants" start heating the air and also storing heat in the ground for night time. Only when the ground is humid, the HAC decreases, and more humid air from the ocean can enter the land. When the humid conditions end, solar radiation restarts the HAC. As the continents progress towards desertification, the wet events become less frequent but more violent due to humidity pressure.

Cities are similar to deserts from the climatic point of view, thus the global desertification trend.

4.5.3 The Avalanche Solar Heating (ASH)

Studies like [144] show the connection of deforestation to desertification, that I better explain next.

Direction and intensity of wind are important in deforestation drying. HAC develops with rather weak marine breeze. Extreme forest drying took centuries in Athens (Greece) due to the prevailing wind blowing from the northern forests towards the southern city. However, all vegetation dried up as houses encroached into the hilly woodlands, leading to the recent Athens fires.

On the other side of the Mediterranean, the Sahara coast is oppositely oriented to Athens, making Sahara optimal for discussing the avalanche solar heating (ASH), as first described in [52].

At around 1,000BC, North Africa had mild climate with forests and savannas, making it ideal for settlement by Phoenicians, who established the Carthage civilisation around 800BC. They thrived in lands 500 km from the Mediterranean, but rivalled Rome at around 200BC, and were destroyed.

At 100BC, the fields cultivated by the Romans were near the ruins of Carthage, with patches of forest, lakes, and savannah (diagram A in Figure 54). The wind that most of the time blows from the Mediterranean Sea was heated in contact with the ruins of Carthage that were too small to create a real HAC, while nearby was the crops' evapotranspiration, so climate was not much altered.

After the Roman conquest, grain production had to be increase for tribute, leading to more forest clearing for increased crop area. Romans also deforested for larger coastal settlements, leading to significantly increased solar heating, so a coastal HAC began to dry the inland areas in days with weak marine wind. Over the next few hundred years, the dryness accentuated until the coastline mainly became a solar air heater.

However, in days with strong marine wind as in diagram B of Figure 54, the HAC is blown inlands and mixes with the marine air, reducing its humidity, so that hot air dries the inland vegetation. The newly dried-up vegetation cannot evaporate water; thus, it is heated by solar radiation and becomes in its turn another solar air heater, increasing further the temperature of the wind. The avalanche solar air heater (ASH) is a feedback effect between wind and solar heat, so it increases in power as it advances on the land and collects more energy from the ground it had scorched.

As known from solar heating technology, the efficiency of a solar air heater increases with the rate of airflow, extracting more heat from the hot ground as it speeds up. The hot air layers rise due to their high buoyancy, creating suction underneath that entrains more air and increases the ground wind. The ground wind, which is hotter due to its contact with the ground through the Coanda effect moves forward, drying up the vegetation beyond the city. If there is a source of ignition, a violent fire fuelled by the hot wind could start in the dried vegetation.

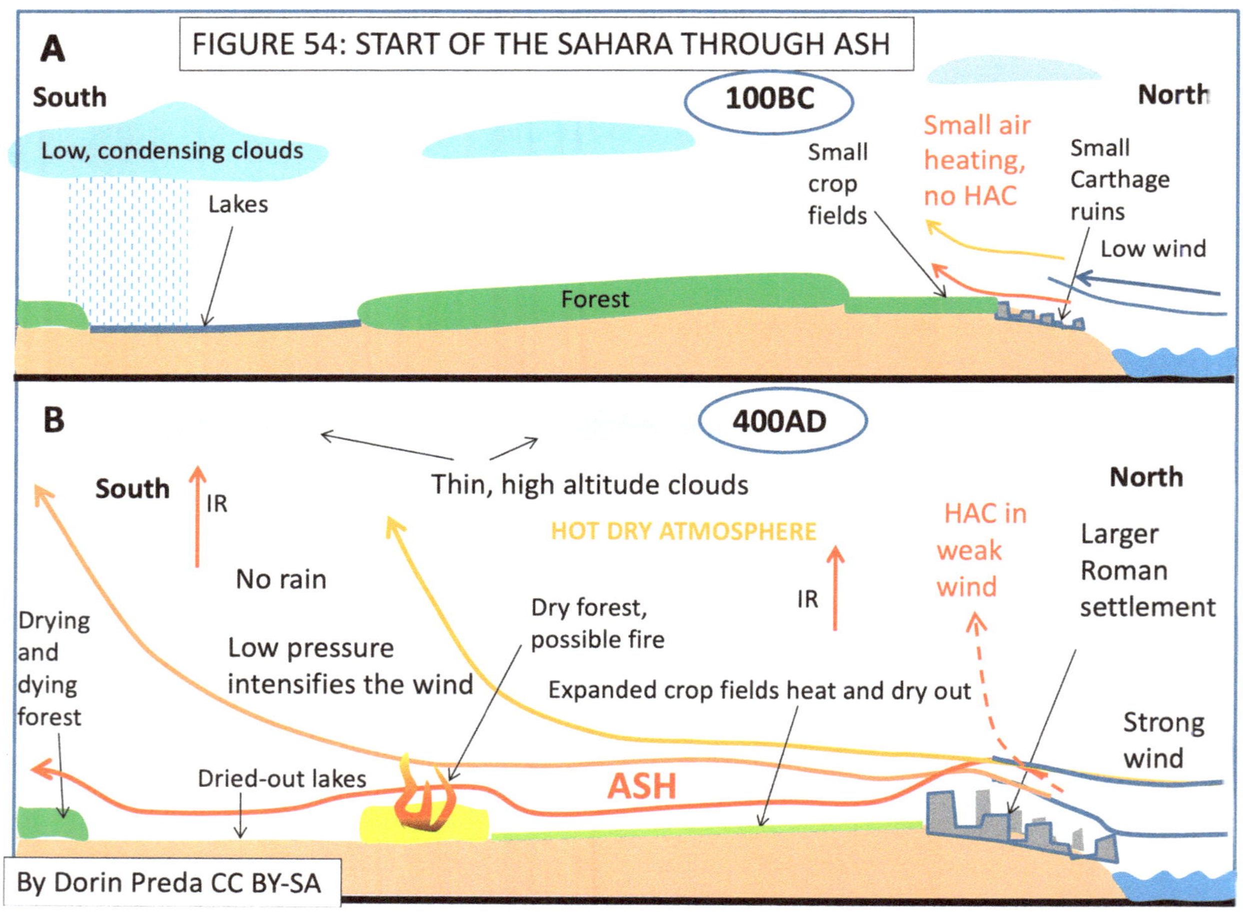

FIGURE 54: START OF THE SAHARA THROUGH ASH

As further inland areas dried up and became infertile, the Romans abandoned them, and nomad pastoralists with goats started using the lands. With the help of goats, the vegetation vanished quickly, and the bare ground dried up under strong sunshine, overheating more air, establishing an almost continuous ASH blowing inland and drying everything, including the lakes.

After the Romans initiated desertification of the Sahara, the ASH continued by its own, resulting in regional overheating (desertification) and permanent climate change.

4.5.4 The Power of ASH

All the old cradles of humanity are now deserts, although people initially settled only in richly vegetated areas. This transformation suggests that the ASH that followed devegetation is more powerful than people, who had to abandon their hard-worked cities and crop-fields.

In Roman times, there were more clouds and precipitation in the Carthage area due to evapotranspiration from the land, resulting in less solar radiation than today; it could have been 6,000MJ/m^2*year instead of the present 7, 000MJ/m^2*year.

Evapotranspiration rates could have been 1,000mm for forest, 600mm for grassland/crops, 200mm for bare ground and 100mm for the ancient city due to some vegetation present.

EH for City: 100kg x 2.2MJ/kg = 220MJ/year; 220 x 100/6000 = 3.7% = 5% (rounded)

EH for Bare ground: 200kg x 2.2MJ/kg = 440MJ/year; 440 x 100/6000 = 7.3% = 5% (rounded)

EH for Grassland: 600kg x 2.2MJ/kg = 1320MJ/year; 1320 x 100/6000 = 22% = 20% (rounded)

EH for Forest: 1000kg x 2.2MJ/kg = 2200MJ/year; 2200 x 100/6000 = 37% = 35% (rounded)

City:

- R Reflected solar radiation is 20%
- EH is 5%
- IR is 20%
- SH is 10%

- P is 0, as it is less than 1%
- AH results as 100 – 20 – 5 – 20 – 15 = 45%

Bare ground:

- R Reflected solar radiation is 20%
- EH is 5%
- IR is 20%
- SH is 15%
- P is 0, as it is less than 1%
- AH results as 100 – 20 – 5 – 20 – 15 = 40%

Grassland:

- R Reflected solar radiation is 25%
- EH estimated to evaporate 500 kg water/year is 20%
- IR is 15%
- SH is 15%
- P is 5% (for a grazed grassland or two crops per year)
- AH results as 100 – 25 – 20 – 15 – 15 – 5 = 20%

Forest:

- R Reflected solar radiation is 15% (deciduous forest)
- EH for forest evapotranspiration of 1100 kg water/year is 35%
- IR is 10%
- SH is 20%
- P is 5%

- AH results as 100 – 15 – 35 – 20 – 10 – 5 = 15%

In the first phase, there was a forest clearing for the city (see A in Figure 55).

Let us do all calculations per linear kilometre of coastline.

A Replacement of 1km inland of coastal forest by the city is: 1,000m x 1,000m x 6,000MJ/m^2 x 50% = 3x10^{9}MJ/year.

B1 Replacement of 3km inland of coastal forest by grassland (crop fields): 3,000m x 1,000m x 6,000MJ/m^2 x 5% = 1x10^{9}MJ/year.

Therefore, in the first stage of development, the additional solar heating per km of coastline was 4x10^{9}MJ/year. That heat was not enough to start an ASH, but it did create a HAC that began to starve the inland areas of humidity.

FIGURE 55 The Increasing Power of ASH

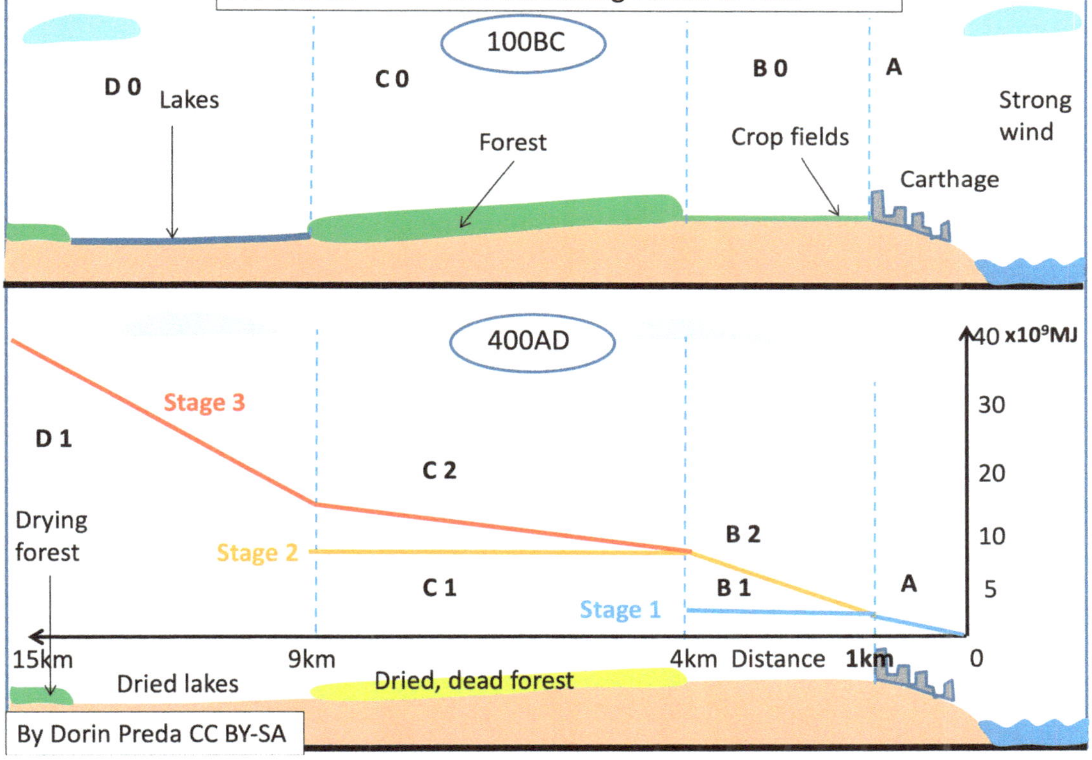

By Dorin Preda CC BY-SA

<u>In the second stage, after the drying of their crop fields, the Romans left (Figure 55 stage 400AD). Carthage which continued to be used by the pastoralists, so it did not revert to forest, but continued to dry the inland areas.</u> The former crop fields turned to bare ground because of the goats of the pastoralists, while the forest next to the past crop fields was half-dried and evaporated only half of the initial water, so it is equivalent to turning to grassland. Since there was less evapotranspiration

water going to the clouds, their extent decreased; therefore, solar radiation increased to 7,000MJ/m^2*year:

B1 to B2: 3,000m x 1,000m x 7,000MJ/m^2 x 25% = 5.25x10^{9}MJ/year.

C 0 to C1: 5,000m x 1,000m x 7,000MJ/m^2 x 5% = 1.75x10^{9}MJ/year.

Therefore, in the second stage, the additional solar heater per kilometre of coastline has increased by 7x10^{9}MJ/year. That heat was enough to behave as an ASH whenever there was significant wind from the North (which is the dominant direction). Thus, there was additional and more powerful drying of further inland areas.

A couple of hundred years later, the new ASH has dried everything on 10km further from stage 2, changing the entire landscape to bare ground (incipient desert):

C1 to C2: 5,000m x 1,000m x 7,000MJ/m^2 x 25% = 8.75x10^{9}MJ/year.

D 0 to D1: 10,000m x 1,000m x 7,000MJ/m^2 x 30% = 21x10^{9}MJ/year.

This way, the third stage added about 30x10^{9}MJ/year of solar heat, which gives a total of about 40x10^{9}MJ/year additional solar heating due to ASH. Said power corresponds to each kilometre in the East-West direction. With the increased capacity, the Sahara ASH started to create its own weather and wind; it's progressing to the South accelerated.

Once started, ASH naturally progresses by drying the vegetation in front of it with the excess solar heat produced by eliminating the previous vegetation. ASH is a "land cancer". Vegetation opposes entropy, while ASH thrives on it.

Before humans, North Africa never had an ASH and I assume that it was greener during the glaciations because the colder belt that brought condensation and precipitations was further south than at present. Sahara probably was never cold and never an advanced desert as it is now.

Many recent studies on the past of Sahara (https://en.wikipedia.org/wiki/African_humid_period), admit that it must have been more humid than now, but there are multiple inadvertences in their interpretations that do not hold. They assume that during the last glacial maximum of 14,000 years ago the Sahara was even hotter and larger than at present, which is illogical firstly because the cold and humid belt was in fact closer to the Equator, and secondly due to the aquatic animals and elephant herds in North Africa. Their least logical assumption is that the cooling that supposedly allowed the African humid period of 12,000 years ago was due to the increased albedo due to vegetation growth following increased rain due to the rise in albedo [79], that is just circular reasoning. Another inadequate conclusion of those studies is that evaporation would be a weaker cooler than the difference in albedo; this is another exclusive albedo intrusion into Earth Sciences. They also assumed that the maximum rate of precipitation in Sahara history was 500mm per year, ignoring that at the huge lakes that existed in Sahara, including the Mega-Chad which was as large as the Caspian Sea, in that climate would evaporate about 2,000mm per year, while the catchment itself would evaporate more than 500mm per year.

It is obvious that in fact, Northern Africa was a savannah with patches of forests and lakes, possible some patches of deserts sometimes, but never a complete sandy desert. I assume that vegetation clearing in the Horn of Africa and the Arabian Peninsula by the proto-Egyptians has triggered there a powerful HAC that remains in force to this day together with that initiated by the Romans in North Africa, so no more humid air reached the Sahara, which suffered avalanche desertification (ASH).

As shown in the lowest part of Figure 55, each stage uses the heat of the previous step to dry the vegetation at the present location. The power of ASH increases proportionally with the length of the territory it has already ravaged. At the end of my calculation, the power of the ASH was $40x10^9$MJ/year*km^2 after it progressed only about 15km from the coast.

Most power rises as hot air, only about 10% stays near the ground in the 10m turbulent layer with the land and its vegetation which means $4x10^9$MJ/year*km^2 applied on 1km width of coastline.

The quantity of water evaporated per year is M=$4x10^9$MJ/2.2MJ= $1.8x10^9$kg/year=57kg/s per km width, or 57grams per metre width and second. If the bulk humidity of vegetation in front of the ASH is 5.7 kg/m^3, it means that ASH can advance with a speed of 1mm per second. While this may not appear much, it means 30 metres per day, or 10km per year potential expansion of ASH.

The Sahara Desert has advanced to the south by 200km in the last century, which means only 2km/year. It results that ASH has enough energy (heat) to continue expanding the Sahara, whatever other climate conditions appear, including a glaciation. The desert reached its irreversible tipping point maybe one thousand years ago.

To be observed that the above calculation is for the power Sahara obtained on its first 17km of inland depth. Now it is 1,800km deep (N – S), so its total power is immense. However, most of that heat rises and heats the global atmosphere, while only a fraction heats the ground as ASH.

Just as a curiosity, let us calculate its additional solar heating power compared to the non-desert stage it was before the Roman times. The average solar radiation is 7,000MJ/m^2*year; we assumed the average evapotranspiration before Romans was 700mm/year, while now it is less than 100mm/year. Reflectance was 20%, and now it is 30%.

Past EH = 700kg x 2.2MJ/kg = 1540MJ x 100/7,000MJ/m^2 = 22%

Present EH = 100kg x 2.2MJ/kg = 220MJ x 100/7,000MJ/m^2 = 3%

Past AH = SR100 – R20 – IR15 – SH15 – P5 – EH22 = 23%

Present AH = SR100 – R30 – IR15 – SH15 – EH3 = 37%, thus an increase of 14%

The area of the Sahara is now $9x10^6$$Km^2$, thus the additional solar heat is $9x10^6$$Km^2$ x $10^6$$m^2$ x 7,000MJ/m^2*year x 14% = $8820x10^{12}$MJ/year = $9x10^{15}$MJ/year

The world energy (heat plus electric) consumption is now $6x10^{14}$MJ/year.

Consequently, the additional solar heating produced by the Sahara becoming a desert has a power 15 times larger than all the power produced by our society.

4.6 The Avalanche Forest Fires and Global Warming

A direct effect of the dryness produced by HAC and ASH is forest fires, which are known to create significant global warming at least through the vast amounts of carbon dioxide they generate [140]. Additional warming occurs less due to heat of vegetation burning, but mostly through an increase in solar heating at ground level because of the loss of vegetation's evaporative cooling.

Prior to human intervention, forests did not experience catastrophic fires that burned entire forests, as evidenced by the presence of very old trees. Lightning typically consumes one tree at a time, as natural forests are too wet to burn. Unfortunately, the local climate on many continents has dried out, reaching a stage when fires become firestorms, increasing local winds and therefore becoming self-propagating like avalanches (positive feedback).

The underlying causes of extreme forest dryness and violent fires include:

- HAC and ASH caused by coastal vegetation clearing which significantly dries out the inland forests.

- Fragmentation of forests for pastures, settlements, roads, etc. (section 3.8).
- Extraction of groundwater from forests, lowering the water table below tree roots.
- Regular human-started fires, reducing above-ground humidity and tree density.
- Global warming.

As coastal vegetation disappears and nothing is left to burn, soil erosion transforms these lands especially in tropical areas, into deserts. In the past ten years, intense forest fires contributed minimally to global warming while burning, but are expected to result in much larger solar heating after burning. To illustrate the effect of these fires on the heat balance, we calculate through method at point 4.1 with values for latitude of 40^0:

- Solar radiation 6,200MJ/m^2*year
- Evapotranspiration of burnt forest/grassland 200mm/year
- Evapotranspiration of grassland/crops 500mm/year
- Evapotranspiration of forests 1,100mm/year

For burnt forest/grassland:

- R Reflected solar radiation: 20%
- EH estimated to evaporate 200 kg water/year is 5%
- IR is 16%
- SH is 14%
- P is 0, as it is less than 1%
- AH results as 100 – 20 – 5 – 16 – 14 = 45%

Grassland:

- R Reflected solar radiation is 25%
- EH estimated to evaporate 500 kg water/year is 20%
- IR is 16%
- SH is 14%

- P is 5% (for a grazed grassland or two crops per year)
- AH results as 100 – 25 – 20 – 16 – 14 – 5 = 20%

Forest:

- R Reflected solar radiation is 15% (deciduous forest)
- EH for forest evapotranspiration of 1100 kg water/year is 40%
- IR is 16%
- SH is 14%
- P is 5%
- AH results as 100 – 15 – 40 – 16 – 14 – 5 = 10%

The EH for Burnt Forest/Grassland: 200kg x 2.2MJ/kg = 440MJ/year; 440 x 100/6200 = 7% ~ 5%.

The EH for Grassland: 500kg x 2.2MJ/kg = 1100MJ/year; 1100 x 100/6200 = 17.7% ~ 20%.

The EH calculation for Forest: 1100kg x 2.2MJ/kg = 2420MJ/year; 2420 x 100/6200 = 39% ~ 40%.

For a forest of 1 km^2, the additional annual solar heating because of burning it down is: 1,000m x 1,000m x 6,200MJ/m^2 x (45% - 10%) = 2.17x10^9MJ/ km^2*year.

For a grassland of 1 km^2, the additional annual solar heating because of burning it is: 1,000m x 1,000m x 6,200MJ/m^2 x (45% - 20%) = 1.55x10^9MJ/ km^2*year.

I do the same comparison with the fossil fuel power plants at 4.7. As a result, the same solar heating as one power plant is produced by deforestation through fire of 19/2.17 = 9km^2 of forest. Similarly, for burnt grassland, we calculate 19/1.55 = 12km^2.

Therefore, each 9km^2 of burnt forest, or 12km^2 of burnt grasslands equals one "power plant" of additional solar heat unleashed on the land until the forest grows back.

Each of the past forest fires all over the world have burned hundreds of square kilometres at a time, thus resulting in real, additional solar heating equivalent to tens of "power plants".

In the past ten years, due to increasing dryness of the continents and human actions, forest fires have increased greatly in size and number. Each mega-fire which burned about a hundred thousand square kilometres generate the equivalent heat of 10,000 "power plants". If that country had about 50 real power plants, that megafire is thus equivalent to increasing 200 times that country's global warming contribution for many years until those forests grow back.

The above is equivalent to global warming and is due to the additional solar heating at ground level because of evapotranspiration loss. The following add up:

1. the significant loss of clouds reflecting cooling because of the loss of forest
2. the heating due to the actual burning of the forests and grasslands
3. the atmospheric CO_2 produced through burning and its global warming effect
4. the loss of human lives and property,
5. the pollution and air quality degradation
6. the loss of trillion of animal lives in the forest fires and subsequent habitat loss
7. the climate change effects: the HAC and the ASH with their drying of surrounding areas
8. the intense soil erosion and loss of nutrients due to the fire ash taken by wind and water.

All the above would not happen if there would be no ignition (multiple simultaneous ignitions).

For point 1 above, we can compare the heat produced by fire to the solar heating induced by the fire. If the quantity of timber was 10kg/m^2, since the heat of combustion of timber is about 10MJ/kg, it results that the fire has generated 100MJ/m^2. However, as seen above, every year, the area of the former forest will produce 2,200MJ/m^2 of additional solar heat. Thus, 2,200MJ/m^2 x 100 years/ (100MJ/m^2) = 2,200 (we round it to 2,000). In each of the 100 years necessary for the burnt forest to grow back, it will produce about 2,000 times more solar heating than the heat it generated through burning.

The "solar heating" is not literally produced by the burnt forest, but it is solar heat allowed to heat the ground and the air because it is not blocked by the clouds of the forest and by the evaporating water. Burning is not only used initially for clearing a parcel of forest, but annually burns the stalks, leaves and roots of the cultivated plants as is said "to return the nutrients to the soil".

Additionally, preventive annual burning is used in the idea that it would avoid larger fires, or cultural burning just because of tradition. However, we should ask:

- What role has ground burning in diminishing the forest fires which are now started high in the canopy and propagate through the canopy which burns first and only afterwards the fire descends to the ground?
- Why would we want to return their nutrients to the soil while the creatures are alive?
- If they are not burned to go up in smoke, but left in the living creatures and soil, wouldn't those nutrients really be left in their best place?
- What is more useful: living vegetation and soil or ash flying in the wind or to the ocean?

An insufficient understanding of natural processes during our entire history, resulted in improper development and management of the coastal zones all over the world, leading to loss of inland vegetation, fires, droughts, and desertification of areas which had abundant plant and animal life before human settlement.

It is possible to avoid and reverse these adverse effects, as shown in Chapters 5,6 & 7.

5 Chapter Five: The Natural and the Human Environments

5.1 The Conquest

More than 1.5 million years ago we were insignificant mammals in Africa, afraid of most other creatures that were more powerful than us. We were living in an almost entirely forested world teeming with all kinds of life forms interconnected in wonderful but scary ways to us.

The Earth was a vast, mighty, magical place dominated on its surface by a living layer we call Nature, where the prime rule was that of life: creatures did all they could to survive. None of them succeeded individually, but only as species. Even species did not last forever, as they gradually changed and evolved, leaving some destitute branches to disappear, while others changed and evolved in something more complex, more wonderful.

The Earth itself was evolving from innate matter to intelligent living beings. The beauty of Nature was its diversity and vibrant vitality, which gave meaning to the previous billion years of senseless evolution of the Universe.

However, the smartness of living beings culminated with our cunningness, and one day about 1.5 million years ago we invented a way to master fire. That totally changed the game, as our wisdom did not match fire's colossal power.

Maybe the legend of Prometheus is true, and a mad individual from an advanced civilisation gave humans too much power too soon. Typically, creatures develop new capabilities as a result of their increased mental capacity; chimpanzees still cannot learn how to make fire, although some scientists try hard on public money to teach them how. The chimpanzees are the decent link.

However, we learned how to use fire and we began to use it all the time, overjoyed at the other creatures running terrified, a thousand dying only for one of them to end up as deliciously cooked food in front of us. The impenetrable forests succumbed to flames creating wide arenas where we could play our domineering games unimpeded.

We just do not know how many habitats and species we have exterminated, because we inherited from our illiterate ancestors just the huge deserts from where any relic or proof was long ago eroded with all the soil, leaving bare rock or sand behind.

Earth and its creatures are totally changed, and their evolution has been substantially altered.

5.2 The Human Way

Just Observation

The very lives of wild creatures are significantly disturbed by us, even in instances in which we just want to observe them. For example, in a recent Nature documentary, a South American puma was chased by a documentary film with their cameras and drones. So as soon as the puma was closing in on a llama, the film team moved closer and deployed their drone above the potential prey to film the kill first-hand. Obviously, the llama got scared and ran away. Once. Twice. Thrice. Unfortunately, the documentary team did not give up, although there were other close-up images of killing filmed by others. In the end, after getting injured by a llama because of the documentary team, the puma got a llama, but it might have thought that next time it will not get llamas, but humans.

The above example is nothing compared to what our biologists and other wildlife conservationists do to the creatures of the Earth, all perfectly legitimately and in the name of our Science Goddess. A hundred years ago they just used to kill, dissect or do experiments on animals in their labs. However, about 50 years ago the biologists decided that everybody should do this, so they made it compulsory for teenagers to dissect live frogs in the loving classroom environment. Thus, everybody learned the human way.

As masters of this world, we have taken the right of life and death on all the other creatures. Some might say that it is acceptable to be careless versus species with billions of individuals not considered useful to us or not under threat. But how do we protect the rare species?

There still is a flightless bird in New Zealand

There are less than 50 such birds left in the world. Biologists monitor them intensively, to find out why their population decreases. They have affixed radio transmitters on all the birds, so they can easily locate them. Since the biologists are paid to show the concern of the authorities for the wellbeing of these birds, there is continuous monitoring of the birds. The biologists periodically find with the radio receiver the location of the underground bird burrows, then grab the bird and drag it out. While the birds try frantically to escape, they scientifically weigh it, measure all its sizes, take all sorts of analyses including blood, genetic samples, etc. The biologists consider the blood analysis as being important, so they do it periodically to each and every bird. There are a lot of surveys and analyses; after all, the biologists need to justify their salaries, and we need to show concern for disappearing animal species. We really love these birds to their death!

There once was an iconic creature: the mighty rhinoceros

It was the embodiment of crankiness; one can find funny cards at the newsagents joking about the irascible mood of this creature. The large herbivore males are known for their grumpiness, which is imprinted deeply in them by Nature because they have to be protective of their females and offspring against the large predators.

This is their attitude and way of life, and should be understood by everybody. They just want to eat grass and be left alone by everybody! And who does not leave them alone, will get charged by the male, because this is why it is a male; this is an intrinsic part of its nature, as everybody (except the biologists) knows.

Because of the rhino male's strong shows of masculinity, some Asian urban myths have created a very lucrative market for impotent people, which threatens to drive rhinoceroses to extinction. Thus, after 2005, the biologists and conservationists "came to the rescue", and decided to capture and isolate all the rhinos to protect them from poachers.

Conservationists organised chases of rhinos in the reservations, to capture and transport them to more secluded places. They tranquilised them, but overdosing killed some of the animals. When the remaining males woke up in small cages, they just exercised their well-known aggressiveness and injured themselves to death by hitting the steel cages. A few less potent males survived and arrived in the other location where they are kept isolated in small yards or pens for the rest of their lives.

The biologists waited for the rhinos to mate naturally, but to their amazement, they did not. So, they tried to drug them to induce mating, with still no result. They also tried artificial insemination by hand without any initial result. The biologists ignore that these are not domesticated animals. Of course, they have to sedate them to do this, inducing massive stress on the animals, and risk of death due to chemicals, that actually happened.

Therefore, three subspecies of rhinos are extinct, one recently. It is not clear if the biologists and conservationists understand that there is a strong interrelation between life conditions, behaviour and male hormones; a male produces good sperm only if it is allowed to behave like a male.

Also, a female would not accept a dominated rhinoceros male, or the hand of a veterinarian for mating. This feature of fierce independence is prominent in rhinoceros. The biologists and conservationists destroyed rhinos through psychological warfare.

Figure 56: The Proud Rhinoceros

https://spectator.sme.sk/c/20028251/rhino-undergoes-artificial-insemination.html

Teams of conservationists worked hard and received generous donations for many years for "trying to save" the endangered rhinoceros, but were they doing the right thing?

The rhinoceros is much more than a large herbivore; it is the concept of freedom, power and independence! A rhinoceros in a pen is a contradiction of terms; it is similar to a fish glued on a table with a hose squirting some water onto its bronchi, while the biologist waits for that fish to naturally spawn...

The captive rhino market should be resolved by understanding that freedom and independence is the essence of rhinoceros, thus the notion of captive rhinoceros is illogical; this would result in forbidding captivity financing and the use of tranquilisers. The horn market problem can be resolved by properly punishing the poachers and the buyers.

Actually, to survive, animals just need to be protected from all humans.

5.3 So Where to Now?

The Earth used to belong to Nature; we conquered the planet and took control, but we do not seem to have a plan, and it seems that things are going wrong:

1) We did some good things (only for us), but we also made mistakes that undermine our interests
2) Do we know what we want for Earth?
3) What do we want for ourselves?
4) How do we get there?
5) Should we preserve Nature?

If we apply some common sense, openness and fairness, we can suggest solutions to the above:

1) It is possible to withdraw from the wrong way, and the safest way is back on the same path.
2) We might want to reverse climatic changes and revitalise Earth (Chapter 7)
3) Improve our wellbeing (Chapters 6 and 7)
4) By using only logical, ethical, scientifically proven and economical methods
5) This is more difficult to justify, because we appear not to care much about animals and plants, while in reality we depend on them. Therefore, we should just think about what we hold most dearly: the image of us. We like to believe that we are Homo sapiens (we have given this name to ourselves), which means "the wise man". And we also believe that we would be fair. As such, we could say: "<u>It would be wise to give the other creatures a fair go. So let them live!</u>"

5.4 The Opposing Environments

Strangely, many people think that they would like to live in Nature, but they do not understand that we have evolved opposite to Nature:

Natural conditions	Human requirements
Rain very much needed	No rain on us
Sunlight (sometimes scorching)	Not too much light
Night darkness	Night light
Wind, storms	No wind
-30^0C to $+50^0$C	+ 18^0C to $+23^0$C
Relative humidity 5% to 100%	Relative humidity 40% to 60%
Irregular ground surface	Flat floor
Mud, stones, water on the ground	Nothing on the floor
All kinds of insects	No insects
All smells imaginable	Only pleasant and not too much
Reptiles, tigers, scorpions	No dangerous creatures

Therefore, the natural and human environments are opposite! We would save our time and effort, and would avoid harming creatures if we admit that our environments should be separate.

Consequently, the notion of human "green living" is absurd. There can be cities with a lot of trees, gardens, zoos and aquariums, but no combination between real wild Nature and human civilisation is possible. The wilderness that adapts to human lifestyle is not natural anymore; it is just "domesticated" and does not evolve naturally, but according to our will.

There were recent attempts of creating small human settlements into forests in the idea of allowing both the survival of Nature and human development. For a short time, the residents enjoyed the appearance of Nature surrounding them. However, dangerous fauna for humans (wolves, snakes) was exterminated. The shy creatures just left that area (deer). As the forest was divided by roads, animal causalities on roads increased, and the surviving creatures had to alter their life cycles to the point of slow demise. Those ecosystems were disrupted and devolved. And then there are the devastating forest fires that have certified that the whole concept was wrong and could not work.

We should renounce the "pastoral life" idea that tried to integrate humans into the natural environment. We can save money and time if we focus on our life, rather than on an impossible marriage with Nature. Our separation from Nature would also be very beneficial for Nature, which only flourishes if it is left by itself, without any human intervention.

There are many modern designs of tall buildings spread in a large park, having vegetation on their roofs and even on their walls, and connected with roads cut through that park (see Figure 57).

Such a setup evaporates more water, keeps city temperature low, and reduces pollution, which is good and should be encouraged. However, those patches of vegetation certainly are not ecosystems and will not be able to survive by themselves; this is not Nature preservation. We should not fool ourselves into believing that this way we accomplish our role of preserving Nature because it is not true!

FIGURE 57: Tall buildings with green envelopes

Credit: Singapore Management University

<u>To really preserve natural biodiversity and allow its evolution, we need to create very large natural reserves without any human intervention.</u> Once we fully understand that we have to separate the human and the natural environments, so their occupants each obtain what they need, what follows is just engineering.

5.5 Nature Reserves

It is imperative to soon setup large Nature Reserves in all significant environments, for all biotopes, because other land developments now encroach rapidly into the existing natural reserves. If the Nature Reserves are not immediately locked out of developmental pressure, the remaining free patches of land would be too small to support unimpeded life of those plants and animals.

Some of the ideas below are already being applied in different countries, but on a small scale and inconsistently. Consistency, international uniformity and control through UNESCO is desired.

Ideally, we would leave the ocean to Nature, stop polluting it and stop excessive fishing.

For Nature to survive on land, it needs vast, undisturbed reserves with no human intervention.

Unlike the current forest reserves governed by local councils, the location and size of each nature reserve should be put aside through National Conservation Schemes that can only be modified by the National Legislative Body in agreement with the entire population (referendum). The UNESCO accord would also be helpful. This way, new land developments could not gradually encroach into a nature reserve.

The rules governing nature reserves and their geographic boundaries should be included in the National Constitution, to ensure they cannot be easily changed when the government changes.

The administrative bodies of the nature reserves should be governed only by the constitutional rules, independently of national governments. The reason is that a nature reserve should be there for thousands of years. All nature reserves should be registered with UNESCO.

A nature reserve has to be very large in size (at least 100km^2), ecologically balanced (with at least one natural river or lake) and un-interrupted by roads. If it must be crossed, a tunnel or suspended highway/railway would be preferred over wildlife corridor overpasses (Figure 58).

There should be no intervention in the lives of animals and plants in the reserves; the whole idea is to let them develop undisturbed. Biologists should not interfere in any way; they should never enter a nature reserve, especially not to analyse creatures or collect specimens for study. Smaller botanical gardens and zoological parks should be used for this purpose.

FIGURE 58 Overpass Wildlife Corridor

Credit: Hsiu Mei Tseng

Fully enclosed visitor passageways with one-way glass viewing points can be installed for tourists, biologists and wildlife researchers, but they should be elevated above the ground not to disturb land creatures. Video cameras would be a preferred alternative.

The single artificial element should be water sprinklers installed at a height on trees to provide water in case of excessive drought, preventing forest fires (especially in Eucalyptus forests maintained for koalas). If there is no permanent lake, large rainwater tanks established for this purpose will ensure that water is available whenever needed. However, with proper and ample reforestation, as suggested in Chapter 7, the climate will become more humid, reducing the risk of forest fires.

No low-altitude helicopter or drone flights should be allowed above or within a nature reserve.

Unfortunately, most of the land was already occupied by sprawling human settlements and other land developments, thus we have to find a solution for Nature preservation on Earth. This is possible as shown in the next chapter, by rethinking our settlements strategy, resulting in benefits for human society, besides reserving minimal areas for natural reserves.

Another important issue is the tendency for exaggeration of each party in this tug-of-war between conservationists and land developers; they each take excessive steps and decrease agreement. Conservationists now stop the establishment of large water reservoirs (which are good for restoring climate and protect against forest fires) by claiming that they damage the ecosystem of some little birds. Obviously, the birds and even land animals would not be dramatically harmed, as they can move as the water of the reservoir very slowly rises. Therefore, balance and reason are necessary in dealing with environmental issues (section 7.8). We have to choose wisely between two harms.

Correct planning for natural reserves could have great advantage for land developers also, who's actions are at present impeded by erratic land locks in poorly justified categories of ownership, protection, etc. It would be beneficial if proper and sufficient Nature Reserves would be soon setup, while the rest of the land would be clearly marked for sooner or later development.

6 Chapter Six: The URB

We have occupied most of the land with our settlements and our agricultural, natural resources, industrial and infrastructure developments, clearing more and more of what was the natural environment. However, this development was mostly chaotic and resulted in many inefficiencies.

The coastal zone was especially under pressure and suffered the most intense damage, inducing desertification of inland areas due to HAC and ASH. This chapter aims to introduce a solution that allows a large population on the coast with high living conditions, while also providing room for vegetation to protect from HAC and ASH and to allow conservation of the natural environment.

6.1 Past Settlements and Land Developments

The usual land development stages of a plot of land involves demolishing old buildings and constructing new ones, as the place transforms from rural to urban:

- Single country house, with the large surrounding plot of land being used for harvesting
- Single town house, with the large surrounding land being used for a suburban garden
- Subdivision of the land for a few single-story townhouses; the large garden disappears
- Demolition of the single storey for multistory townhouses
- Demolition of the multistory townhouses for a city block of flats or offices

This transformation may take 100 years; thus, each generation demolishes and rebuilds one stage, incurring large expense and significant work. If money is borrowed for the expense, the mortgage extends over the entire 20 years of each step, meaning that the expense is doubled due to interest.

At the township level there is also a lot of work in reshaping roads, redesigning public structures, and retracing cables and pipes. Actually, the entire township is a continuous workplace that frequently blocks traffic and annoys neighbours.

This ongoing construction work is the most expensive, time and resource-consuming human activity. Unfortunately, almost nothing is recycled from demolished buildings, meaning that we continuously transform natural resources into waste.

This chapter aims for an improved type of land development that avoids the above inefficiencies, unnecessary expenses and waste. Ideally, it should be applied on undeveloped land, but it can also be included between the above steps on large plots of already developed land. Finally, there are cases where this new type of development just cannot be applied, for example due to pre-existing cultural heritage buildings, or too-expensive existing land developments. In these areas, our focus should be on improving the thermodynamics and water cycle of that area, as outlined in Chapter 7.

6.2 The Future of Settlements and Land Developments

Human settlements have always been influenced by the needs of their inhabitants and their natural settings. In subtropical climates, the first shelters were simple roofs made of tree branches and leaves to shield from rain. In colder climates, people-built walls for their huts, while in extremely cold or hot climates, natural caves were preferred for their thermal properties.

The first prehistoric city at Catalhoyuk in modern Turkey, dating back 9,500 years, had a unique characteristic of streets on the roofs of houses, with entrances leading downward. This design was a response to the presence of lions in the surrounding jungle, highlighting security as a top priority.

Over time, the characteristics of human settlements have evolved based on technological advances and changing requirements. With the rapid technological progress in the past century, there has been a shift in lifestyle needs, prompting a redefinition of ideal future settlements.

Key features of the new settlements include a transition from heavy to medium-light industry and services, prioritising quality over quantity, providing good entertainment, enabling work from home, and emphasizing information production and consumption. The infrastructure should feature improved transportation and be structured like a small city to meet all essential needs within close proximity, similar to a "15-minute city", but more compact in space.

These new settlements should prioritize self-sufficiency, reduced transportation needs, and enhanced local services, like education and healthcare. The ideal size for the new settlement is estimated to be around 20,000 residents, although smaller settlements may be viable depending on local conditions. My suggested new settlements would function as small cities due to their integrative characteristics; they could be referred to as "Unitary Rational Buildings" or "URB".

An important issue is the cost of land, which is soaring all over the world, thus more than half of the value of the dwelling is in the land it occupies. The cost of a house in developed countries has reached about half the life income of an average employee. Therefore, it is necessary to build in inexpensive multi-storey style. To ensure URBs success, location selection should be based on geographical features, soil type, water availability, workforce requirements, and natural resources. They should be as compact as possible to minimise transportation needs, well-connected to information and power networks, and designed for self-sufficiency based on their population size.

In essence, observing the diminishing gap between rural and urban lifestyles, I propose building in compact city-style from the start, to avoid the need for future demolition and reconstruction.

In conclusion, the future of land development lies in establishing URBs that meet the future needs of residents, while promoting quality and efficiency, for a sustainable world.

6.3 New Settlements versus HAC and ASH

As shown, in the past our coastlines have been covered with very wide strips of concrete, tiled roofs, and asphalt, leading to intense solar heating. These hot air curtains (HAC) prevent humid marine air from reaching inland areas that dried out, while discarding excessive rainfall on the coast.

To avoid HAC and ASH, I suggest to build relatively small and compact self-contained URBs on coastlines, surrounded by large forests, orchards, plantations or agrivoltaics fields (see Figure 59).

FIGURE 59: URBs on the coast

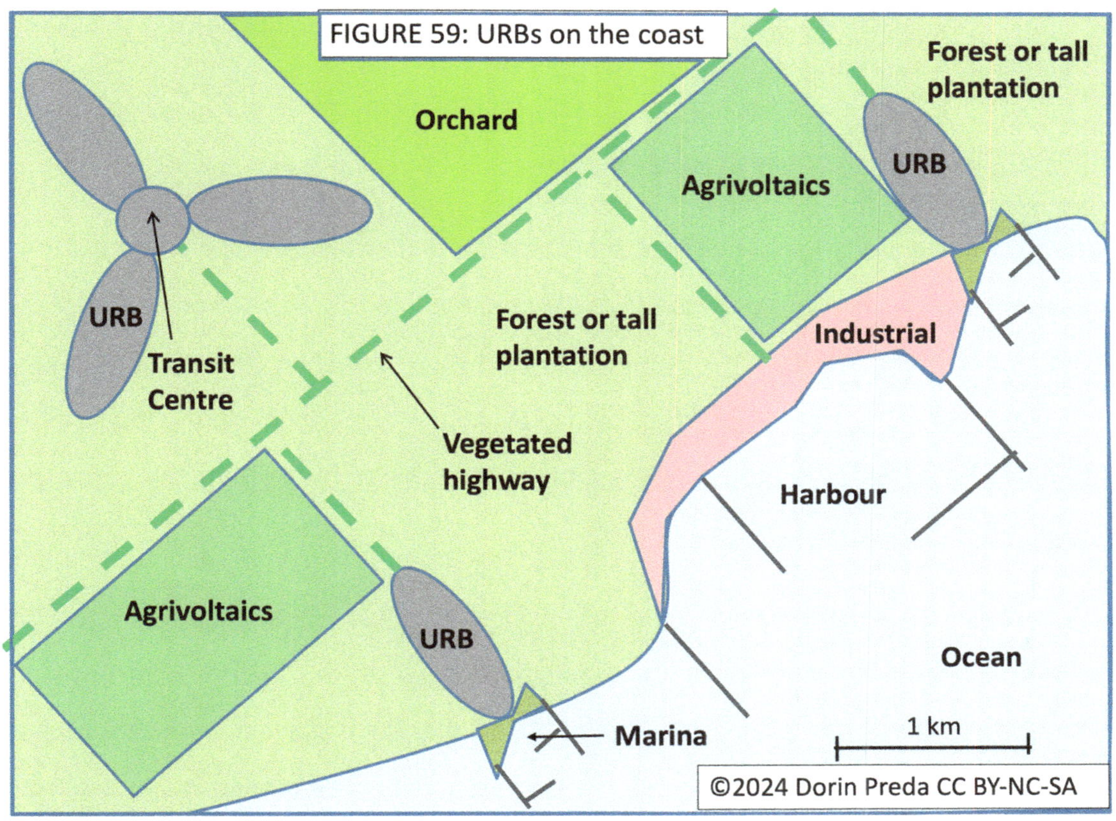

URBs longer than 1 km would have too much floor area per person, so more populous URBs could be created by merging smaller URBs into a common transit centre, like in the above figure.

A URB on a smaller land area than current settlements would not produce a HAC, especially if its roof collects solar radiation without overheating. When vegetation occupies more land than URBs, climate will be dominated by the former, preventing formation of HAC or ASH on the humid coast.

By "tree plantation" I mean useful trees or alternation of fruit trees and grassland crops (intercropping). Agrivoltaics can provide both fresh produce and power to the nearby URBs.

Large trees should be planted on the sides and even on the centreline of the highways to avoid HAC and create the "vegetated highways" shown in figure 59, that are also more driver-friendly.

URBs are high-density settlements that accommodate both a large population and ample green spaces on the coast.

6.4 Mars versus Earth Deserts

Due to a lack of planning, former cities were built, used and abandoned. They were repurposed, demolished, or left to become deserts. The reality is that we deplete lands and move onto new ones, leaving wastelands behind because we do not know how to create closed cycles, like Nature.

Instead of investing in research for a sustainable way of life, we continue to promote and support wasteful consumerism, which is leading to the downfall of our civilisation and of Nature.

We only understand open cycles: use something, transform it into waste and dump it. Now we are not just content with using-ruining-dumping cities, but we want to do the same to planets! Some suggest moving to Mars because "we cannot repair Earth"! They likely plan to move to Mars and ruin it as well! This is a very dangerous idea, because it creates a wrong mentality among uninformed people, which, if supported by leaders, would create the acceptance of totally ruining Earth, when in fact it is realistically salvageable.

To send 1 tonne to the Moon, we need about 50 tonnes of liquid oxygen and hydrogen, and even more for Mars. These pose significant environmental costs to Earth, since we currently use fossil fuels to produce rocket fuels. Given the high energy demands of space colonisation, we should focus on developing a renewable energy industry on Earth before venturing into space.

We have directed our valuable resources towards unrealistic plans to migrating to Mars, when we have not even achieved a self-sustained habitat on Earth. The Biosphere 2 experiment failed 30 years ago: https://en.wikipedia.org/wiki/Biosphere_2. We have yet to figure out how to create a self-sustained airtight habitat where people can live, grow food and recycle everything.

The Biosphere 2 building is still available, so we first need to find out how to live indefinitely in a perfectly airtight enclosure with no matter exchange with the exterior. If successful for 3-4 years, we can afterwards attempt to establish a permanent base on the Moon, which is not only tightly enclosed like Biosphere 2, but is also exposed to cosmic radiation, like Mars. After gaining experience on the Moon, we can expand to Mars and other planets if successful. We should observe that, wherever we want to live in a maintainable way for more than 500 years, on Earth, Moon or Mars, we must develop self-sustainable, closed-cycle lifestyles to ensure the future of humanity.

The main challenges on the Moon or Mars are breathable air, and protection from cosmic radiation. On Earth, we do not face cosmic radiation, and we have an abundance of air. Therefore, creating a self-sustained human habitat in the Sahara, where temperatures are milder than on the Moon or Mars, would be more feasible. The Sahara has much more water than the Moon or Mars, and we can wear regular clothes instead of space suits.

By revitalising the Sahara through revegetation and bringing back rainfall, we can transform it into a paradise where people from northern regions would happily migrate. Restoring vegetation in the Sahara, would lower temperatures to very pleasant year-round averages, making it more appealing than cold and misty Europe. Chapter 7 below outlines how we could revegetate Sahara.

Settlements that do not produce solar heating, like the URB (this chapter) or those described in section 3.8 within large forests or plantations are the initial step towards a more rational and enjoyable way of living. This contrasts with the sprawling, irrational developments that have plagued Earth with wasteful settlements.

Investing in Earth rehabilitation would yield profits within the lifetimes of the investors, unlike investments in massive rockets or a Mars colony, which may take hundreds of years to pay off (if successful).

Additionally, by first restoring the natural world, we will be rewarded with its beauty and balance.

6.5 Advantages of Apartments Over Houses

1. Our idea of Nature close to home often involves a well-maintained lawn or a garden with ornamental bushes and flowers, which are not natural ecosystems. This green space can easily be created on a few square metres apartment terrace and have the same pleasant effect as the hundreds of square metres around a house. Additionally, it is much easier to maintain, as most people do not enjoy spending their weekends mowing a large lawn, trimming bushes or cleaning out gutters filled with dried leaves.
2. There is a strong push in advertising for large houses, as they bring in more profit for real estate businesses and banks. The idea of a bigger house may seem more desirable in commercials, but once purchased, people often realise how difficult and expensive it is to maintain. Many people end-up working most of their lives to pay off the mortgage for a house, leading to financial strain and family issues. It is wiser to live a more comfortable and peaceful life in an apartment rather than being burdened by the debt of a large house.

FIGURE 60 Privacy Problems

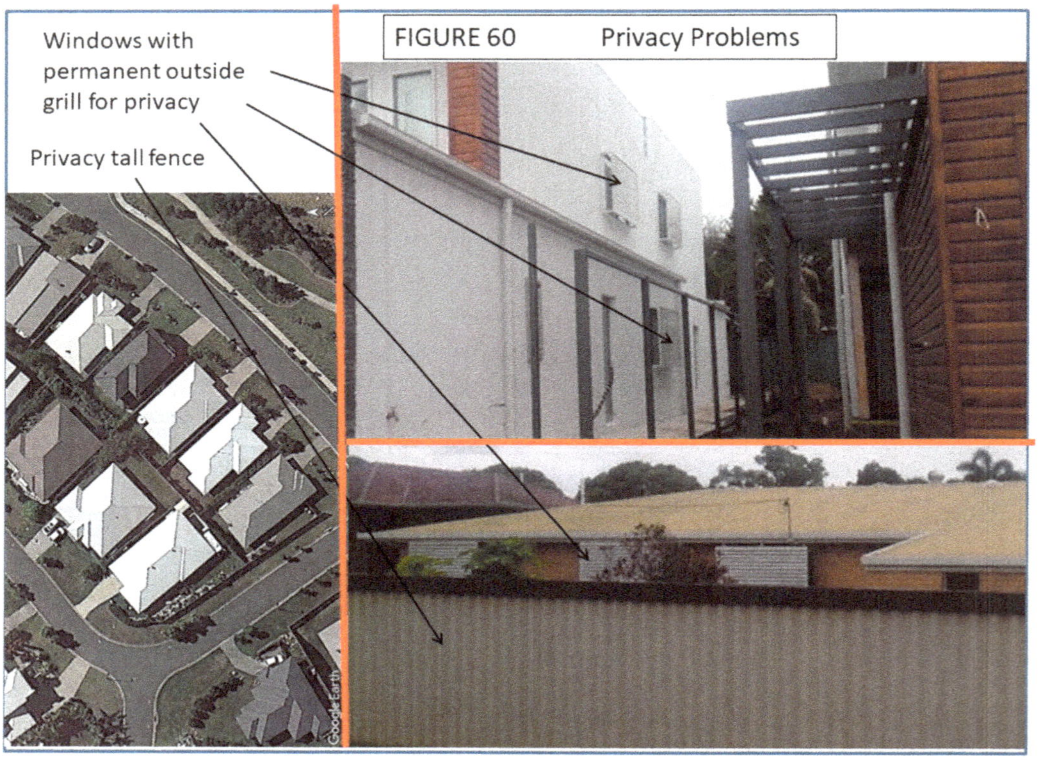

3. As shown on the left in Figure 60, modern houses tend to take up all available land, losing the traditional concept of a house and resembling more apartments with very small or non-existent courtyards.

4. Despite having many windows, people often cannot enjoy them due to privacy concerns or lack of a pleasant view, leading to installation of permanent blinds or grills to their windows.

5. Modern life has become more static than centuries ago, with people shifting their focus from the outdoors to indoors. With increased isolation and more individuals working from home, the home office in an apartment has become more relevant than the barns, vegetable garden, or the home workshop found in older, large houses.

6.6 URB Requirements

Societal: - Ensure a city style, civilised life

- Protection against floods, earthquakes, fires, hail, storms, pests.
- Education centres: kindergarten, library, primary and high schools
- Social services: hospital, infrastructure, firefighting, police
- Service centres, shopping centres, supermarkets
- Spaces for restaurants, bars, cafés, entertainment centres, leisure and sports
- Efficient transport and excellent telecommunications

Economic: low annual distributed capital cost and minimal running and maintenance costs.

Individual comfort: good air quality and temperature, low noise, minimal transport, privacy, safety, high-quality low-cost services, and local employment opportunities.

Environmental: separation of the human and natural environments, no pollution, no groundwater extraction, URBs provide compost to the fields, allowing protection of the natural environment.

6.7 URB Description

6.7.1 Structure:

- For a population of 20,000, the terrain size would be 1000m x 300m, approximately 0.3km^2.
- To provide a large floor area for a small land area, the URB should have about 20 stories, which is feasible with average engineering anywhere.
- Since streets, parking spaces and spaces between buildings of a usual city take more than half of its area, we eliminate them to increase density. This way, all tall buildings of a city merge into a single very large structure: the URB, as seen in Figure 61.

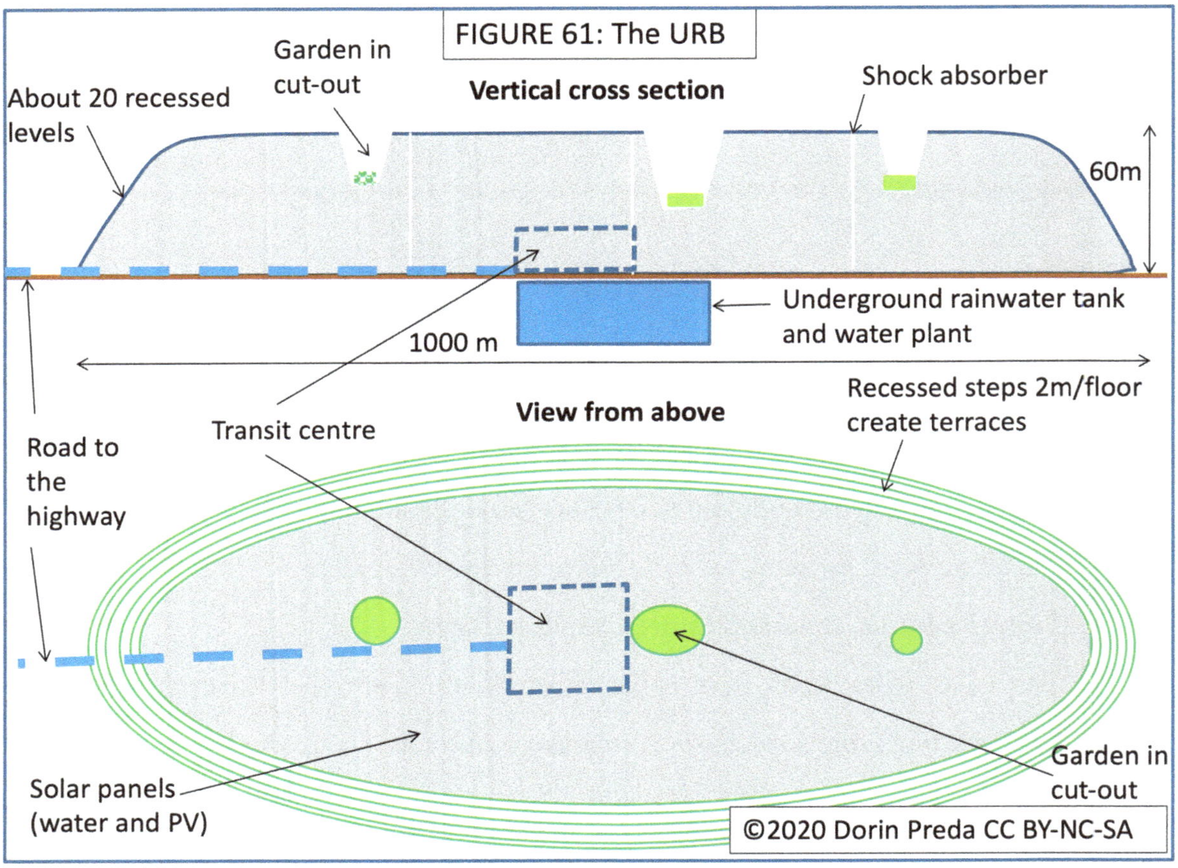

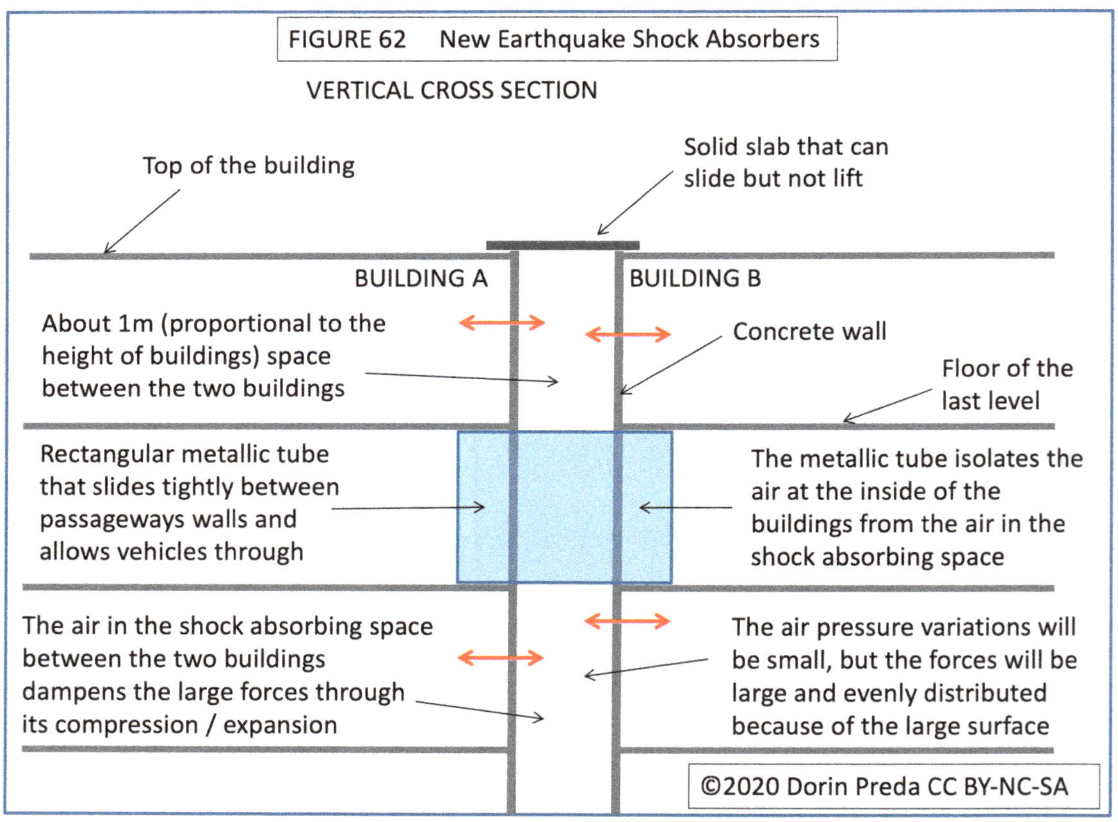

- Although it is about 60m tall, the URB is very stable because it is very wide. Each floor is 3m high including an engineering space of 0.5m at the top.

- The URB offers a lot of space (about 200m^2 per inhabitant including the public spaces); it can be varied by increasing/decreasing the width of the URB.
- In some places, there can be cut-outs in floors, allowing fresh air and light to reach gardens.
- A large rainwater tank and a water plant would be placed underground.
- URB has a structural frame made out of high strength and long-lasting materials, while the walls could be made from less resilient but fireproofed materials.
- My new type shock absorbers between building segments prevent earthquake damage (see Figure 62), and have very high resilience, while the cost is hundreds of times smaller than other large shock absorbers of the same effect. The air pressure variation in the shock absorber would be small, but the resulting damping forces are large due to the large surface area. There are rectangular metal tubes connecting the passageways from two adjacent buildings and separating the air in the passageway from that in the shock absorber.

6.7.2 Functionality

- The URB should contain everything a community of about 20,000 people needs, except agriculture and heavy industry. Besides the apartments for the inhabitants, the URB also contains offices, business places including light industries and services, kindergartens, primary and high schools, hospital, fire station, commercial centres, service centres, police station, entertainment centres, indoor sport centres, etc (see Figure 63).
- There are single-way and double-way passageways for pedal and light electric vehicles; lifts and ramps.
- Natural light penetrates in through reflective air vents in the roof.
- The side of the URB would be tilted, with each level extending about 2m further outside than the level above it, to allow vertical perspective for a terrace.
- Small gardens would bring additional light deep inside the URB.

FIGURE 63: Vertical Cross Section through URB

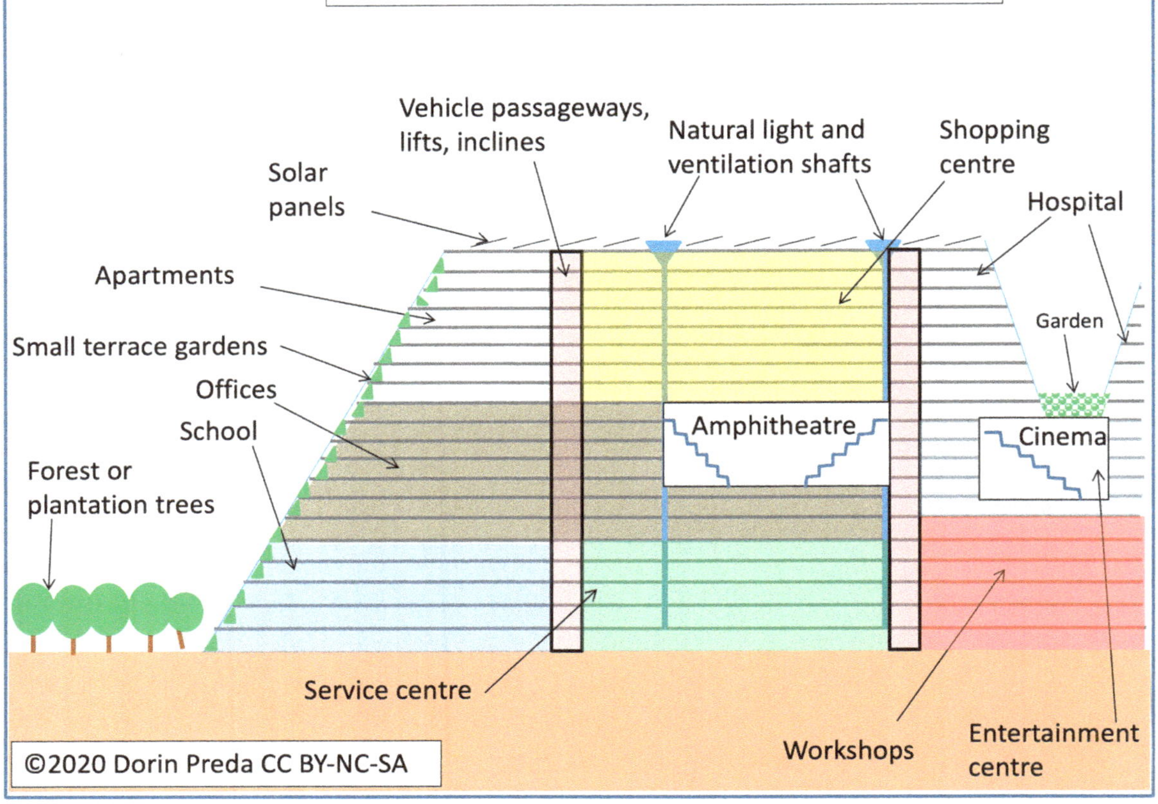

FIGURE 64A URB Apartments

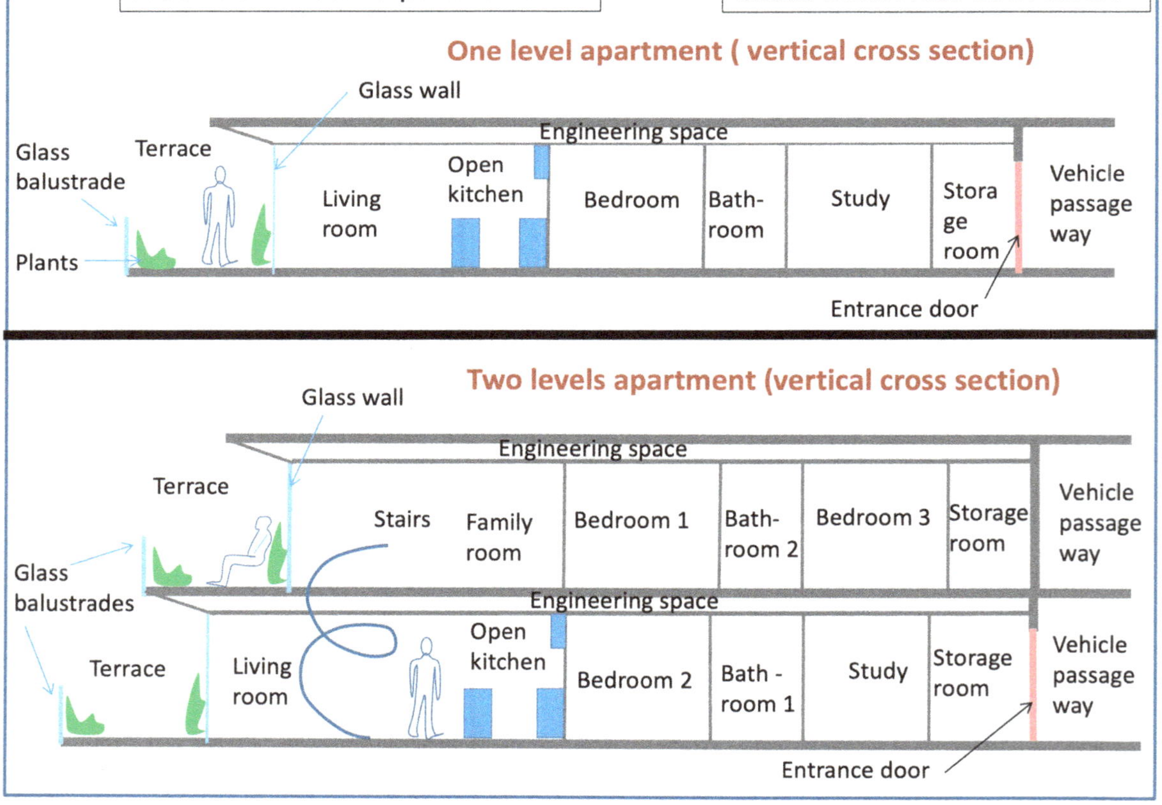

- My new "Principle of optimised façade" states that a room or other URB space should only be placed on the outside envelope if it needs a window. Living rooms need windows, but not the bathrooms, bedrooms, storage rooms, etc. Schools, offices and hospitals need windows, but not the workshops, cinemas, warehouses or supermarkets which do not have windows at present either. The streets (passageways), and factories, do not need windows.
- The outside envelope of the URB is unchangeable because it is made up only of windows, while the passageways are on the inside of the URB. This way, there are no streets that penetrate this envelope for people to build new houses around the URB. Therefore, a URB cannot grow, which solves the uncontrolled expansion (sprawling) problem of other settlements.
- Each apartment (see Figure 64A) has natural sunlight and clean air, and a wonderful view of the surrounding forest/plantation, much better than existing suburban houses.
- The URB apartments are delimited by their structural floors, ceilings and walls with the neighbours. A false ceiling will create with the structural ceiling an engineering space of 0.5m which houses the pipes, air conduits and wiring necessary for that apartment
- Therefore, the structure of an apartment and the placement of the internal walls can be decided by the dweller. Still, it would be logical to place the living room towards the glass wall, which offers a beautiful forest view. The dweller could actually decide where to place the external glass wall, creating either a larger living room with a small terrace outside, or a smaller living room with a more spacious outside terrace.
- There can be small apartments, large apartments, and even two-level apartments.
- Since a URB can last hundreds of years, there will be many residents using an apartment over time; each of them could change the position of the internal walls and rooms functions.
- URBs can be made out of concrete reinforced with basalt or carbon fibre instead of steel, as basalt/carbon do not rust and are more environmentally friendly and last longer than steel.
- Therefore, the outside envelopes of URBs are made only of structurally sound and environmentally resistant glass, long lasting, very easy to maintain, which confers an elegant outside aspect to URBs. The green patches of ornamental plants on URB terraces would visually connect to the vegetal surroundings, as suggested in Figure 64B.

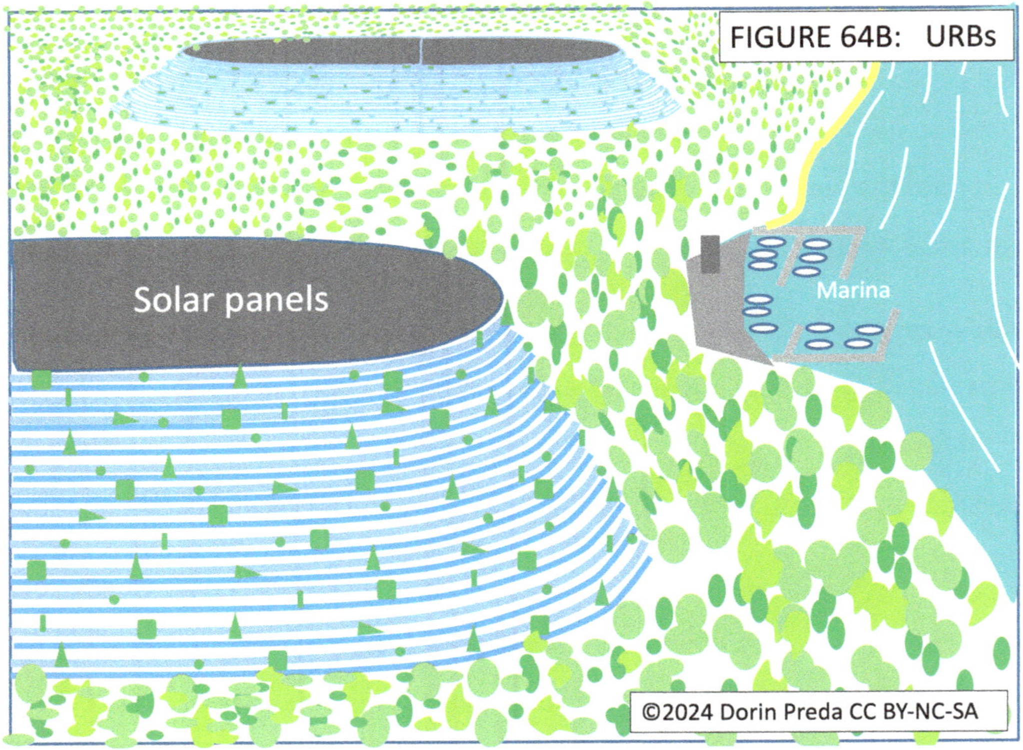

6.7.3 Advantages of URBs over Present Houses

- Nicer view for all residents (only the trees, mountains or seashore, with no other houses or streets, or back walls of another building, or the neighbour's clothes drier.
- Perfect privacy (nobody can look inside).
- Silence, since the engineering spaces provide sound insulation; the noise made by one neighbour travels only towards the forest, while the forest is not noisy like a street.
- Safe from fatal fall, since there is another terrace 3m below.
- Cleaner air, as outside is only forest (or plantation), and no streets with cars or trucks
- Much lower capital cost, since the ever-lasing walls, structural floors and ceilings are shared with the other apartments which will last hundreds of years (low capital cost/year).
- Much lower maintenance, as there is no large garden or dependences to clean up and maintain, no outside walls or roof or gutters to clean, no need to repaint (it is only glass).
- Much lower running cost, as heating and cooling costs will be very low due to the very small interface area (about $15m^2$ of façade per apartment).

- Due to the much higher density, it is possible to collect separately the liquid and solid dejections and process them in fertiliser and compost for agriculture.
- Higher security, as there are no fences that could be jumped over.
- For a URB of 20,000 inhabitants, 20 floors, 1000m length and 300m width, to each person it would correspond: 9m^{2}of windows, 200m^2 of floor space (private and public spaces), and only 20m^2 of building envelope (this illustrates the minimal heating/cooling needs). The advantages and economic performance are obvious when compared to a present city: 20m^2 of windows, 1000m^2 (including public spaces) and 2,000m^2 of building envelope per person. The URB dweller can pay only 1/100 of the heating and cooling expense of present cities. If URBS were to become widely used in a country, the energy consumption of that country would be reduced tens of times, with multiple benefits.

6.7.4 The Roof

- A "green roof" is environmentally better than bare concrete, but tree roots can penetrate the waterproofing membranes, making it very difficult to locate and repair a breach.
- The roof could be partially covered with solar water heaters to provide the URB's daily requirement of 500 tonnes of hot water from vacuum-type solar water heaters that do not heat the air more than the leaves of the forest.
- The rest of the roof could be covered with solar PV panels to produce electric power.
- By using solar collectors and solar panels on the roof (not green roof), it is also possible to collect rainwater from the roof for the water treatment plant in the basement.

6.7.5 Air Conditioning

In a warm climate, a URB surrounded by large vegetation will need its air cooled and de-humidified. This task is much better and more economically done by the centralised air conditioning plants of the URB than with small individual units.

The air conditioning systems have to exhaust heat, which is usually done by dumping hot air that is not desired due to its HAC effect. Therefore, the solution is to have the condensers of the heat pumps cooled by water through partial evaporation in cooling towers (see Figure 65). This way the URB behaves like a forest, ejecting water vapours, not hot air.

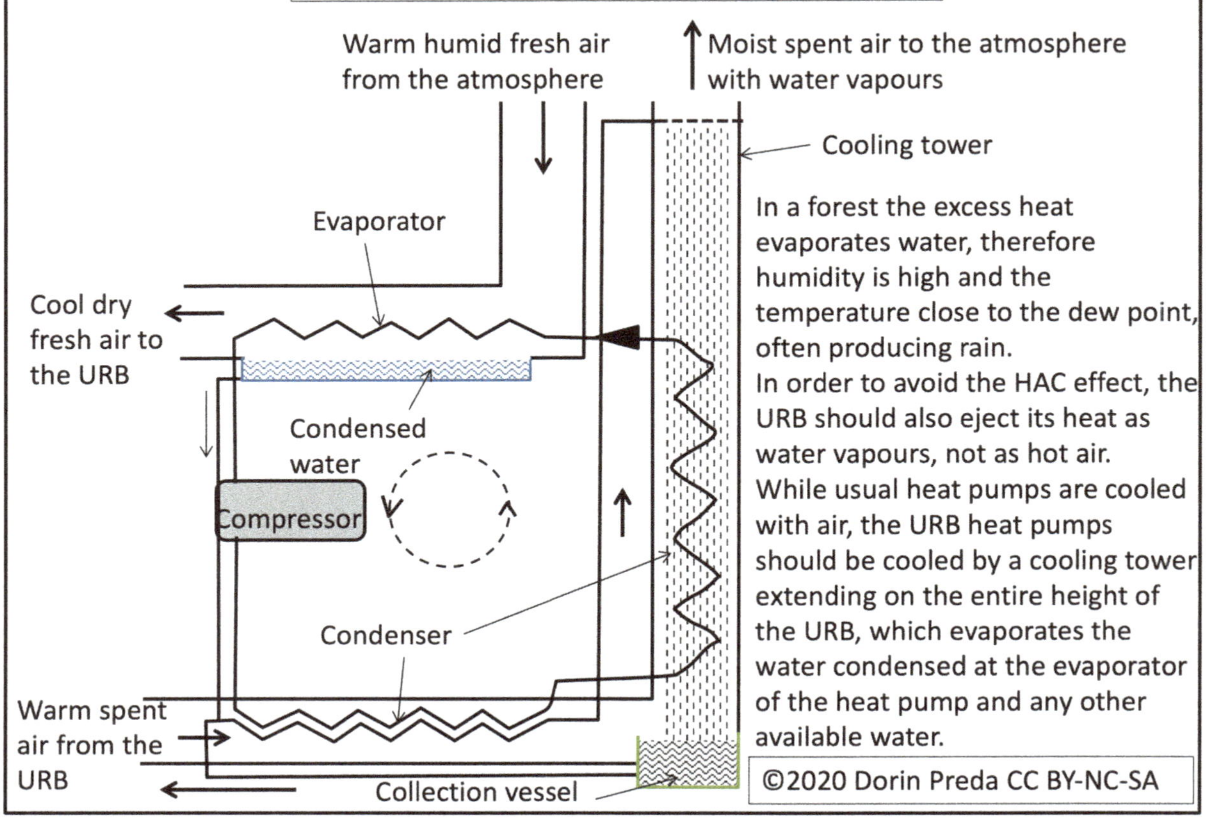

FIGURE 65: Water Vapours Heat Exhaust

Alternatively, if the cooling load is not too high (as in the subtropical climate), the excess heat of the URB could be used to heat with a heat pump the hot water, which will be needed anyway.

6.7.6 The URB Transport System

Outside the URB there is the usual transport system with cars, buses, trains, trucks, either on the highway and/or rail which connect the URB with airports, seaports and other URBs.

Due to the short distances inside a URB, and the need of exercise for people, it would be preferable that all healthy people just walk and eventually use hand trolleys for shopping.

However, for people who do not want or cannot walk, my transport solution **inside** the URB is:

- Chairs in houses, cars, trucks, trains, and supermarket trolleys of the URB are replaced with Personal Electric Vehicles (PEV), and the entire URB is adapted to make use of them everywhere. Examples of already existing PEVs: [31], [32] and top of Figure 66.
- Each person could have his/her PEV, which fits his/her body and is the personal vehicle to use in passageways (PW). It would also be his/her chair everywhere: at home, at the office, in concert halls and cinemas, schools, in conference rooms, restaurants, etc.

FIGURE 66 : A Foldable PEV and two FEV models

FIGURE 67: URB Transit Centre **Vertical cross section**

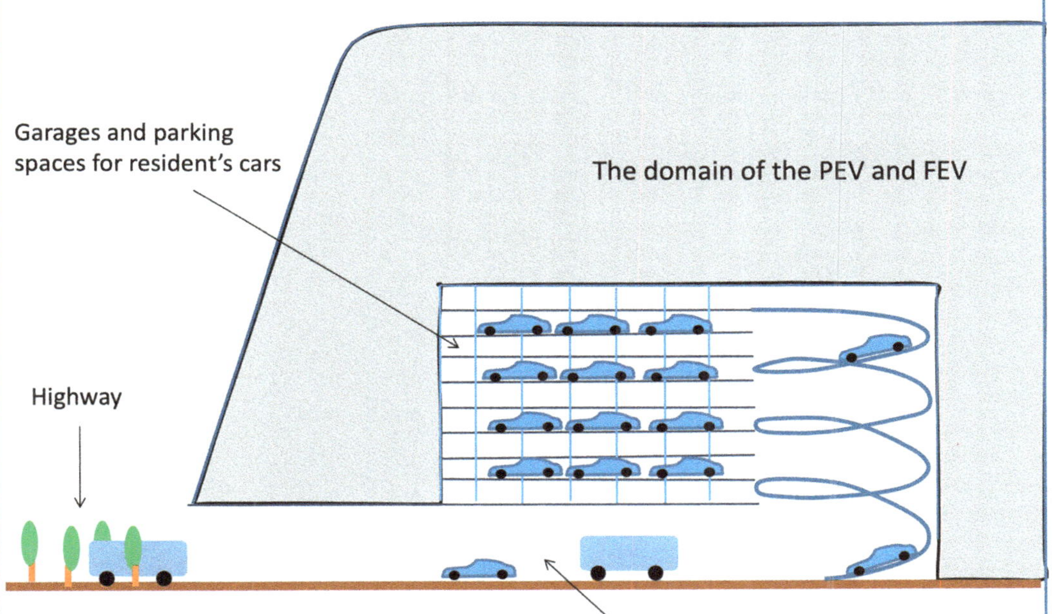

- The PEV should be controlled with a joystick on its armrest (see Figure 66); the bike or steering wheel type is not compatible with desks and tables.
- In cinemas, theatres, conference rooms, restaurants, there should be no chairs, but only platforms and inclines (ramps) for PEVs

- PEVs do not need parking spaces because when not used they just take the place of chairs in apartments.
- In supermarkets people could walk behind their PEV and push it in front of them, using the PEV as a trolley; after paying at the cashier, they put the groceries in a large foldable bag behind the seat of their PEV and then drive it home.
- PEVs only need lanes 1m wide on the passageway (PW) of the URB, not 3m wide as cars.
- A one-way PW would be 4m wide and would have two lanes for travel plus one lane on each side for manoeuvring.
- A two-way PW would be 6m wide and would have four lanes, two per direction of travel plus one lane on each side for manoeuvring.
- Pedestrians could use the manoeuvring lanes.
- Inside an URB, a PEV driving at 15km/h can get anywhere in less than 10 minutes.
- Since the maximum speed is 15km/h, there will be fewer accidents and no casualties
- People on PEVs can travel between URBs in dedicated buses that have platforms instead of seats, so they can ride their own PEV on the bus.
- Foldable PEVs (see Figure 66) could be carried in the luggage compartment of coaches, or in the trunk of cars.
- PEVs are smaller than a car which usually carries a single person during weekdays; at work PEVs take the place of the desk chair.
- PEVs have much smaller capital costs than cars
- PEVs need far less energy than a car and can be recharged at home, school, work, etc.
- Cleaner than fuel powered cars, PEVs are less expensive than electric cars
- Couriers can deliver large objects with FEVs (Freight Electric Vehicles) of Figure 66.
- The FEV would drive on two PW lanes, as they are wider than a PEV.
- Many PEVs and FEVs already exist; they are a mature technology.

- All the rooms, spaces and passageways (PW) of a URB are adapted to the PEV – FEV system, resulting in large advantages compared to the present car-truck transport system
- PEVs do not require driving licenses and can be used by children; and they eliminate the need for school buses
- Can be used by disabled people, saving in caretaker car drivers.
- If a PEV malfunctions, the person can just walk behind it on the manoeuvring lane of PW and push it to a nearby repair shop; no towing services are needed.
- All places in the URB may have foldable chairs to offer to people who for one reason or another do not use a PEV.
- People who chose not to use PEVs can just walk the short distances of URBs.
- The Transit Centre (see Figure 67) is situated at ground level in the URB. It has bus stations, train stations, freight exchange platforms, and all necessary facilities to connect the PEV - FEV system inside the URB with the transport system of the outside world.
- The cars of the URB inhabitants would be parked in their garages at the Transit Centre and would rarely be needed, since the URB provides most of life's requirements. When they want to go outside the URB, people would ride their PEV to their garage at the Transit Centre where they park their PEV in the place of their car, then drive their car out of the URB.
- People who do not live in a URB but come to visit, would arrive in a car, bus or train to the Transit Centre where they can just hire a PEV; it is easy and quick to learn how to drive a PEV; no licence is needed. Alternatively, people could just walk the short URB distances.

6.8 The Economics of URBs

- There will be an initial considerable effort and expense to build each URB.
- An URB can also be designed with the main common infrastructure (water treatment plant, transport centre, etc) at one of its ends, such that initially only about one quarter of an URB is built and able to house the first residents. Subsequently it is extended according to

population requirements, to increase its capacity to maximum. This way the initial capital is smaller and the start-up is more rapid.

- The compactness (shared walls and floors) ensures a minimal capital cost per square metre.
- Since it can last for hundreds of years, the capital cost per year can be much smaller than in present cities, which are continually changed (demolished and rebuilt).
- There is a much smaller social cost due to less energy and timey lost with demolishing a building and rebuilding old type buildings.
- Minimal running costs for utility services result from its efficiency, scale and integration.
- URBs minimise exchanges with the outside, as most services can be provided by its residents.
- URBs placed in remote areas like country URBs, or mining URBs, attract people with their civilised comfortable and inexpensive lifestyle.
- In one financial version, a group of investors registers a company that builds a URB and subsequently rents the apartments and the other spaces to users. The investment will last for hundreds of years, bringing consistent benefits with minimal risks.
- Renting would be more convenient than ownership, as people would have more independence, not lose time and money with selling/buying, and can easily move between non-URB settlements and URBs, or from one URB to another, or change apartments/commercial spaces in the same URB.
- Thus, the optimal investment is not buying URB apartments or commercial spaces, but buying shares of the company that builds and administers the URB.
- Given the comprehensive structure of the URB, all types of jobs could be available locally.
- Since life in a URB will be less expensive, the average work requirement of the inhabitants will reduce, allowing more leisure or other activities.

7 Chapter Seven: Climate Change and Climate Repair

Besides the greenhouse gases emissions, lesser causes of climate change mentioned by climate science are aerosols and the land use and land cover change (LULCC), as in [84], [73].

Chapters 2, 3, and 4 have shown that humans altered and deforested most of Earth and produced most of its deserts, while chapters 5, 6 and 7 present solutions to restore and adapt it to our needs.

Coastal areas were highly naturally productive because they received regular rainfall, promoting vegetation and the best soils. However, people also favoured coasts for their cool, pleasant weather, abundance of fish and easy sea transport. As a result, people settled on the shores and covered the best soils with tiles, cobbled stone and later with asphalt and concrete.

Scientific progress brought the industrial era 250 years ago, and industrial agriculture 100 years ago. Life improved, populations increased, and coastal settlements expanded through booming real estate. Development of coastlines accelerated with modern architectural designs boasting larger buildings, spacious plazas, roads and immense parking spaces, at the expense of coastal vegetation.

Two hundred years ago, 1 billion people in developed countries lived two to a room; now 4 billion people each have two rooms. Houses take up the largest part of coastal land, but there also are many public buildings, which by the time increase in number and occupy larger and larger areas.

Our large coastal cities are surrounded by sprawling services and industrial areas, where wide asphalt highways and driveways feed a never-ending sea of flat roofs devoid of vegetation.

Consequently, cities have increased tenfold, and almost all coastlines have been deforested and transformed into powerful solar air heaters. The resulting HACs dry out the tropical inland areas, while the ASH transforms them into deserts, displaces people and wreak havoc over the world.

Most of our work is dedicated to increasing our human environment at the expense of the natural one. As a large reaction corresponds to a large action, climatic changes become increasingly powerful and now threaten to destroy our proudly built human environment.

Naturally, revegetation, soil restoration and carbon sequestration in trees would result in climate repair if we would regain our correct understanding of the natural world, and properly assess the relative magnitude of the different causes of climate change.

7.1 Climate Change Effects, Causes, Mechanism and Prevention Measures

7.1.1 Comparison of Anthropogenic Global Warming and Radiative Forcing Effects as in [52]:

1) **Carbon gases since 1750** and other human-induced gases have created (according to IPCC) a radiative forcing equivalent to 2.5W/m^2, and a land temperature rise of 0.8^{0}C [16].

2) **Deforestation since 1750** as mentioned in Section 2.6.2 generated 2.4W/m^2 from correct albedo and around 0.9W/m^2 from water, resulting in a deforestation effect of 3.3W/m^2.

3) **Total anthropogenic effect since 1750** (1 + 2) results in: 2.5W/m^2 + 3.3 W/m^2 = 5.8W/m^2.

4) **Deforestation by humans** estimated in section 4.1 to have produced about 20W/m^2 might have resulted in a warming of around 3^{0}C, as shown in Section 3.6.2.

5) **Total anthropogenic effect** (1 + 4) results in: 2.5W/m^2 + 20W/m2 = 22.5W/m^2 and 4^{0}C.

6) **The remaining forests** now cool us by around 14W/m^2, that is interestingly non-linear.

7.1.2 Causes of Anthropogenic Climate Change

- 1,000,000 BCE: Invention of artificial fire and forest thinning for hunting in Africa, after 70,000 BCE in Australia, 50,000 in India, decreased precipitation and started desertification
- 12,000 BCE: Deforestation for the newly invented pastoralism in North and Southern Africa, Eurasian steppe, and Gobi, have stopped the tree humidity pumps in those sensitive lands.
- 10,000 BCE: Invention of agriculture, global effects of deforestation after around 8,000BCE.
- 5,000 BCE: Deforestation of coastal zones started the first HACs in the Middle East and China
- 100 CE: Deforestation and soil exhaustion by the Romans started the ASH of Sahara Desert.
- 1800: Industrial era carbon gases emissions; the railways facilitated increased deforestation
- 1900: Increase in human population and specific consumption resulted in excessive carbon emissions and increased deforestation for settlements, agriculture and extractive industry.
- Present: Consumerism, pro-fossil fuels and pro-deforestation interests dominate society.
- Human induced and ignited forest fires produce more carbon emissions and deforestation.
- Political competition between countries increases energy waste and decreases agreement.
- Prevalence of short time micro-social benefits over long term macro-social interests.
- Insufficient independence of scientists and international organisational deficiencies.
- Insufficient scientific acceptance 2020-2024 of the devegetation role in climate change [52].

7.1.3 The Mechanism of Climate Change

Essentially, the artificial climate change mechanism is:

1. Before humans, almost all land, especially the tropical areas, was covered with forests. Tropical forests had more powerful tree-ocean humidity pumps, which produced larger rainfall. The medium-high latitude lands were cooler and drier because they received less humidity from the tropical ocean, which discharged its heat in the tropical humidity pumps. The large quantity of water on land and in the low troposphere created an intense water cycle that kept relative humidity and dew point high, so the low thick clouds readily condensed all vapours, while the air above clouds was drier, thus the greenhouse effect was low. Since almost all land was covered with forests, the average climate was similar to that in a large forest now: temperature differences were lower, the wind and vertical convection were low, rain was more frequent, but the rate of rainfall was smaller, and there were more clouds and rain in the cooler low latitudes and less rain in the cooler high latitudes. Lower overall temperatures and no air overheating means that there were less high thin clouds.

2. Humans repeatedly started bushfires for hunting in the tropical forests and savannah they first inhabited and stopped forest regeneration by burning the saplings and stopped full maturation of grasslands. Gradually, the tropical forests, precipitation and cloud cover decreased, thus solar radiation, air temperature and dryness increased from 100,000 to 10,000 years ago, producing high **land degradation** and **alternating dryness cycles**.

3. Some people stayed in dry places and adapted, others moved to wetter subtropical areas, invented pastoralism and agriculture, and cleared most forests, this way stopping the regional tree humidity pumps, decreasing precipitation and cloud cover. Since the low latitude areas receive most solar radiation, the entire Earth started to overheat due to more solar radiation reaching the ground, thus the **planetary temperature increases** in an ongoing process, due to continuously increasing intertropical deforestation to present.
4. The lack of precipitation due to deforestations meant that there is less water available for evapotranspiration, which is the single way to remove solar heat from the ground, that is the largest cause of global warming, as explained in Section 3.6.2. The solar heated ground emits the infrared radiation that is amplified by the GHG and also heats the air, dries it and creates the hot thermals. Therefore, there is an **increase of the above ground air temperature** (the measure of global warming) due to deforestation and GHG, which also

explains **heatwaves**. The prolonged dryness and the agricultural tillage stress have increased soil exhausting and resulted in **soil loss** through **increased erosion** and ultimately **desertification** in all past cradles of civilisations.

5. In the past 200 years, carbon dioxide from burning fossil fuels and from deforestation exceeded the ocean's sequestration capacity, added to other greenhouse gases and amplified the effect of the solar heating of point 4, resulting in the modern global warming.

6. Deforestation has replaced evapotranspiration cooling with air heating, increasing the temperature and density inversions of the atmosphere, that increases convection, creates **high thin clouds**, produces **tornadoes** and amplifies **storms** that propel humid air at high altitude, resulting in **hail.** The increased storms produce **dangerously large waves** on water bodies and **coastal erosion**. The clear air turbulence created by strong air convection may be a **threat for airplanes**.

7. In low wind, deforestation HACs of the tropical coastlines produces **droughts** of its inlands by stopping the humidity pumps, diverting away humid marine wind, which **overheats** and **floods** the higher latitude regions. This is also responsible for **melting glaciers, polar ice, and the Siberian permafrost**. The droughts of the tropical lands have removed above-soil water, while excessive pumping for agriculture has removed groundwater, this continental water loss added to water from melting land ice and produced **sea level rise**.

8. The evaporating intertropical ocean is over-pressured by the nearby HACs, closing access to those regional lands, this rise in partial pressure usually producing **extensive floods** in non-HAC intertropical regions, or sometimes bursts through HACs to produce **flash-flooding**. Ocean evaporation is also increased by the hot air of the deforested land and by the excess solar radiation from the lack of clouds above deserts, starting **hurricanes** on the West of Sahara, and producing **excessive precipitation** elsewhere, that may trigger **landslides**.

9. In high wind conditions, ASH extracts more heat from the ground, its hot air dries up vegetation, which loses water and heats more air, that dries more vegetation in an avalanche (positive feedback) which results in **devegetation** and **soil erosion,** in **dust and sand storms**, **desertification**, and sometimes **forest fires**.

10. The above-land hot convective currents produced by deforestation propel more water vapours to the top of the troposphere, increasing the high thin clouds and the greenhouse

effect in parallel with the increased carbon dioxide and methane emissions. The avalanche (positive feedback) climate change effects are symptoms of a destabilised water cycle.

Unfortunately, fossil fuels burning, deforestation and their carbon emissions increase.

An explanation at global scale of the above phenomena must consider that a dry atmosphere working only on sensible heat transfers might result in more than 200°C difference between the Poles and Equator. The powerful latent heat transfer of water creates evaporation pressure mainly at low latitudes and condensation vacuum mainly at high latitudes. This process overcomes flow resistances, despite the troposphere only being 1/1000 of the distance between a Pole and Equator. Therefore, the atmosphere is a gravity-confined, partly-evacuated heat pipe with water as its most effective working fluid.

At tropical latitudes the descending Hadley humid warm air and the Ferrel cold air converge and mix with the low altitude oceanic humidity, resulting in normal precipitation in the cool, cloudy and vegetated lands of South-eastern Asia, Central and South America, many islands and in North Africa before human intervention. However, in the present cloud-free deforested regions of North Africa, the Middle East and Australia, the descending air is overheated by infrared radiation and thermals from the deserts below, preventing rain formation. This is illustrated in Figure 68.

FIGURE 68: Latitudinal Heat Cells and Cloud Migration

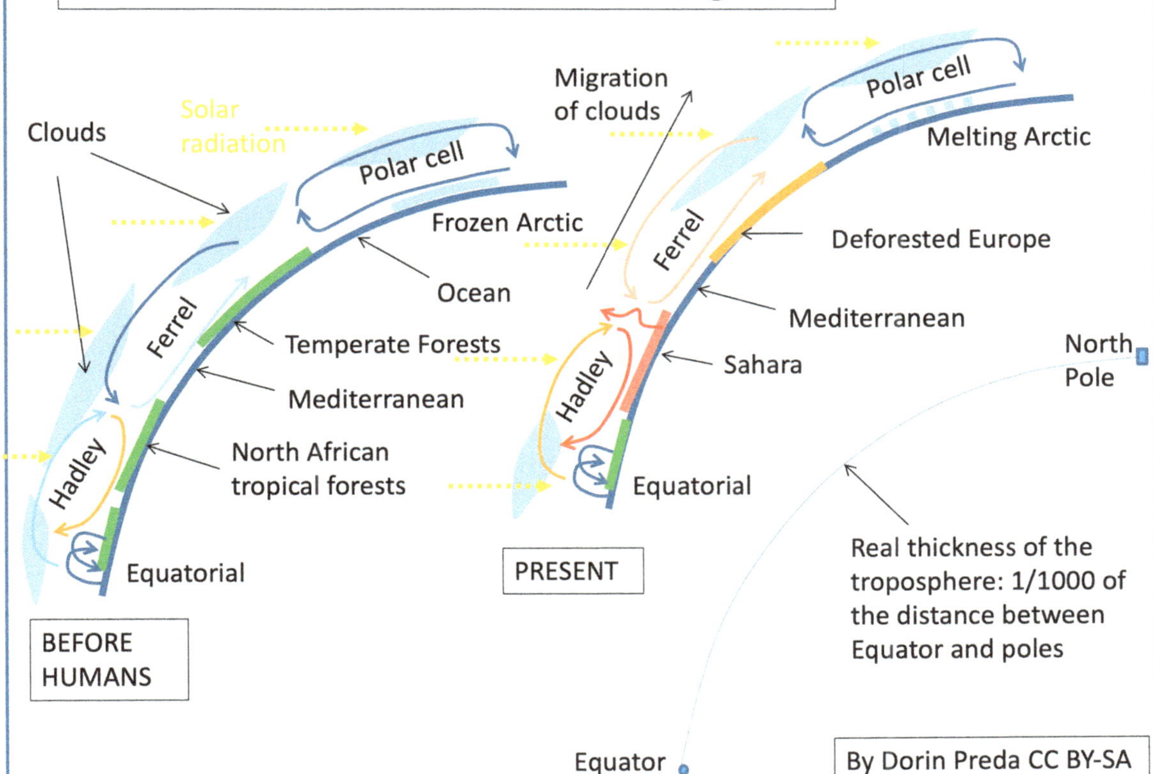

By Dorin Preda CC BY-SA

After humans easily devegetated the tropical areas that were prone to overheating due to their geographical position, statistically less clouds formed there and probably moved to higher latitudes. However, since tropical areas receive more solar radiation than lower latitudes, the net global reflectivity has decreased, leading to an increase in the planetary temperature and in the actual solar radiation absorbed by Earth. Unfortunately, deforestation of the intertropical area accelerates at present, further increasing planetary and above-ground air temperatures.

The low horizontal branch of the Hadley cell adds deforested tropical land heat to the intertropical ocean surface, increasing evaporation and Atlantic hurricanes, as shown in Figure 69.

Most evaporation occurs in the intertropical ocean, with the humidity being discarded in the closest sinks lacking HACs, like the forested tropical and equatorial lands. The humid ocean wind is usually unable to penetrate the coastal HACs of the deforested tropical regions, leading to additional vapour pressure in the warm oceans, that can result in periodical floods in the unprotected equatorial lands, a phenomenon that has become more frequent with rising ocean temperatures.

Since most tropical lands are blocked by HACs, more humidity is pushed towards higher latitudes, where it condenses, overheating and flooding those areas [90]. The hottest and lowest humidity air masses can only reach condensation in the Arctic and Antarctic, where they melt the ice. Tropical deforestation has global consequences.

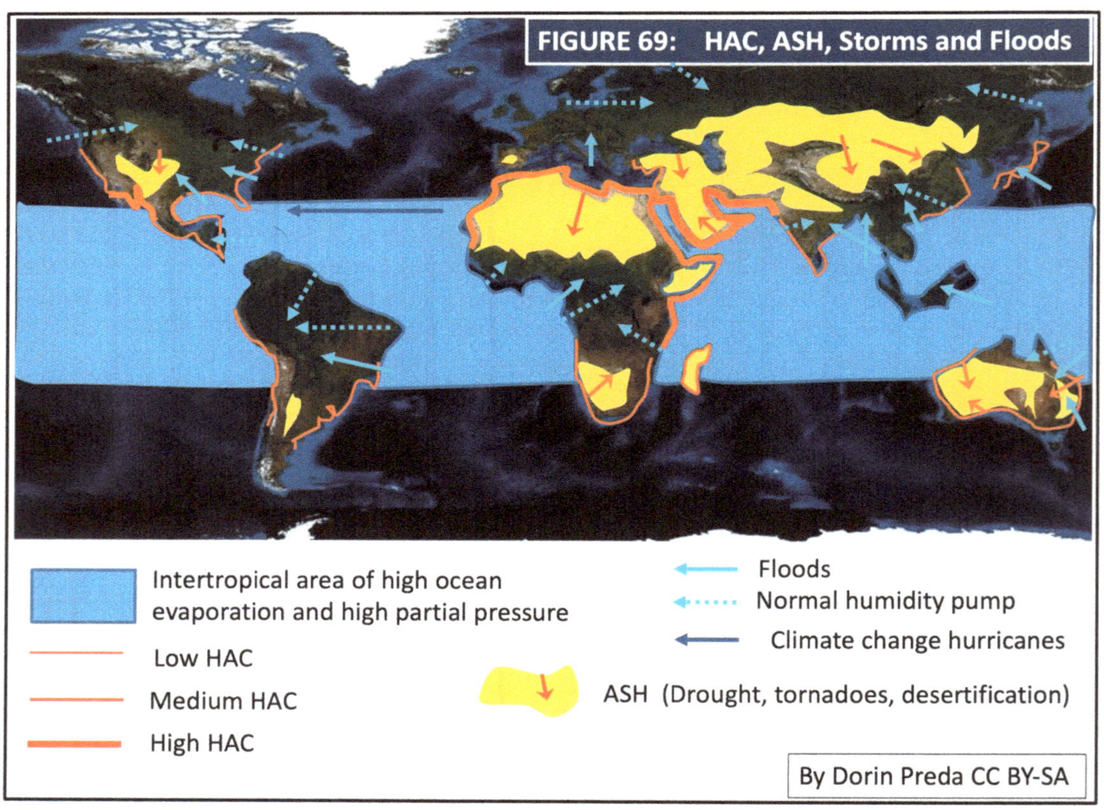

ASH overheats tropical land air, while HAC prevents mixing with ocean humidity, leading to no precipitation in the desert despite that air contains water vapours, just increases the GHE and creates high thin clouds instead of the beneficial low thick clouds.

However, if a strong storm penetrates a HAC and cools the ground with rainwater, it cancels the HAC and through condensation it creates local atmospheric low pressure. This allows ocean evaporation pressure to push more humid air into that pressure low (similarly to a humidity pump), resulting in increased precipitation and rapid flooding in a feedback process. These events are becoming more frequent and are a symptom of excessively high vapour pressure over the tropical ocean exacerbated by HACs blocking the tropical land that previously helped regulate pressure through daily rain in rainforests.

Initially, almost all land was covered with forests, which actively created precipitation conditions through cloud reflective cooling, evaporative cooling, low infrared emission and the generation of organic carbon aerosols [69]. The much cloudier tropical forests kept temperatures low not only in those regions, but also in higher latitudes that were receiving less precipitation than at present. More clouds led to better cooling of the Earth. Currently, the situation is reversed, with overheating both in tropical areas and at higher latitudes (thus higher global temperature) due to reduced precipitation in tropical lands following their deforestation. This is illustrated in Figure 70.

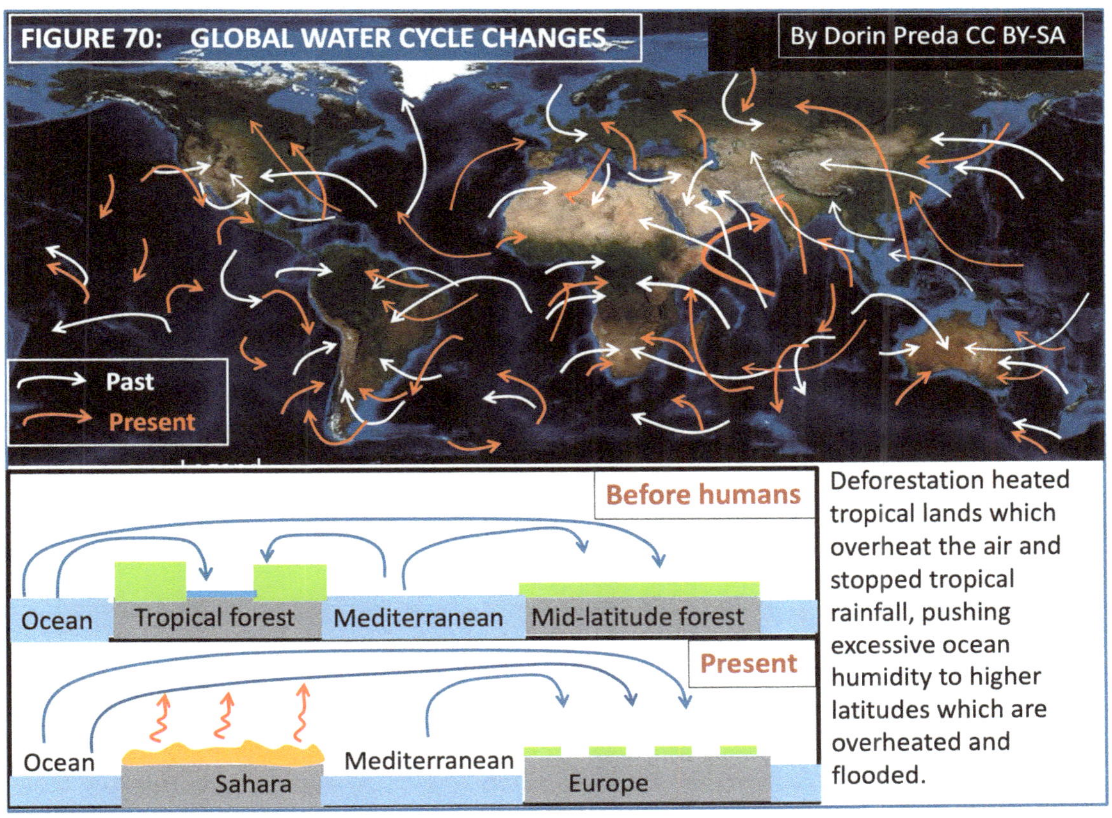

FIGURE 70: GLOBAL WATER CYCLE CHANGES

By Dorin Preda CC BY-SA

I argue that climate change effects are disruptions of the natural water cycle as indicated in Figure 72, mainly caused by deforestation of tropical and subtropical regions. The additional ground infrared radiation due to solar heating of deforestation is amplified by carbon emissions and by the water vapours sent to high altitude by deforestation thermals, further increasing global warming.

As we know from heat pipes, pressure automatically increases in closed systems to overcome flow resistance. The Earth acts as a heat pipe pressurised by gravity, with the ocean as its main evaporator and the high altitudes and high latitudes as condensers. Deforestation of tropical lands has transformed them from condensers into over-heaters, because they cannot retain water.

Heat fluxes, pressures, temperatures and areas of heat exchange are interdependent. A decrease in condenser area due to deforestation leads to increased temperature, pressure and velocity of water vapours.

My experience with heat pipes confirms that the heat transfer efficiency is inversely proportional to the distance between the evaporator and the condenser and to the temperature difference. If we increase the distance between the evaporator and the condenser (replacing tropical condensation with high latitude condensation), we increase flow resistance, thus temperature automatically increases to create more pressure to overcome the increased flow resistance. This additional effect is confirmed by the present increase in Earth average temperature.

Additionally, I observed in partially evacuated heat pipes (which have both air and water) that when efficiency is high and the difference in temperature is low (the Earth before humans), water vapours stay close to the evaporator, do not rise much in the condenser, and they push the dry air up. This is similar to the observations of low altitude clouds (even at tree level) in environments with high relative humidity. Conversely, the high temperatures produced through deforestation and desertification raise pressure and send water high into the troposphere, where it has a strong greenhouse effect, but cannot condense. Therefore, we should allow large vegetation (thus a lot of water) on the tropical land, because it will keep temperatures low and humidity close to the ground.

7.1.4 Methods of Climate Change Prevention:

- Reduce fossil fuel dependence and consumption; reduce energy waste.
- Stop net deforestation, de-vegetation and groundwater extraction from areas near forests.
- Prevent forest fires, "controlled" burning, slash and burn, and cultural burning (see note).

- Keep goats in their barns, and eliminate large feral herbivores (goats, horses, camels).
- Stop pastoralism and grain cropping in areas with less than 500mm/year total water input.
- Limit urban sprawling, and favour URBs and similar compact styles.

Note: *An issue arises when people illegally start forest fires and blame fire prevention measures for allowing combustible material to build-up. The current sparse forests are drier and more flammable. To address these issues, we can either re-create dense and humid forests that are less prone to fires, or we should plant useful trees that will be protected by their owners.*

7.2 Principles and Systems of Climate Repair and Strategy

Global warming can be assessed in terms of planetary temperature (derived from albedo) or above-ground air temperature (the measurements of the meteorological station), the latter being more credible, as shown in Section 2.7. Consider the impact of devegetation around meteorological stations in the past 200 years. Climate primarily affects the lower troposphere, emphasizing the importance of ground-level research. While satellite observations are useful for some data collection, and conducting comparative areal studies, they should be validated more effectively.

7.2.1 Temporal Approach

Climate change is viewed as undesirable, indicating a preference for the past state. In the past there were more forests and water on land, thus computer simulations are wrong suggesting that forests and water would cause global warming. The key difference with the past is that deforestation has replaced the cooling evapotranspiration of vegetation with solar heating, which is amplified by excessive carbon gases emissions. Through deforestation, we have changed Earth's powerful water cycle from friend to foe. Significant revegetation would restore the water cycle and also absorb in trees the present atmospheric carbon excess, reverting the climate change.

7.2.2 Spatial Approach

The real benchmark for global warming is the above-ground air temperature measured at meteorological stations.

Some argue that a local climate is not relevant for the global climate. However, I suggest the axiom: "Global climate is the result of all local climates."

This means that acting locally we can obtain global results. For example, if a few kilometres of land around all meteorological stations were revegetated as they were about 100 - 200 years ago, the measured average land temperatures would drop, giving _the appearance_ that global warming has disappeared in a reversed "urban heat island" effect. However, if we revegetate most of Earth, global warming would _truly_ disappear.

7.2.3 The Method

At global scale, revegetating intertropical areas will attract local precipitation of water evaporated from nearby ocean, preventing those water vapours from overheating high latitude areas, and will bring the solar radiation reflecting clouds closer to Equator, thus decreasing planetary temperature.

On a local scale, revegetation would bring water closer to the ground, decreasing the greenhouse effect and replacing warming high thin clouds with cool low thick clouds, as suggested in Figure 71.

FIGURE 71: Revegetation Mechanism

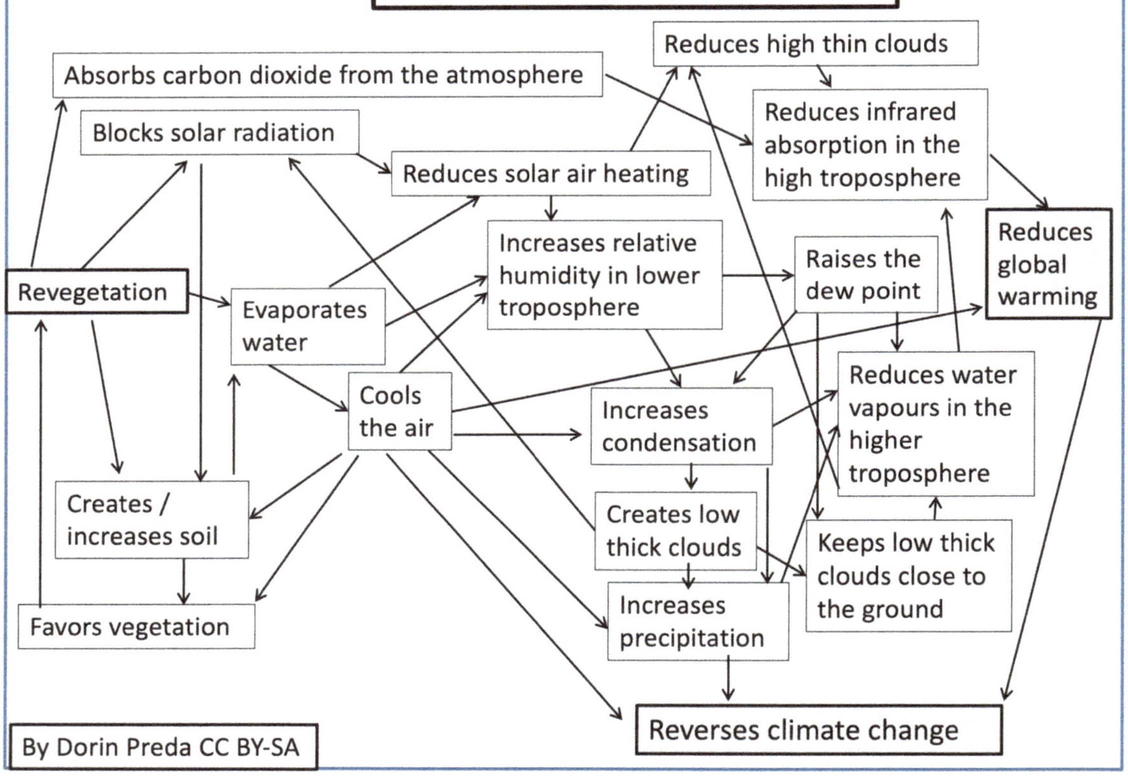

Water is the most potent working fluid on Earth. It depends on our land-use whether too little water warms the climate, or a lot of it cools it down. Reforestation is the obvious solution that was proposed and retracted by some scientists [55], due to the strong opposition of most climatologists to reforestation. They reject both reforestation and afforestation with multiple reasons similar to those in Section 2.1.3, despite China and other countries achieving good results with their tree planting programs.

I suggest revegetation (not necessarily reforestation) with useful trees, because it is the most realistic method to repair the climate through the water cycle, while it also reduces CO_2 in the atmosphere, in parallel with emissions reductions. To achieve a green island effect, it must be high plant density under a canopy of useful trees, preferably privately owned, for protection and care.

We should not accept geoengineering solutions because they are damaging and risky.

Unlike our ancestors, who deforested the Earth, we now understand enough about climatic processes to realise that humans have destabilised the climate through widespread land changes.

FIGURE 72 : Climate Change and Repair

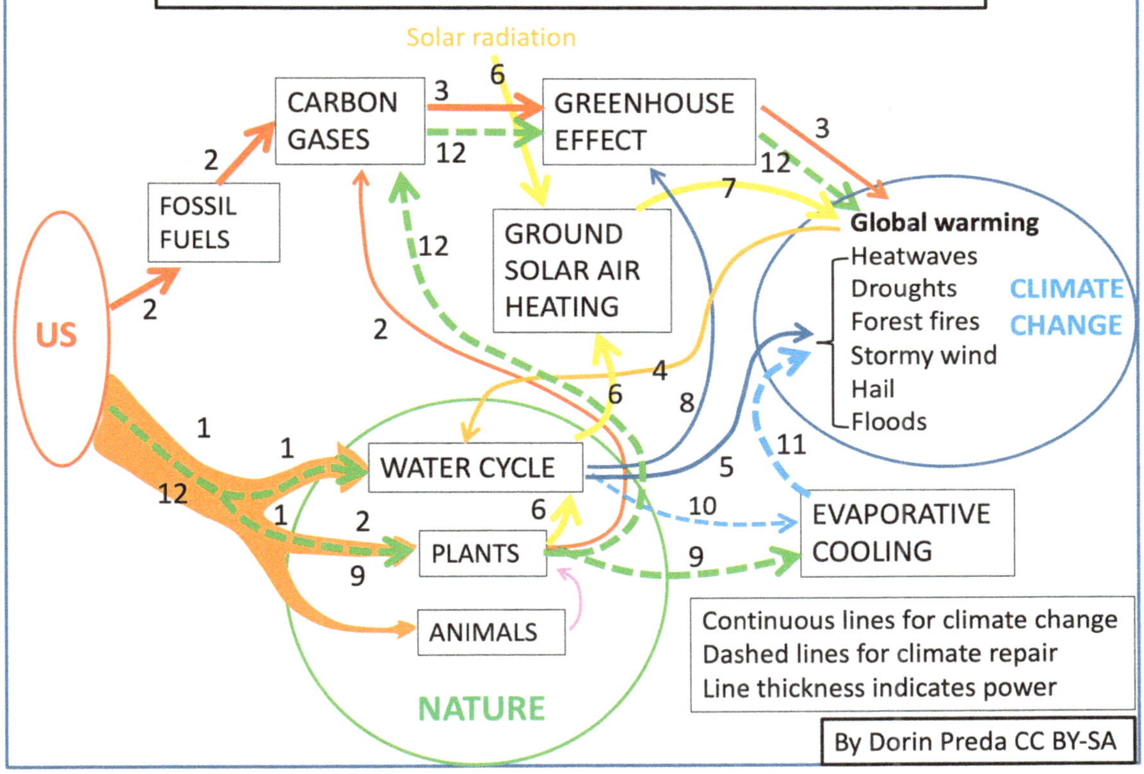

The upper global warming path in Figure 72 can be replaced with the revegetation lower path, to restore climate while absorbing carbon gases emissions as detailed below:

1. Humans cleared forests for development and altered the very powerful water cycle
2. Deforestations and burning of fossil fuels have produced excessive carbon gases
3. Excessive carbon gases have increased the greenhouse effect and global warming
4. Global warming and deforestations damaged water cycle, which is the main energy actor

5. The damaged water cycle resulted in the major climate change effects
6. Deforestations have reduced evaporation on land and increased solar heating
7. Excessive solar heating now replaces evaporation and produces global warming
8. Hot air pushes water at high altitude, where it increases the greenhouse effect
9. We must restore evaporation through revegetation to decrease solar heating
10. We can also decrease solar heating in cities with artificial water evaporation
11. Restoring evaporation and the water cycle will stop the effects of climate change.
12. Revegetation will also sequester carbon gases and thus decrease global warming.

We cannot identically revegetate all coastlines because most of our cities are already there, but solutions of improvement will be suggested further. We must revegetate wasteland deserts, to repair the water cycle and the climate, decrease fossil fuel use, and sequester carbon in trees.

There are practical experiences and programs like the Farmer Managed Natural Regeneration (https://fmnrhub.com.au/), which support the above theory and proves that returning the Earth as much as possible to its initial vegetated state, improves the land and the climate.

7.2.4 Reforestation and the Greenhouse Effect

The argument that "Reforestation increases the greenhouse effect" is stated by some scientists who oppose reforestation claiming that it would increase humidity, leading to an increase of the greenhouse effect and Earth's temperature.

While vegetation does increase relative humidity, it also decreases temperature through evapotranspiration under the reflecting clouds. However, above forests humidity is low in the upper troposphere due to water condensing into the low thick clouds below.

I argue that the greenhouse effect depends on absolute humidity, not on relative humidity. As air temperature increases due to global warming, absolute humidity is high from increased ocean evaporation, even if relative humidity is decreased by the high temperature.

In forested areas, relative humidity might be high due to the cool air, but absolute humidity could be low due to low dew point and frequent condensation, resulting in low greenhouse effect, while in a nearby region without vegetation hot air may have low relative humidity but high absolute humidity, generating a larger greenhouse effect. Additionally, forests humidity is in the lowest part of the troposphere, while approximately 90% of the greenhouse effect as stated in numerical experiments [67] and satellite measurements [68] would take place in the upper troposphere. Therefore, there is no direct correlation between vegetation and the greenhouse effect.

Satellite observations show that in hot dry areas, the infrared emission, thus greenhouse effect, are larger than in cloud covered areas, as seen in Figure 35A.

The illustration on the left in Figure 73 shows that in the past there was more vegetation on land, therefore more clouds. The low above-ground air temperature was maintained by evapotranspiration taking latent heat to the clouds where it was transformed into sensible heat of air and emitted upwards as infrared radiation into an atmosphere of low absolute humidity, as water already condensed into clouds. This resulted in a low greenhouse effect above the clouds, characteristic of a fully vegetated planet.

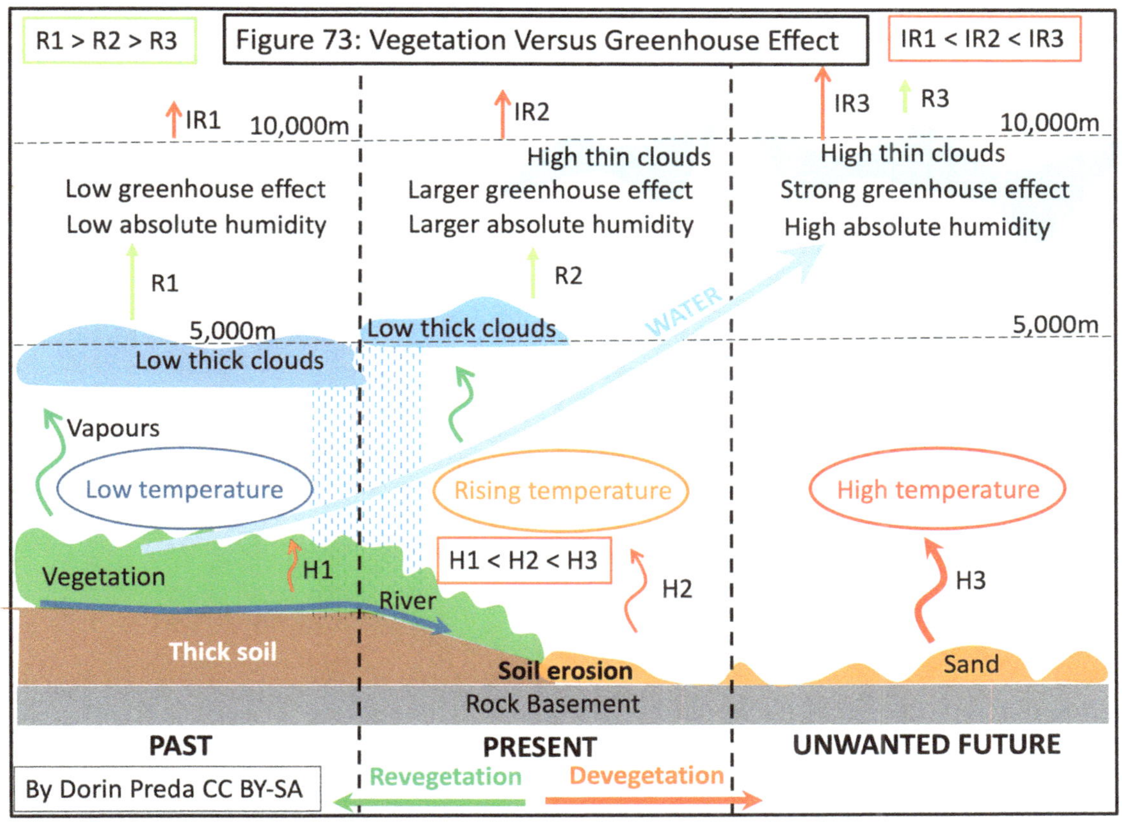

Figure 73: Vegetation Versus Greenhouse Effect

Air temperature under the clouds was low, because a significant part of solar radiation was reflected as R1 by clouds, and because evapotranspiration had used all the remaining heat H1, thus there was no air heating. Since H1 was low, the greenhouse effect was low. The intense precipitation maintained a large quantity of water on the ground as permanent rivers and swamps into which the present fossil fuels formed.

At present, as shown in the middle panel of Figure 73, rising air temperatures are due to deforestation, which decreased evapotranspiration, thus solar energy is used for the larger H2 to heat the air and emit infrared upwards, decreasing the relative humidity and resulting in fewer low thick clouds, thus a lower reflection R2 of solar radiation, and a higher IR2. The deserted tropical areas generate powerful hot currents which on one hand take their significant absolute humidity to high altitude, increasing the greenhouse effect, and on the other reach the top of the troposphere, where their vapours freeze into high thin clouds, that further increase the greenhouse effect. As air temperature is higher, the low thick clouds are also higher, as seen in the middle panel.

Therefore, at present, due to lower vegetation, there is less water on the land, but more at the top of the troposphere, where most of the greenhouse effect takes place.

Deserts, as shown on the right in Figure 73, do not have vegetation water to evaporate, thus solar radiation is transformed into a large H3 (infrared radiation and heat of the air) decreasing relative humidity, evaporating the low thick clouds, such that the very low reflection R3 results in a high IR3 and planetary temperature. Despite generating more absolute humidity through ocean evaporation due to high air temperatures, vapours are entrained high in the atmosphere by strong dry convection currents of the desert, creating high thin clouds, which along with the high absolute humidity and high infrared of the desert, generate a strong greenhouse effect, that further raise air temperature in an unwanted avalanche (positive feedback) effect.

The intense water cycle easily precipitates water between vegetation and the low thick clouds, while in deforested areas the hotter air prevents precipitation but increases the greenhouse effect.

To stop temperature increases and climate change, we must return the Earth to its past vegetated state. By keeping water vapours close to the ground with precipitation conditions induced by revegetation, we can significantly decrease the greenhouse effect and mitigate climate change effects, additionally decreasing carbon emissions to stabilise the system, as illustrated in Figure 72.

7.2.5 The System

We will observe the climatic adjustments in the higher troposphere, but the common climate processes mainly occur below 5,000m that is also where we can directly act. This altitude also marks the upper limit of human settlements (our area of interest) and encompasses all meteorological stations serving as reference points for global warming (see Figure 74).

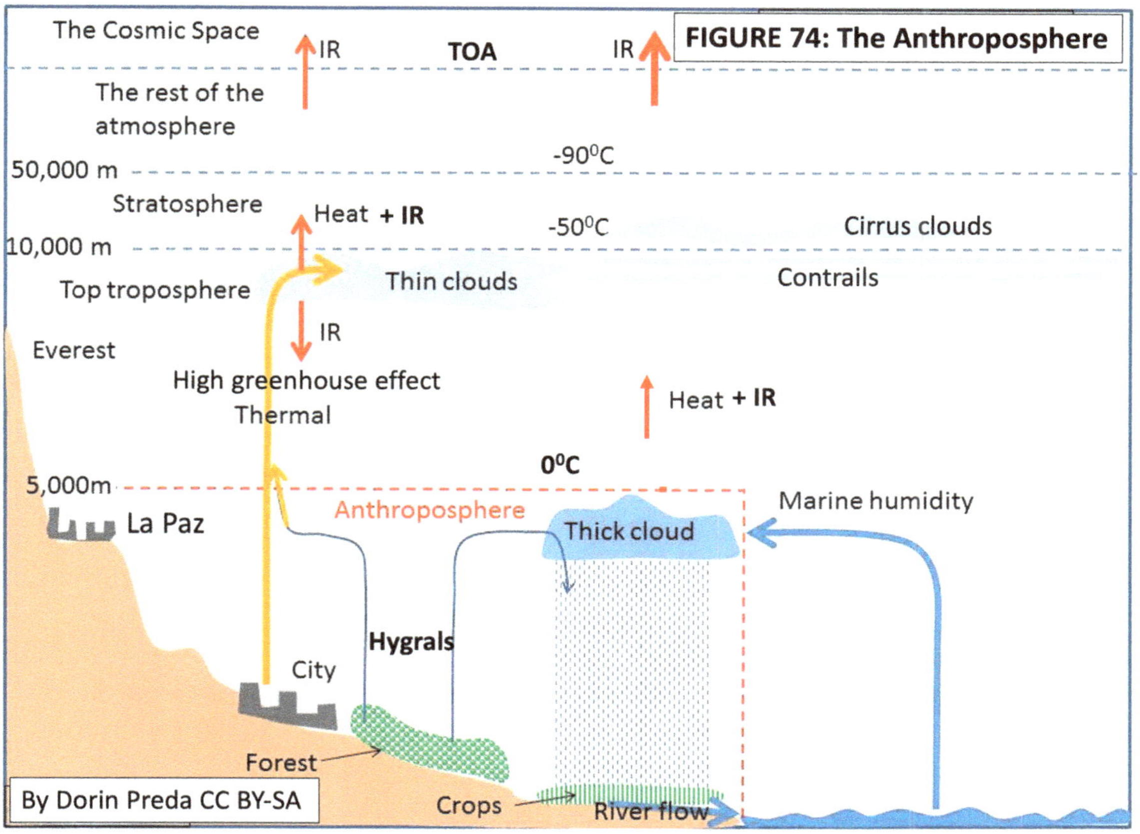

FIGURE 74: The Anthroposphere

Most importantly, 5,000m is the maximum altitude for low thick clouds, which play a crucial role in regulating a healthy climate. While high thin clouds and storm clouds exist at higher altitude, they are either less common or not part of a cool and balanced climate. Most heat is dissipated in the troposphere above 5000m, mostly through the radiation of air and greenhouse gases, as well as through conduction and convection with the atmosphere above.

I define our system as the land areas below 5,000m altitude, which I refer to as the Anthroposphere. However, unlike other definitions of the "anthroposphere", I include the natural world of the land (because we rely on it and we can manage it), while excluding the oceans, which do not require direct intervention (refer to Figure 74). We have to restore temperatures of the anthroposphere, which will in turn normalize the top troposphere and the oceans. This approach is justified, as our past actions have solely impacted the anthroposphere, leading to the current climate crisis.

There are four mechanisms through which energy is released from the top limit of the anthroposphere:

a) *Reflection by clouds* or the ground, with low thick clouds being the best reflectors.

b) *Latent heat exchange* through the higher clouds, especially high thin clouds (undesirable).

c) *Sensible heat exchange,* which currently predominates in tropical regions (undesirable).

d) *Infrared radiation* from the ground, clouds and atmosphere.

7.2.6 More Details on Clouds

Large volumes of humid air, which we could term "hygrals", ascend from the ground and cool adiabatically to the dew point, where condensation forms visible low altitude thick clouds. Even when these clouds do not immediately produce rain, they play a crucial role in reflecting solar radiation, absorbing and emitting infrared radiation and maintaining the energy balance.

The low thick clouds are the most effective reflectors and radiators of electromagnetic waves, as well as potent carriers of latent heat. At the Equator, they transport significant heat without causing air movement (the doldrums), while in the Sahara sand storms resulting from convective currents working on sensible heat can strip tree bark without carrying as much heat (see Figure 37).

Vapours act as heat sponges, absorbing heat to cool and humidify the air, thus resolving warming and dryness. The denser a cloud, the more powerful its effects, but also the shorter its lifespan. In conditions of relatively low air temperature and high absolute humidity, the dew point (the temperature at which water condenses from air) is elevated. Misty equatorial forests are the best examples, where air does not have to rise to reach its dew point and make fog.

Low thick clouds efficiently gather vapours from the air, which are then released as rain, leaving behind dry air with a reduced greenhouse effect. These clouds also capture dust particles that aid in droplets formation. This was the predominant type of clouds of the early Earth, characterised by abundant water and vegetation that formed frequent swamps, accumulating dead trees that eventually transformed into coal.

Another type of clouds that transport heat rapidly and reduces atmospheric water and greenhouse effect are **large storm clouds,** like cumulonimbus, although they are infrequent, and often result in hail due to the mixing of liquid water and ice at high altitudes. These clouds also clear the air of humidity, leaving behind dry, clear air that facilitates the outflow of infrared radiation. Thus, maintaining water at low altitude in the troposphere with low thick clouds is crucial.

Unfortunately, the prevalent **high thin clouds** in today's skies, particularly in problematic tropical regions, are unable to produce rain due to their composition of minute ice crystals. When heated, these clouds sublimate rather than form droplets. These clouds forming at 7,000m – 15,000m are undesirable due to their strong greenhouse effect.

The low thick clouds covering the forests are reduced due to the latitudinal cloud shift (Section 2.6.3), while the subsequent rise in temperature favours formation of high thin clouds which reflect less solar radiation and have a higher greenhouse effect. Thus, there still are clouds after deforestation, but they are of the kind that increase global warming.

It is imperative to prevent the formation of high thin clouds, such as airplane contrails and cirrus clouds generated by hot thermals from intense heating sources like deserts, cities and industrial zones. The high-altitude haze serves as a testament to our alteration of the atmosphere.

7.2.7 Our diet has varied throughout history, as we transitioned from hunter-gatherers to agriculturalists and pastoralists. Despite fruit and vegetables being healthier, the diet shifted from fruit to cereal, and from white meat to red meat and milk products.

The problem with fruit and vegetables is their seasonal availability in all areas except the equatorial forests. Even there, the quantity is limited; competition probably forced towards diversification. As early humans migrated to the rest of the world, they changed their diet to make their lives more dependable on their work than on the variability of nature. However, the introduced changes depended on human's low technological stage at that time, and gradually created larger and larger detrimental effects on their health, lifestyle and on Nature, that gradually increased to this present critical stage. **The fact is that our change in diet resulted in climate change.**

Primate and early human diet consisted mainly of fruit and a smaller quantity of roots and vegetables, sometimes supplemented by white meat of small animals, birds and fish they could catch. Since this was their diet for millions of years, their (and our) metabolism was adapted to it.

Small diet variations for short time (like chewing grass seeds) helped them survive when fruit was not available, but their digestive system did not allow it permanently. However, with the advent of stone tools they succeeded to crush the seeds and by using fire they succeeded to make seeds more edible. Thus, acceptance and demand for grains grew, leading to invention of agriculture and wide deforestations for crop fields. Despite that mechanical ease of digestion, our metabolism remained unadapted to cereals, and many people still suffer from celiac disease, while carbohydrate excess leads to malnutrition, obesity, heart disease, diabetes and cancer in the rest of the people.

Most pastoralists remained roaming populations following with their grazing animals the fresh grass growth, while some of them settled and partially adapted the diet of their animals to cultivated grasses. Humans started to continuously consume animal milk, which is well assimilable only in the early stages of life by that species of mammals, not by humans. It is known that human milk is different of herbivores milk, which has too much fat and calcium and different enzymes. Consuming milk and derived animal products by adult humans led to disease due to lactose intolerance, obesity, cancers and heart disease produced by the excessive, different mammal fat.

However, the greatest change was the switch from the vegetarian diet supplemented with little white meat, to a predominantly red meat diet, or the blood diet of some nomad populations. The red meat of grazing animals is connected to cancer [95] and to human aggression [96], suggesting that this is an unnatural food for people, who evolved from apes, not from felines.

Many people die of red meat cancers, of fat-caused heart disease, or are allergic to gluten from cereals and milk lactose because our bodies did not adapt to these foods in the past 10,000 years.

In the process of changing the Earth to produce the new foods for our new diet, we deforested 2/3 of the land and changed it into pastures and cereal crop fields, leading to extinction of many species and degrading Nature. Pastures is known to now occupy half of the present agricultural land (https://ourworldindata.org/grapher/cropland-pasture-per-person). Deforestation for pastures presently continues, leading to desertification in Sahel, Eurasian Steppe, Australia, N and S America.

Deforestation and intensive agriculture with tilling and goats is known to produce desertification, resulting not only in unusable land, but in climate change due to destabilising the water cycle.

However, the most significant observation is not only that the change in diet was unhealthy for people and destructive for Nature, but that it is totally uneconomical, as shown by the following brief analysis, that leads to major possible improvements. This also leads to reflection onto the power of human habits, unjustified traditions and unused reason.

The information in Table 3 was collected from a multitude of sources on the internet and includes high variability. It was processed to show the two main parameters in land development for agriculture: specific producer income and energy of that food per hectare. There are five categories in suggestive colours: grain cereals, vegetables, fruit, grazer red meat and omnivorous white meat. Product vitamins and other nutritional values were difficult to represent due to complexity of that information, but the conclusion is that cereals are the least nutritious, having less vitamins.

The figures for animals and birds include meat, milk, eggs, and wool. The tonnes/ha figures for grazing animals were derived from their grass food per hectare. Pigs and poultry production were also estimated from their food in the fruit and vegetables categories, considering a rate of assimilation of ½ for chicken and 1/3 for pigs and rabbits. Therefore, production of poultry per hectare that results from consuming a significant proportion of cereals is smaller than that of pig/rabbit, which can result from fruit and vegetables of larger production per hectare. However, other food chains can be considered, for example fruit-worms-poultry.

TABLE 3 Income and Energy from Agricultural Products

Annual Production	Tonnes /ha	Producer price [$/t]	Producer income [$/ha]	Energy [kcal/kg]	Energy [Million kcal/ha]
Wheat	4	210	840	3,300	13
Rice	10	300	3,000	2,800	28
Maize	8	200	1,600	3,500	28
Barley	4	200	800	3,100	12
Rye	4	250	1,000	3,190	13
Potatoes	40	200	8,000	770	31
Tomatoes	60	500	30,000	180	11
Sugar beet	60	100	6,000	200	12
Carrots	30	300	9,000	410	12
Apples	50	700	35,000	520	26
Oranges	40	700	28,000	470	19
Banana	60	200	12,000	890	53
Dates	10	1,500	15,000	2,770	28
Cassava	20	400	8,000	1,600	32
Jackfruit	80	250	20,000	940	75
Palm oil	6	900	5,400	8,850	53
Cattle	0.9	1,400	1,260	1,300	1
Sheep/Goat	0.6	1,700	1,020	1,450	1
Pig, rabbit	10	1,300	13,000	1,700	17
Poultry	5	1,800	9,000	1,800	9

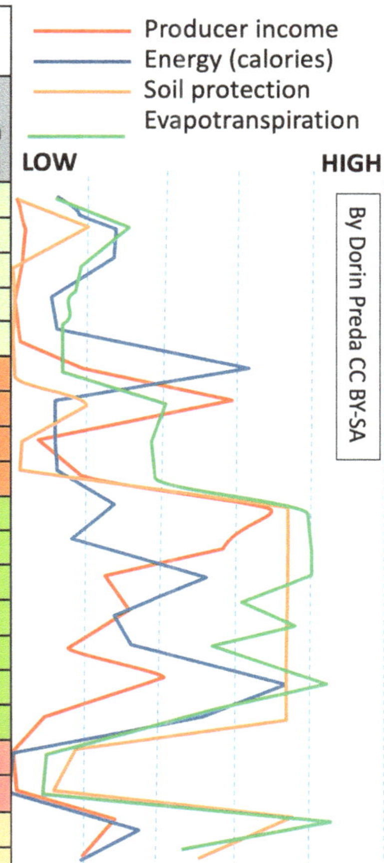

The price and energy content of the entire animal were considered, not only those of their meat.

High evapotranspiration means low solar heating, thus low climate damage. Soil protection (the inverse of soil erosion produced through tillage) and evapotranspiration for the animal categories are those of the plants consumed by those animals.

Grains (except rice) are the least productive and least nutritious; their tillage damages the soil and they have rather low evapotranspiration. Rice is more productive, protects the soil much better, has less carbohydrates and has higher evapotranspiration because it grows mainly in water.

Fruit brings the highest financial and energy benefits, is nutritious for people and animals and is packed with vitamins. Their trees produce most evapotranspiration for climate protection and are the best at soil protection.

It is obvious that grazing animals are the least productive despite producing 14.5% of global carbon emissions [94] and their pastures ruined the climate through the needed deforestation. Especially in subtropical and tropical areas, the wrong practice of total deforestation for grazing did not increase the production of grass, but decreased it through precipitation decrease and soil erosion. Entire continents were ruined by the people's appetite for red meat, which is the easiest to obtain through the primitive practices of hunting or pastoralism. The market price for red meat shows the irrational nature of human habits: the price rises because of the decrease in productivity of those lands, and because of the access to fashion food by newly rich people. Conversely, it is also absurd for the least rich people of Sahel to have a red meat diet, when this is the most expensive, the least land productive and increases desertification through their goats. It would be much more beneficial for those people to plant fruit trees.

By replacing red meat with white meat, the meat production can increase at least 10fold, thus there would be no need for pastures, for which most land used by present agriculture was deforested.

One hectare of potatoes and sugar beet with apple tree lines can produce at least 30t legumes and 10 t of apples, that can be fed to pigs to produce at least 10t of pigs, compared to 4t of wheat/ha.

The seeds of fruit, vegetables and legumes should be separated from their pulp, which would be used as food for people and animals, while the seeds could be fed to chicken to reduce to minimum their requirement of grains.

Pigs, rabbits, poultry, fish and similar animals have by far the largest financial and caloric output, their white meat healthier than red meat, and the orchards that produce their food protect soils.

Omnivorous and vegetarian animals like pig, rabbit and fish could be fed mostly fruit from trees and vegetables growing under those trees, but no grains. Those animals would eat the fruits irrespective of their appearance, so insecticides would not be needed to stop the insects which are food also for those animals. Food chains like fruit-insects-birds or predatory fish can be considered.

A diet of 50% fruit, 30% vegetables, 5% cereals, and 15% white meat is probably the healthiest for people. Many cultures have recipes for dishes that use fruit and vegetables instead of cereals and legumes. Dried fruit make very healthy snacks.

Useful trees adapted to all climates can be planted everywhere replacing "useless" trees and increasing the economic value of gardens, towns, crop fields, pastures and even forests.

Fruit trees and other useful trees ownership is the single realistic measure of protection against forest fires in countries which now experience multiple simultaneous forest fires.

Preservation was the single advantage of grains over fruit and vegetables for primitive people. However, at present we have many methods of preservation for fruit, vegetables, meat. I present in Section 7.3.4.6 a few suggestions that can help the transition from cereals to fruit and vegetables. Today we also have the option of fast crop transport to remote areas where fresh fruit and vegetables do not grow in necessary quantities (cities). This does not stop aeroponics cultivation of vegetables on flat roofs, that is a good source of fresh product and evapotranspiration cooling.

These measures would reduce by more than 90% the need for cereals, which are vitamin-less crops of smallest yield and small income and are responsible for most world soil loss through tillage. Therefore, most pastures and grain crop-fields should be replaced with orchards or vegetables fields with tree lines, **resulting in more than doubling the global agricultural production.**

Advantages of fruit and vegetables versus cereals:

- Around 10 times higher producer income per hectare
- Require less mechanical work and use of fossil fuels
- Produce around 2 times more energy as calories per hectare
- Larger evapotranspiration of trees compered to cereals, thus improved climate
- The soils are protected against erosion by roots, through shade from the sun and do not require tilling or other destructive intervention; the soil is enhanced by branches and leaves.
- Unlike cereals, trees can grow in poor soil, rocky areas and in hilly terrain.
- Trees do not require pesticides; weeds grow, die, decay in place and return minerals to soil.
- When fruit and vegetables are used as animal feed, insecticides are not needed; it is possible to choose only the unblemished fruit of a tree for people, while feeding the rest to animals.
- Unlike grasses, trees sequester much more carbon and can hold it for centuries.
- There are modern methods of preserving fruit and vegetables; there is no grain advantage.
- Fruits have much more vitamins than grains and have fructose, that is a healthy sugar
- Fruit and vegetables do not induce celiac disease, diabetes, obesity or heart disease.
- Unlike grains, valuable vegetables can be grown on flat roofs in aeroponics systems.

- Trees can protect smaller useful plants underneath from excessive sunshine or wind.
- Tornadoes can form in cereal crop fields, but not in orchards.

Advantages of white meat animals over red meat grazers:

- There are only 5 main species of edible grazing animals, but hundreds of non-grazing ones.
- Red meat was associated with cancers in humans and detrimental behaviour of people.
- Income per hectare from white meat animals is 10 times larger than from red meat animals.
- Energy production per hectare of white meat animals is around 10 times larger than grazers
- Pastures needed for red meat animals resulted in about 1/3 of world deforestation areas.
- Pastures cannot protect soils as well as trees from erosion and solar radiation degradation
- White meat animals, birds and fish can eat fruit, vegetables or similar animals/insects, in more complex and biodiverse chains closer to Nature than the monoculture of pastures.
- Pastures many times require pesticides and fertilisers.
- Trees evapotranspiration (thus climatic performance) is three times that of pastures.
- Pastures cannot store carbon like the fruit trees needed for white meat animals.
- Tornadoes can form in deforested pastures, but not in orchards.

Such improvements in agriculture would be driven by market forces, would improve population health and wellbeing, would protect soils and would significantly contribute to climate restoration.

The above is in line with the studies on adaptation of agriculture to changing climate conditions [26], to which it adds the effect of correcting climate through transformed agriculture.

7.3 The Avalanche Greening and Cooling (AGC)

AGC is large-scale revegetation that once properly started self-propagates like an avalanche (positive feedback) naturally, or can be substantially accelerated artificially to mitigate climate change, including air temperature and carbon gases in the anthroposphere. AGC is one of the main focuses of this book because it is the best method to reverse climate change. AGC involves revegetation compensated with evaporative cooling (EC) in areas where full revegetation may not be feasible, for example in existing urban areas.

AGC is the opposite of ASH, despite also advancing like an avalanche. Unlike ASH, which is a quick and powerful process due to its entropic nature, AGC is an anti-entropic, slower and less forceful process that initially greened Earth eons ago. With internationally committed help, AGC could start re-greening essential regions in just a few decades and make a significant climate improvement.

There have been many previous suggestions and initiatives for revegetation to restore Earth's climate, such as the Great Green Walls of Africa and China. However, contrary to them, I propose revegetation in a downwind direction, as shown in Figure 85. This approach allows to halt HAC and gradually deprive ASH of energy and sand, while not exposing the new plants to its hot attacking front and sandstorms.

Like ASH, AGC also progresses with the prevailing winds of the region and it could be artificially accelerated to the rate of possible ground coverage.

Additional locally adapted measures will optimise AGC and enhance its effectiveness.

The revegetation and water tools presented in 2020 in [52] address these issues, as detailed below.

7.3.1 AGC Tools

7.3.1.1 The Water Cycle is the primary tool for climate repair, because:

- It is the only tool with enough power to counter-balance solar radiation and air heating through cloud reflection and evapotranspiration.
- It has the highest heat transfer capacity, and can transport a lot of heat at a low, constant temperature from the Equator to the Poles through the very thin troposphere.
- Climate change effects (droughts, storms, heatwaves, hail, fires, and floods) are symptoms of a disrupted water cycle.
- Water has exceptional infrared absorption and emission capacity.
- The ocean, the main heat storer, is part of the water cycle.
- Water is the medium of life, of which we are a part.
- Only the 7.3.1.1.1 - 7.3.1.1.3 and 7.3.1.3.4 tools are strictly necessary; the rest are optional.

7.3.1.1.1 **Runoff Water** that is excess surficial water from rain preferably should be stored on site in the groundwater or in reservoirs before it flows away because no rainwater should return to the sea. Collection of water is through terracing where available, or through V or crescent-shaped retaining shallow walls pointing downhill, that prevent soil erosion and temporarily retain rainwater giving it time to seep into the ground. At the lowest point and inside of the retaining walls a relatively shallow borehole with gravel wall to improve ground flow will allow better seepage when charging and eventually pumping out the water when discharging. Tree roots can withdraw this water from a depth of a few meters. Depending on the nature of the local ground, the borehole can be deeper, to offer more storage capacity. An average of rainwater storage of 1,500mm (above ground) of which 1,000mm is realistically available for pumping out is enough to assure water independence. In case of a low porosity ground of 0.15, this requires a 10m deep borehole, while in a good porosity ground 5m depth would suffice. There are cases (especially in hilly terrain) when the ground structure cannot retain the groundwater, irrespective of the quantity of water delivered to the borehole. In such cases, storage of water in reservoirs is necessary, preference being given to deep reservoirs which occupy less ground area. They can be artificial ponds or dams that inevitably lose water through evaporation and beneficially humidify the surrounding area and provide vapours for cloud formation, or there may be many covered small reservoirs.

7.3.1.1.2 River Water and Reservoirs

Large reservoirs like dams and pumping stations are necessary in the first stages of the AGC to retain all freshwater on land and prevent its loss to the sea before the natural humidity pump (section 7.3.1.3.4) properly starts. River and creek water is the most accessible source of freshwater that can be extracted with pumps from the closest point of a river and pumped in the nearby reservoir. Another purpose of the large water reservoirs and pumping network is for extracting and supplying water during rare, small natural droughts and floods. However, a rehabilitated climate will likely reduce the occurrence of large droughts or floods. Returning water to the land also protects the marine ecosystem, especially the coral reefs, from eroded soils and fertilisers. Solutions exist to maintain the bio-environmental role of discharged freshwater in deltas and estuaries even if most river water is redirected to the land.

7.3.1.1.3 **Boreholes Along the Shoreline** prevent surficial runoff, extract groundwater that would otherwise be lost to the sea, and pump it into reservoirs. This groundwater which may contain fertilisers, must not contaminate the marine environment, but can be utilised for AGC vegetation.

7.3.1.1.4 **Evaporative Cooling (EC),** is necessary to compensate for the lack of evapotranspiration where vegetation cannot be used or is not as dense as it should be. In the fields, EC can be just irrigation water on crops to compensate for the difference in evapotranspiration compared to trees. In urban or rural areas without vegetation or with insufficient vegetation, EC could be water sprinklers or humidified canopies made of porous materials that also work as sunshades. Automatic dispensers would keep the canopies wet without allowing dripping or runoff that would be a waste. This way the optimum quantity of water is automatically evaporated, proportionally with solar radiation.

7.3.1.1.5 **Rainwater Collectors** like in Figure 75A can provide water for revegetation if river water is not available in the initial stages of AGC, until the humidity pump starts.

FIGURE 75 : Marine methods (vertical cross sections)

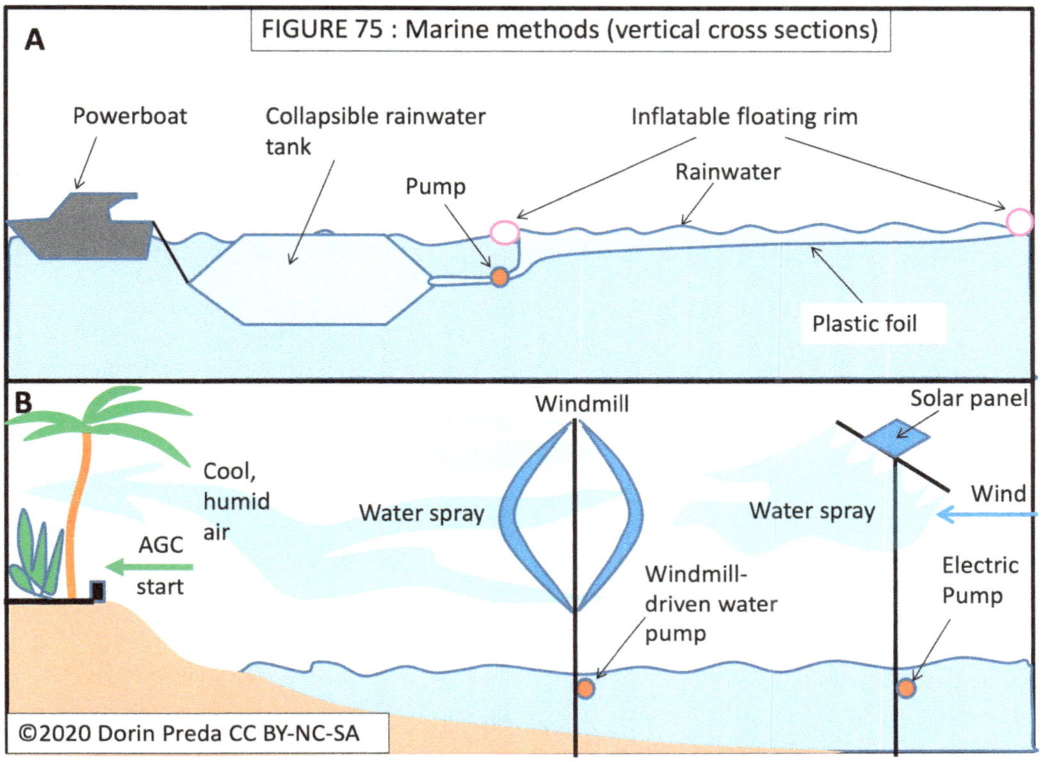

My method of collecting rainwater from the sea uses a large rain collector made of plastic foil surrounded by an inflatable rim that float on the sea. The collected rainwater may be pumped into a collapsible floating tank and transported to the shore. This method can also be used in lagoons of atolls and islands with freshwater deficits.

7.3.1.1.6 **Air Humidifiers on the Sea** as shown in Figure 75 B help increase air humidity near the shore in front of the AGC start-up line, by using wind or solar energy to pump water and create a fine spray that humidifies the air. These devices should be placed at least 100m from the shore to prevent salty spray from reaching the land. To avoid seawater filtration for their nozzles, they could instead use fast spinning winged wheels onto which unfiltered seawater flows and is dispersed. Different other models of humidifiers can be employed, including passive evaporative filaments, meshes or pads humidified with seawater with pumps or through their own movement in the wind, but they will have to be resilient in storms. Marine air average relative humidity in Tripoli and in

Brisbane is around 60% annually and 70% in summer. These humidifiers could raise low level relative humidity by around 10%, so the subsequent onshore conditions coupled with the natural humidity variations can induce rainfall.

7.3.1.1.7 **Air Humidifiers on the Land** consist of evaporative pads as illustrated in Figure 76.

FIGURE 76 Seawater Evaporative Pads

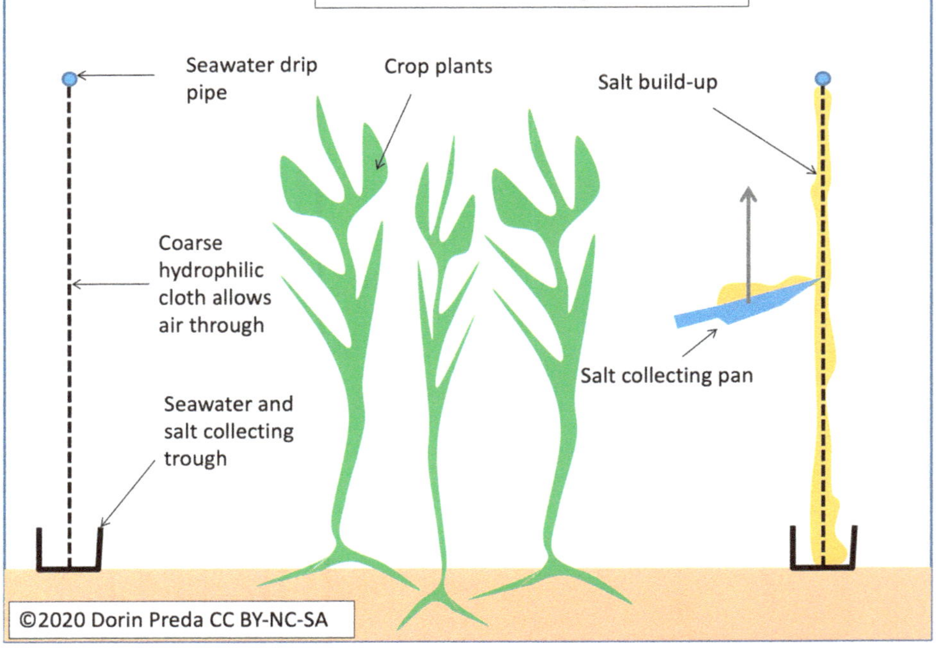

These cloth pads above long troughs are humidified with seawater pumped using solar panels to cool and humidify the air. They promote plant growth, provide shade and wind protection to young plants near the ground, and allow taller plants to reach full sunlight. If the rate of seawater flow is large enough, no salt would be formed. The salt produced when water flow is too low can be collected and sold.

7.3.1.1.8 **Fog Collectors and Air Wells** of many types exist and can be employed if there is not enough river water available to help start-up the AGC in a hilly region.

7.3.1.1.9 **The Rain Tower** as in diagram A of Figure 77, is one of my design solutions for providing the water necessary for the start-up of AGC if there is no river water available. The tower would be constructed on the seashore in front of the AGC, to collect the humid marine air pushed by the wind and allow it to rise vertically, expand adiabatically and achieve precipitation if the tower is tall enough (a few hundred meters). The tower base could extend sideways alongshore in a wider funnel (Venturi) to collect more wind energy, which is the start-up energy. If precipitation conditions are achieved in the tower, they will decrease pressure and attract more marine air in, while the rising in the tower of the warm dry air resulting after precipitation will accelerate the flow. Initially, when precipitation occurs at the top of the tower, rainfall inside the tower will humidify and cool the air below, raising its dew point, so the precipitation area will move downwards. However, this means that the top of the tower will exert more suction due to rising of more warm air, that will accelerate the airflow in the tower and drag precipitation area up, until an equilibrium is established. Rainwater can be collected either at the base of the tower, or with troughs higher in the tower, to provide pressurized rainwater for a small hydro-electric generator. It would be more beneficial to let water rain inside the tower because it enhances the tower's precipitation performance.

FIGURE 77: The Rain Tower, the Rain Trough and the Humidity Turbine

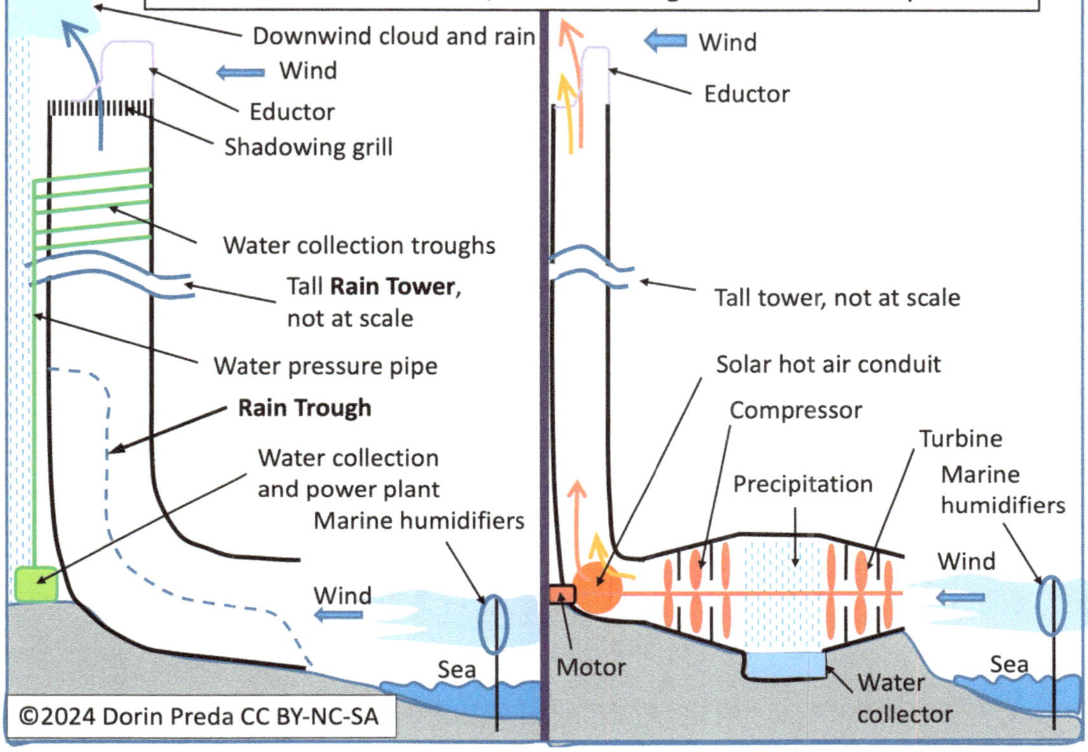

The tower could be made of basalt reinforced concrete; thus, it needs compression, which can be achieved with many diametral cables made of basalt fibre. It is expected that moss will naturally form inside the tower, playing the best hydrophilic role possible on all surfaces like the walls and the diametral cables, promoting precipitation. A grill with black vertical walls at the top of the tower would reinforce it structurally and would prevent sunlight from heating the inside of the tower. Marine humidifiers like in Section 7.3.1.1.6 would increase inlet air humidity and decrease temperature to favour precipitation in the tower. The tower should be painted on the outside with reflective paint. My eductor design (the tower is taller on the windward side by a height equal to its diameter) increases air flow without obstructing it. Depending on meteorological conditions, precipitation will occur in the tower, or higher in the troposphere from where it would rain down on the land of the AGC. A rain tower with a diameter of 100m that extends 300m at its base in the alongshore direction, could deliver around 10,000t of water per day from a 3m/s wind.

7.3.1.1.10 **The Rain Trough** shown in Figure 77A, is another design which could help the start-up of AGC and would work on the same principle like the mountain slopes which naturally enhance precipitation. The Rain Trough is less tall than the Rain Tower and less expensive. It would have a similar curved shape like the base of the tower, but without the wind-side, so its role would be just to direct the strong sea breeze high in the troposphere where it could achieve adiabatic expansion and precipitation conditions. I estimate that such a Rain Trough should be at least 200m wide in the seashore direction and at least 300m tall with 200m curvature radius. It would work in rather strong wind.

7.3.1.1.11 **The Humidity Turbine** depicted in diagram B of Figure 77 is another of my designs that would help if there is no river water available to start the AGC. This mechanical device lowers the pressure of the humid air to make it release its water, then re-pressurizes the dry air which warms up and rises in a tall tower to recover the energy for the process, and is preferably assisted by a solar air heater. The turbine system has to be started with a motor, but in good weather conditions it would subsequently work without the motor, using the latent heat released by water upon condensation, and especially the additional heat from a solar air heater. The device uses a turbine at the inlet which is installed on the same shaft with a compressor and a motor at the outlet. The turbine would be installed on a seashore and receives cool, humid air from some marine humidifiers. The inlet air rushes into the turbine because air is attracted into the precipitation zone which has a lower than atmospheric pressure. As inlet air actuates the turbine and expands, it also cools, allowing water to precipitate in the central zone of the device where pressure is smaller than atmospheric. Water flows in a collector from where it is evacuated with a pump to the outside. The dry air at small pressure in the precipitation zone is then absorbed by the compressor which at start-up is actuated by the motor. The compressed air is warmer than the inlet air because it received the latent heat of water, so by rising in a tall tower together with very hot air from a solar air heater, they help the compressor overcome the inefficiencies of the system, so the motor does not have to work all day. The solar air heater is a very large device, not pictured in Figure 77, except for its red conduit, and it can produce electric power besides helping the humidity turbine. It is similar to the solar updraft tower: https://www.sbp.solar/solar-updraft-tower/?lang=en. In 2002 I patented the first solar updraft tower with permeable (transpired) solar radiation absorber [91], and I increased its working temperature in 2007 with [92]. The tower itself could provide start-up solar heating if its sunny side is transparent and the opposite side is black in colour at the inside. Such a humidity turbine in the subtropical climate, near the sea, could deliver around 7,000t of water per daylight day (without the motor working) if the diameter of the turbine is 40m, and inlet air speed is 30m/s. It could also produce 20,000t in the rest of the day when the motor has to work, but it might be more economical than seawater desalination. This device does not rely on wind for its functioning. Another version, in a windy environment, could replace the motor with a wind turbine.

7.3.1.1.12 Rainwater Availability is mistakenly believed to be a problem in most tropical deserts that are thought not able to support large quantities of vegetation. In fact, a conservative estimate could consider the input of humidity from marine wind blowing with a mild speed of 2m/s across 1m length of shoreline. Let us say that only the bottom 3,000m of the wind column would suffer convection and condensation as the air volume travels deeper inland, this giving a volume of 6,000m3 of air crossing the shoreline per second. At 50% relative humidity and 250C temperature, that volume of air holds 12g/m3 of water, or 72kg for the 6,000m3. If condensation conditions would subsequently occur above that land and only 2/3 of that water transforms in rainfall, and it only rains 1/3 of the day, it results in around 1,400t per day. Let us further assume that there would be only 3 rainy days per week, that means around 270,000t per year. If the depth inland on which this rain spreads is 270km, results an average precipitation from marine input of 1,000mm of rainwater per annum. In a rainforest about 40% of the rain is rainforest input through evapotranspiration, that would give a total rainfall of 1,400mm per year which is basic for a rainforest. However, only 400mm of the 1,500mm evapotranspiration of the AGC was used, thus a good estimate would be that marine humidity input at only 2m/s wind would suffice for around 500km of inland territory of rainforest. In stronger wind days more humidity is available. Additionally, local wind is part of the convectional global circulation that can bring more humidity from the evaporating oceans which are not in close proximity to the AGC area. It is important to remember that the rate of precipitation on the oceans is much larger than on land (see Section 1.10) because most lands are blocked for the marine humidity due to HACs. When these precipitation sinks will open, there will be a decrease of global water vapours partial pressure which will redirect part of the present ocean precipitation towards lands and will also increase ocean evaporation. Given the very large evaporation potential of the ocean, the much smaller area of land than of ocean, and the high mixing of the atmosphere, enough vapours will be available anywhere good precipitation conditions will occur. The southern hemisphere has a much larger ocean to land area, thus capacity of providing water vapours; this will ease revegetation of Australia. New studies could reveal if there is significant difference in the absolute humidity between the two hemispheres and what is the "permeability" to water vapours of the inter-convergence Equatorial zone. Therefore, the problem is not the source of humidity, but to favour rain conditions, as suggested further.

Each region has its own characteristics, thus customised solutions have to be designed and deployed for improved results.

7.3.1.2 **Carbon Dioxide Sequestration** is another tool in the climate change/repair process. Artificial carbon sequestration is difficult and expensive, while the water tool and revegetation provide larger carbon sequestration at lower costs. The large cooling effect of the water tool (as shown in sections 2.5.3 and 2.6.3) can also offset the global warming of carbon dioxide, until its concentration in the atmosphere is reduced through artificial sequestration, or through natural sequestration in trees or the ocean.

7.3.1.3 **Vegetation** is a vital tool that harnesses the power of water and offers additional benefits (fruit and vegetables) besides carbon capture and climate repair.

7.3.1.3.1 **Trees** are very efficient automatic water collectors, storers and evaporators, that also provide shade, soil stabilisation, wind shelter to other plants, carbon storage and organic carbon aerosols for vapour condensation. They are the best and largest evaporators, able to maintain their function during small droughts, while grasses simply dry out and die. Trees do not require water tanks, as they can store some water in their trunks and rely on deep groundwater. In natural forests trees create their optimum environment depending on latitude, ground type, wind and humidity regime, etc. They adapt to those conditions and gradually the forest expands as much as land is available. In an AGC, trees are planted artificially, but it will be a learned science and an art to choose the proper combination of tree species when planting them. The reconstituted forests, the useful tree plantations and the orchards can be as dense as natural forests, thus no EC would be necessary. Cities, however dense vegetation is possibly established, will need EC. Owners should plant as many trees in their gardens as possible, while townships could plant trees anywhere possible around buildings, increase tree density in parks, plant them on the sides of roads and even on the centreline. Tree cover must be very dense, to reduce expensive but necessary EC to eliminate solar heating. It is essential not to install an elevated rim around tree open patches in the asphalt, as rainwater must seep into the ground and reach the tree roots. Cobble stone pavement where possible is preferable to asphalt. Integrating trees open patches into the township stormwater system would also be advantageous. Tree saplings should be planted deep in their holes, to prevent large roots from damaging surrounding asphalt.

7.3.1.3.2 **Lines of Trees** planted in crop fields are already being used to prevent soil erosion, improve evapotranspiration, and decrease solar radiation on crops, known as "alley cropping". This is especially necessary in the intense sunshine tropical areas. In sloping terrain, as depicted in Figure 78, trees should be planted on contour lines in pastures and crop fields, and bund walls or tarpaulins supported by pegs and trees should be installed to retain water for a few hours after rain and allow it to seep into the ground.

FIGURE 78 : Harvester between tree lines

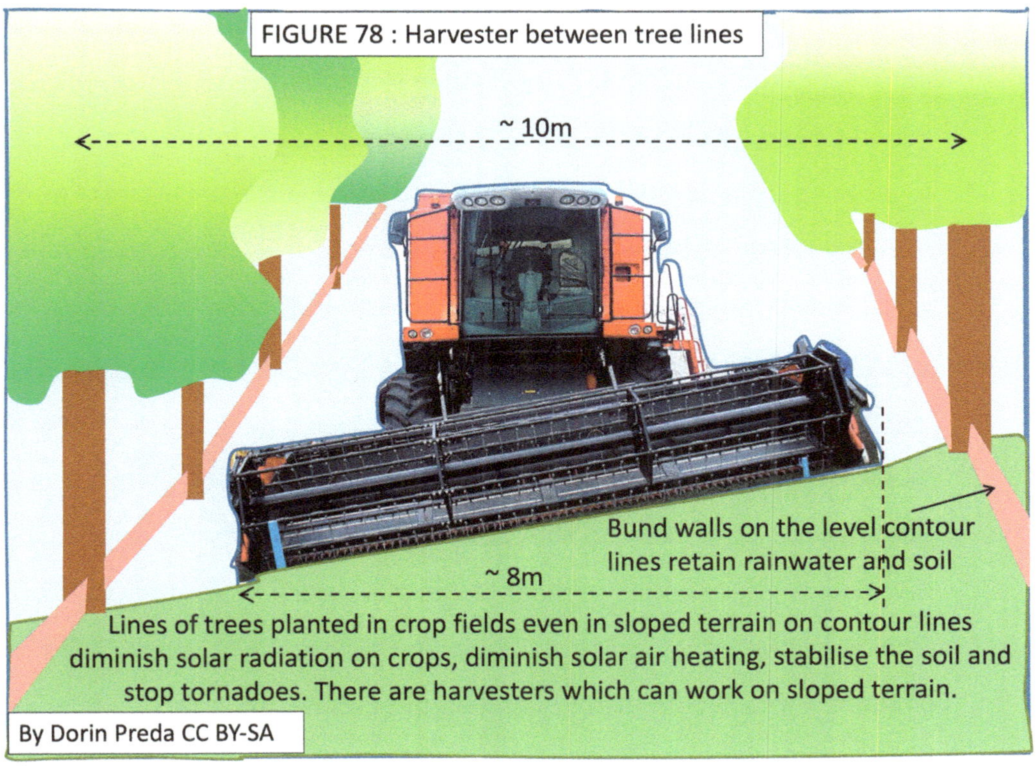

By Dorin Preda CC BY-SA

Over time, the moving soil will form horizontal terraces based on the bund walls. Alternatively, terraces could be directly built at the time of bund walls work, as historically done in Southeast Asia and the Americas.

7.3.1.3.3 **Tree Ownership** is critical, as deforestation occurred because people found more beneficial use for the land than the "useless" trees of native forests. By planting useful trees, they will be valued, cared for and protected by their owners, resolving this way the problem of forest fires.

7.3.1.3.4 **Tree Humidity Pumps** as first presented in [52] and in Section 2.2.2.4 of this book are the main tool of AGC that transfer water from the ocean to land.

The great humidity pumping capacity of the Amazonian Forest is calculated in Section 2.2.2.2 as an example. By re-establishing tree humidity pumps between the tropical ocean and the tropical forests, plantations or orchards, we prevent ocean humidity from being pushed to mid and high latitudes to heat and flood them, as stated in [90] and illustrated in Figure 70. For such a humidity pump to start and deliver more water than the AGC consumes, the minimum width alongshore should probably be at least 20km in the difficult conditions of North Africa, or 5km in the more favourable East Australia case, while the depth inland would be 7km in North Africa and 3km in East Australia. These dimensions are justified by the wind regimes, present rates of precipitation and monthly relative humidity. The reservoirs will collect the seasonal excess water and deliver it to the plants in the dry season, while the annual excess will be used for further inland advancement of the AGC. As the AGC extends inland, the excess water will tend to create permanent rivers and creeks. To prevent the necessity of pumping river water from the river mouth back upstream, preference would be given to collecting that excess water locally in reservoirs and in groundwater boreholes all over the territory.

7.3.1.3.5 **Green Canopies** could be set-up in areas where trees are not feasible, by using strong mesh supported by poles or hung with cables from buildings. Climbing plants could then be grown in these canopies above roofs of buildings (both tilted and flat) parking lots, schoolyards, open warehouses, etc, because there should not be any solar heating. Installing mesh or other suitable structures on the walls and roofs of buildings would allow climbing plants to cover them (refer to Figure 57, bottom left corner). The leaves of these plants would prevent rainwater from reaching the walls, reducing wall decay and heating costs (as drying a porous wall after rain significantly cools it). Many useful climbing plants such as grapes, berries, and even vegetables could help offset the cost of maintenance and could provide food for people or animals.

7.3.1.3.6 **Grasses,** including crop grains, are smaller plants that are not as effective as trees for managing climate for many reasons:

- Grasses evaporate less water than forests, requiring additional use of EC.
- Grasses have a shorter lifespan than trees, and release their carbon quickly.

- Grass-like plantations require more work to sow and harvest compared to trees.
- Crops require tillage, which damages soils, while tree roots protect them.
- After tilling, the soil is left uncovered, allowing sunlight to damage it.
- Grasses even when mature do not provide as much ground protection as trees.
- Agriculture is unsustainable and can ruin in one year soil created over a century.
- Trees lock in about 150 times more carbon per unit of land than grasses.
- Grasslands cannot prevent tornado formation, while forests and orchards can.
- Even a brief drought can kill grasses, turning the land into a solar air heater.
- Claims that grass agriculture (for biofuel) could replace forests are unfounded.
- Thus, trees should be preferred to grasslands and pastures.

7.3.1.3.7 **Organic Carbon Aerosols** produced by trees not only promote vapour condensation and rain in humidity pumps but also have a positive effect on reflecting solar radiation. Re-vegetating the Earth would increase the global concentration of organic carbon aerosols produced by plants, contributing to global cooling without the need of geoengineering.

7.3.1.3.8 **Chemical Plants** are needed to produce concentrated solutions of nutrients tailored to different plant species because of the required high plant density in poor soils.

7.3.1.4 **Soil** is the second largest challenge (after water) in re-vegetating deserts. Unfortunately, careless deforestation has led to partial or total soil erosion over more than half of Earth land areas. Another careless destruction is the practice of burning vegetal residues from slashing, cropland clearance after harvest, city tree clean-up, etc. All vegetal matter and organic residues should be turned into compost and combined with suitable inorganic materials to create the necessary soils for desert areas. Methods exist to use desert sand as a base for plant growth when humidity retention is provided. The fact that Earth's deserts are a result of human mismanagement, not natural climate change, is evident in the fact that intertropical areas of the Americas and East Asia are not deserts. This is because local populations did not burn or cut down their forests, and did not overgraze their lands with goats. They practiced small scale agriculture with great care for the soils, regenerating them with biochar, crop/forest rotation (Central America), and terracing (East Asia). Terracing is the best method for soil stabilization for cropping and should be implemented even in slightly sloped terrain to prevent erosion, promote soil and water retention, and increase soil health. A large plant density is required for AGC, that raises the problem of soil nutrients. Initially, these will be artificially prepared and delivered. However, after real thick soil is formed, it will be able to support that large plant density of different symbiotic species (if they are properly chosen) for the same reason for which the Amazonian Forest can.

7.3.1.5 **Urban aeroponics** is a good solution for flat roofs to provide more evapotranspiration than a green canopy and can provide fresh produce and good education for youngsters

7.3.1.6 **Reflective Surfaces** as simple and inexpensive like painting the roof white can be used especially in urban and rural areas where humidified canopies or green canopies cannot be installed. However, some EC would still be needed because any materials would still heat up.

7.3.2 AGC Description

The past deforestation of Earth has had negative consequences, particularly when soils were lost, leaving us with deserts that contribute to climate overheating and lack economic value. Therefore, it is crucial to setup the first AGCs on coastlines to stop the HACs and allow more marine humidity enter the land areas. The method is feasible on coastlines where the average far offshore wind blows from the sea towards the land. The implementation of AGC will use the tools described in section 7.3.1 above, of which only the 7.3.1.1.1, 7.3.1.1.2, 7.3.1.1.3 and 7.3.1.3.4 of the water tools are strictly necessary.

The general steps of the AGC started on a deserted coastline are:

- Put aside the desired Natural Reserves (Section 5.5) and close all human access to them.
- A total rate of evapotranspiration from vegetation and EC is established such to prevent solar heating in the starting line which continuously occupies that stretch of coastline.
- The new plants evaporate water, humidify and cool the air and emit organic carbon aerosols, promoting cloud condensation, which reduces solar radiation and air heating.
- The HAC is reduced and pushed inland by the increasingly vegetated coast.
- Marine moist wind is further humidified by vegetation evapotranspiration and pushed upwards by the HAC formed at the AGC/desert interface, generating rainfall further inland.
- Inland rainfall lowers barometric pressure, attracting more clouds from the ocean, creating a humidity pump that supports AGC progress.

- Once initiated, AGC can advance downwind at the rate of achievable planting.

7.3.2.1 The length of the AGC front along the coast should be a minimum of 10km (the height of the troposphere) and should be expanded as much as possible over time. If land is available along the coast, the AGC should not be deeper inland than wide. Smaller than 10km patches of AGC on a desert coast can create an island with cool humid air, but have smaller chances to setup a natural humidity pump, thus will probably require water from storage, or even artificial water in the dry season, depending on local conditions. In case of large international AGC programs, the AGC front length should be decided on geography reasons, as shown further. The AGC cannot start in the middle of a desert, unless there is a lot of water available to cool the incoming hot wind, and it cannot become water independent.

7.3.2.2 EC necessity can be calculated or measured. It is calculated from solar radiation; for example, in tropical areas solar radiation is about 6,000MJ/m^2*year, of which approximately 1,200MJ/m^2*year is reflected back by a forest or dense orchard/plantation and around 1,500MJ/m^2*year is stored for night dissipation, thus the remaining 3,300MJ/m^2*year evaporates about 1500mm of water, to cancel air heating. This figure corresponds with the measured evapotranspiration of rainforests, which do not need EC. No gaps in the AGC should exist either on its starting line or further inland as it progresses, so that all areas evaporate at least 1500mm per year with EC help where necessary. However, EC is inferior to vegetation, as it does not emit organic carbon aerosols, and does not extract CO_2 from the atmosphere, instead requiring expensive water tanks and pumping stations. Therefore, emphasis should be put on tree planting and green canopies as much as possible, to reduce the need for EC. Ideally, the EC rate of water flow should be proportional to the instant solar radiation. Additional methods of air humidifying with seawater could be considered (7.3.1.6 &7.3.1.7). The necessary EC can also be inferred from measuring air temperature above the canopy of AGC vegetation at the starting line (on the seashore) and in the inland place of the AGC where EC has to be inferred. If the latter is higher, more EC should be added, until air temperatures are equal, therefore there is no solar heating by AGC areas. A less exact method of inferring the necessary EC would be by measuring solar radiation under the canopy of the AGC, which should be no more than 10% of that above the canopy.

7.3.2.3 Water reservoirs capacity results from the quantity of additional water needed for EC in each area. The large water reservoirs would be uncovered or partially covered with floating solar panels and would retain as much freshwater as possible; their surface evaporation will play a key role through EC. Their capacity should be large enough to prevent any freshwater return to the sea, because all water is needed for the AGC trees and for EC. The large capacity is needed especially on the coast, where the HACs may induce excessive rainfall. The solar panels that float on the lakes during the day and the conventional power plants at night could pump water upwards through the network of creeks, boreholes, and reservoirs situated at different heights on the terrain. As more rainfall will occur, it will allow for further AGC inland expansion without additional artificial water input.

7.3.2.4 Plant density is a key element of success for AGC, because it influences evapotranspiration, EC, soil health and air temperature and humidity. When plants (different sizes and species) are in high density in a place, they benefit from each other's shade, wind protection and evapotranspiration humidity, so their individual evaporation is smaller, and plants thrive. **The reason for forest drying, forest fires, and failed attempts of revegetation is the decreased plant density.** Rushing to cover large areas of AGC with insufficient plant density (size and number) is certitude to failure. One rule is to increase plant density until it blocks at least 90% of average solar radiation at ground level to stop air heating. Another rule is to have enough plant density in any place of the AGC that above canopy temperature is the same with the upwind air temperature on the starting line of the AGC. This can be achieved with shorter plants under tall ones (may be different but compatible species). EC like surface irrigation must keep the soil cool and moist to protect it from heat damage if there is no sufficient plant cover until trees grow. Ideally, by placing saplings and young trees in mobile pots between permanent but still young trees it is possible to have good ground coverage with young plants. As the saplings grow, their pots are enlarged and placed at larger distance, until it is the time to permanently plant them in the ground. The three essential AGC parameters that should be maximised are: the tall, multi-layered canopy, the plant density and the total area of the AGC. As result, the necessary condition is sufficient water with nutrients from chemical plants (7.3.1.3.8).

7.3.2.5 In cities, EC mist diffusers placed as high as possible can disperse fine water droplets that evaporate before reaching the ground. Alternatively, automatically humidified sunshades made of hydrophilic fabric could be stretched aloft to create shade and even evaporation. In cities and rural settlements, humidified sunshades could also be the mechanical support for green canopies, providing together evaporation, shade, fruit/vegetables and organic matter (leaves) for compost production. Proper maintenance requires participation of a suitably educated population which actively supports the wellbeing of their township.

7.3.2.6 Crop-fields and pastures should be planted with useful tree lines, for example a 7-8m wide strip of crop or pasture would be flanked by two lines of trees 8 -10m apart, with vegetables or other useful plants using the land between two trees on a line, as suggested in Figure 78. EC here can consist of irrigation that humidifies the soil to stop solar heating and to protect the soil from decay, while providing water for plants and for evapotranspiration.

7.3.2.7 The AGC advances as far inland as the water reserve from the reservoirs allows the 1500mm evapotranspiration condition. The climate of those areas will become wetter because of the gradually improved coastal evapotranspiration, and the increased precipitation will fill the reservoirs that in turn will allow AGC advance inland. Excess rainfall can be just pumped into the inland desert behind the AGC, to create a water table. Water must not flow out to sea.

7.3.2.8 Coastlines generally tilt towards the sea; thus, there is a highland behind the coastline that can be benefited from. Marine winds are pushed up the sloped coastlines and may reach the dewpoint and create rainfall. This can be enhanced with Rain Troughs. That rain will return downhill as runoff, creeks, and rivers, and all their water would be collected in the reservoirs that are gradually built inland as the AGC advances.

7.3.2.9 The AGC line is moved further inland when the capacity of the reservoirs becomes insufficient because of increased precipitation, increasing the water availability for EC to the new balance. This way AGC could gradually revegetate the land, making it productive for new population that works that land and lives in new, civilised and economical URBs that are environmentally friendly, do not waste land and do not produce solar heating.

7.3.2.10 **Water retention** in AGC lands will decrease sea level if AGC and humidity pumps would be re-established at global scale. Until now, human activity has decreased by a few meters (X) the groundwater and above ground freshwater level on all lands, thus has caused a rise of ocean level by a height of around X/2.5 due to the ratios of land area to ocean. This is an additional cause of the sea level rise, besides glaciers melting. Conversely, if new AGC will be developed in all bare lands of Earth, the humidity pumps will reinstate those groundwater and above land water stores, significantly lowering global sea level. This would be another climate change symptom resolved through Earth revegetation.

7.3.2.11 **The Economics of AGC** are very beneficial in the long run, but AGC requires significant start-up capital investment for population support, for agriculture energy and materials, and for new settlements (like URB) energy and materials. If AGC is started in a desert, the population support can be resolved by granting titles of property to the settlers over the parcels of AGC they will develop, with the condition that they follow the proper development plan. Similar contracts with the dwellers can be setup for financing the construction of the URBs. Other financial tools can be setup for the rest of the AGC enterprise either nationally or internationally, warranted by the high future benefit that will be brought by AGC through improved agricultural land, as depicted in Table 3.

7.3.2.12 **An example of AGC** is shown in Figure 79, as a coastline facing the sea to N (similar to the Sahara), where marine humidifiers like in Section 7.3.1.1.5 float and cool the air, rising its dew point. It is essential for the AGC that the average wind blows from the sea towards a coast. It once had a natural dense forest, but now is a shrubland with the remnants of that forest on the right, and an old village in the centre of the coast. The red right side of the figure is a detail of the smaller red rectangle in the picture on the left, which shows the Stage 1 of the AGC, that extends about 2km inland from the shore.

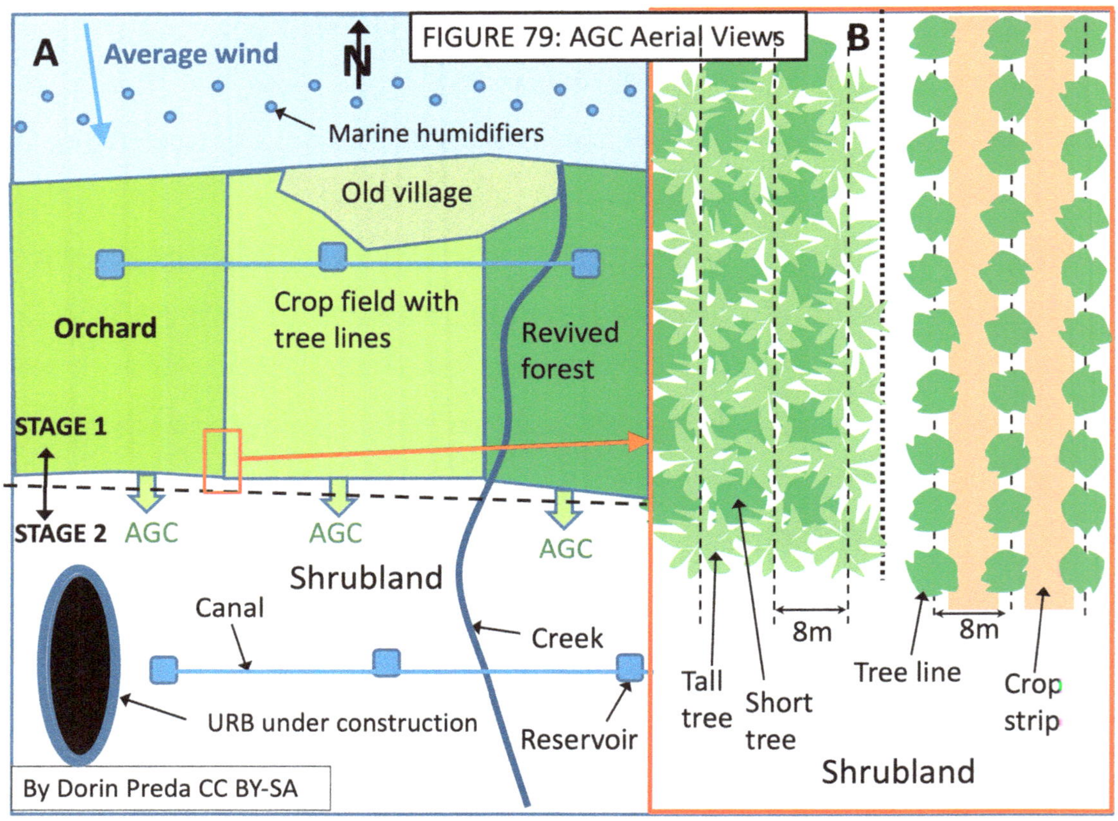

Initially, the AGC is prepared by constructing the large reservoirs and the canals from the creek. In this preparatory step the entire old village should be planted with trees, green canopies or EC such to avoid creating any solar air heating. The reservoirs should be large enough to assure water for 6 months for the planned green areas of Stage 1. After they are filled with water, planting can be started by the agricultural teams, while the construction workers would start work for the URB. As seen in the detail figure on the right, in the orchard there are tall trees lines 8m apart intercalated with short trees lines also 8 m part, such that there is 4 m distance between the tree lines, enough for trucks to drive between the tree lines and under their canopies. Initially, the saplings are planted in large baskets (or plastic bags) with soil that are placed (with forklifts) much closer to one another than 4m, to properly shade the ground and create high plant density. After a few years, when the trees start to overgrow their baskets, they are placed permanently in holes in the ground at 4 m distance one to another. In the centre of the picture is also shown a crop field (for example vegetables) with tree lines that are similarly managed with the orchard, to assure high plant density. In this first stage area actually the young trees for the stage two are also grown for a few years, until they grow enough for permanent planting in the stage two area. Artificially prepared nutrient solutions will be important for a proper start-up of the trees until they are permanently planted in the ground, and even afterwards. Therefore, there will be a slow and gradual process that will give time to the workers to properly setup a permanent infrastructure for a sustainable future. Given the fact that trees live for hundreds of years, the work investment will be worthwhile because it would permanently regenerate an otherwise barren coastline and will be able to support new population in a sustainable way. As a side effect, when such developments will occupy areas of tens of square kilometres, they will improve the local climate (decreased temperatures and increased, regular precipitation), while when such developments would occupy hundreds of thousands of square kilometres, the continental climate will be improved.

7.3.3 Sources of High Thin Clouds

In good weather, airplanes should fly under the water freezing altitude, so their vapours condense as low thick clouds, rather than freeze as high thin clouds. Large industrial plants and factories, power plants, refineries and furnaces should exhaust excess heat as larger quantities of lower temperature air or as humid air to avoid producing high thin clouds.

Flammable gases should not be burned in open flares. The excess heat from industry could be utilised to provide heating to nearby towns or for agricultural processes like drying.

7.3.4 Energy Issues

The energy sector is in a complex and rapidly changing situation, influenced less by professional analysis and more by financial interests, social misinformation and political indecision.

Before the industrial era, most of the energy used by humans came from animal power, wind power for propelling ships and pumping water, passive solar heating and drying and especially burning timber for heating and cooking.

7.3.4.1 Fossil Fuel Energy

Coal and petroleum were used for heating in ancient China, including in small personal coal heaters hidden in their large robes. The Roman architect Vitruvius described the first revolving steam engine, while ancient Greeks used steam power in piston engines.

The large-scale use of coal for mechanical energy began in Western Europe in the early 1800's. Later, oil begun to be used for large scale heating and transport, in parallel with natural gas for lighting and heating applications but less in transportation due to storage difficulty.

Due to its much smaller price than oil and gas, coal was and still is the main source of energy for large electric power plants, which through "co-generation" also sometimes provide useful heat from their cooling water. There remain coal reserves for about 150 years, but often they are not located in the place of use; hence coal has to be transported, often with great pollution, expense, and quality losses. Coal extraction and burning are major contributors to carbon dioxide and methane emissions. Less coal is used for steel making and in the chemical industry.

In the future, fossil fuels could be sold for higher prices to chemical industry, thus it would be in the interest of fossil fuels owners not dump their fossil fuels on today's small prices market.

Fossil fuel power plants (and especially the natural gas ones) began to be used as backups for the growing renewable energy.

7.3.4.2 Nuclear Energy

Nuclear fission is mainly used to generate electric power, and it is nationally dominant only in France. Heating-only reactors also exist. Fission nuclear reactors' main restrictions are their high capital cost and the potential of dangerous accidents. Thorium reactors are promising.

Controlled nuclear fusion hopefully will be achieved to generate pollution-less electric power and heat for near-by users. The aim after achieving sustained reaction will be to make this energy economical and really environmentally friendly.

7.3.4.3 Renewable Energy

Similar to fossil fuels, all renewable energy originates in solar energy. Hydroelectric power was the first used, and it is still the main contributor. Wind power has grown lately by more than 10% annually, and they install more than 1000GW worldwide. Solar energy which initially was used only in thermal applications, now produces large scale electric power, and is set to overtake wind power, despite both are hampered by storage and transmission problems. Besides solar electric power, solar thermal energy could provide heat for buildings and for processes in industry and agriculture, but different barriers exist. Despite large advancements, the proportion of renewable energy from total remains small because of technological, environmental, economic and socio-political reasons.

7.3.4.4 Geothermal Energy

This is not renewable energy, as it accelerates the local solidification of the Earth's crust, inducing discontinuities which can cause or amplify earthquakes. Geothermal energy has environmental repercussions especially due to the use of fracking during their extensive drilling needs. It is economically viable in only a few locations in the world, the rest needing subsidies or capital risk.

7.3.4.5 Requirements

Fossil fuels now provide the vast majority of our energy because renewable energy cannot supply for large scale applications, and especially not for winter heating due to natural and technological reasons. Most of the world's population lives in the Northern hemisphere, in a temperate climate where heating of buildings is unavoidable and impossible to provide from solar energy because those countries have too little solar radiation during winter. A very large wind power industry could provide only a small part of that heating and would be unreliable.

Current renewable energy technology cannot provide for large cities, factories and the transport system, because of real limitations. Improvements in energy generation, transport and storage with good economic and environmental performance are essential for sustainable transformation of the energy sector.

7.3.4.6 Solutions

Preservation of fruit and vegetables can be done economically:

a) Solar dryers like in Figure 80 could be used for inexpensive drying of fruit and vegetables by using solar energy in the intertropical climate. The dryer uses a tall chimney assisted by the fan to make the air flow through the solar air heaters and the produce in the dryer. The DC fan is also actuated by solar radiation; thus, its rate of airflow is proportional to solar radiation, so it is self-tunning with the variable power of the solar panels. Similarly, the chimney will self-tune with the solar air heaters. The solar air heaters may be of the type with permeable solar radiation absorber, which are more inexpensive than those covered with glass. The optimum rate of airflow is 0.01m^3/s for each square metre of solar air heater.

FIGURE 80: Solar Dryer

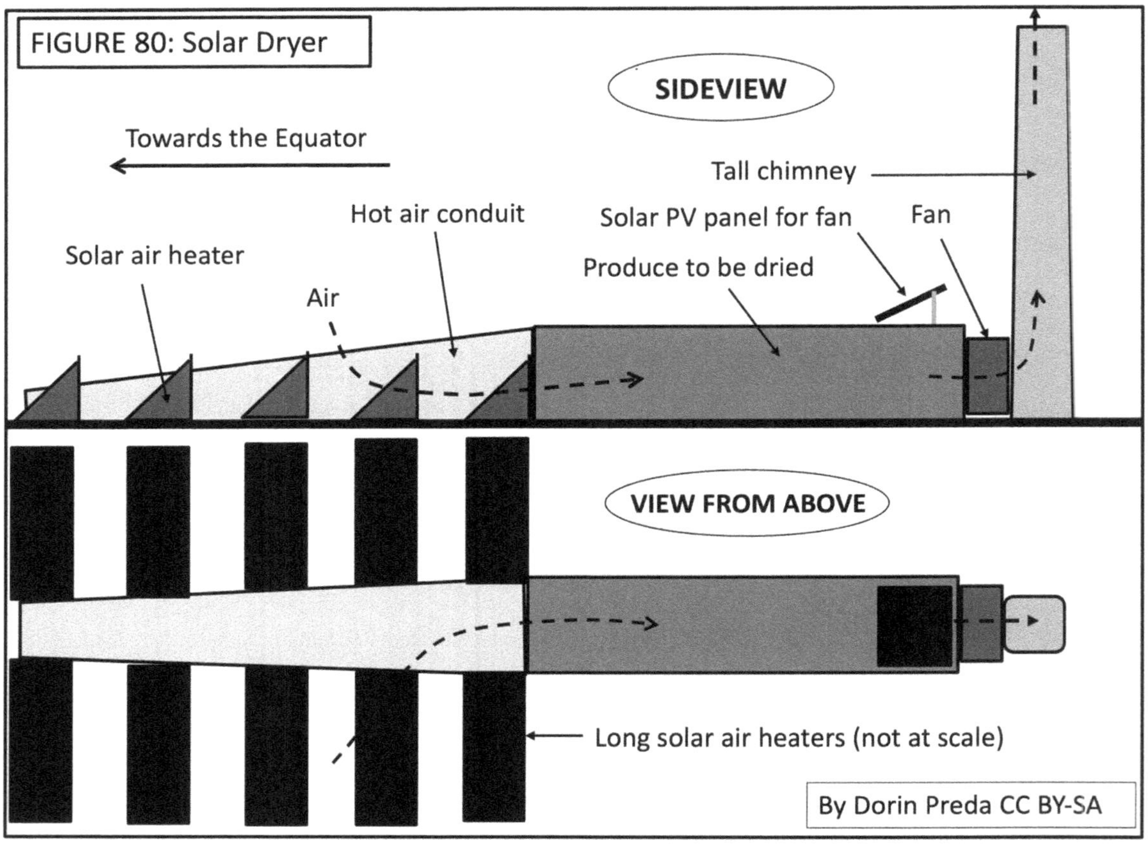

For small household production of under 1,000kg of produce per month, the dryer following the above general layout would require just a prism-shaped solar air heater of 10m total horizontal length and 2.5m tilted width (25m^2 aperture). The produce could be placed on stacked trays that allow sideways airflow, creating a long prism as in Figure 80A. Through on one end of the prism hot air from the solar air heater would be sucked it, while through the opposite end the humid and colder air would be extracted. The stack of trays should be supported above the covered ground and could be covered with a large waterproofed tarpaulin having its sides laid on the ground and airtightly pressed down by long sacks of about 0.1m diameter filled with sand. The fan (or many smaller fans placed in a common frame) actuated preferably at 12V DC should have a total power of around 250W to create a total rate of airflow of $0.25m^3/s$. The chimney should have a cross-section of around $0.1m^2$ (about 0.3m x 0.3m if rectangular or about 0.38m diameter if round) and should be as tall as possible (but not more than 15m). The dryer will work in suction; therefore, attention should be paid to air tightness. Proper hygienic measures and sterilisation should be undertaken for food safety.

b) Fruit and vegetables for animal feed could be preserved in 10 kg and 20 kg stainless steel or aluminium containers with large airtight lids that may be properly sterilised and reused many times.

c) Fresh fruit and vegetables for animal feed can be transported to the animal farm where they can be stored in very large freezers powered by renewable energy (solar, wind, etc). Very good thermal insulation and the large thermal inertia of the very large freezers will assure that they will not be affected by a few days of power interruption due to variability of their renewable energy actuation. Flat panels with phase change materials at the inside of their walls can eliminate radiative fluctuations. Alternatively, or additionally, they can have dual actuation: solar and wind, given the complementarity of these power sources. The solar panels on the roof will provide beneficial shade. The cold air resulting from thawing the fruit and vegetables for consumption can be used to cool the condenser of the refrigeration plant.

7.3.4.7 Conclusions on Energy

- The consumption of fossil fuels continuously increases, despite the talking of reductions.
- The simplest ways to reduce fossil fuels are energy efficiency and rational consumption.
- The fossil fuel industry should co-operate with the renewable energy industry.
- Must invent how to make renewable energy devices with renewable energy, not fossil fuels.
- Must invent countless renewable energy devices specialised for diverse requirements.
- The transition to renewable energy requires realistic planning and significant time.
- URBs can reduce by about 80% the energy requirements for construction, maintenance, heating/cooling and transport compared to present settlements.

7.3.5 The Carbon Budget and the Action Strategy

As shown in section 3.6, in addition to the large emissions of carbon dioxide from burning fossil fuels and deforestation, and methane emissions from various sources, there is the significant issue of increased solar heating due to deforestation. Fortunately, revegetation can help both capture excess atmospheric carbon (in parallel with reducing the emissions), and combat solar heating.

A) The urgency is to stop net deforestation, and to revegetate with useful trees the deforested Amazon ($1x10^6km^2$) and Central Africa ($1x10^6km^2$); deforested areas should be mostly planted with orchards, and tree lines should be planted in minimal vegetable and crop fields.
B) It is also necessary to revegetate about half of the present tropical deserts that are responsible for most of the water cycle imbalance. Global climate would be significantly restored if the following would be revegetated: part of the Middle East ($1x10^6$ Km^2), and part of Sahara ($4x10^6$ Km^2) in the northern hemisphere, and part of Australia ($4x10^6$ Km^2), part of Southern Africa ($1x10^6$ Km^2) and southern South America ($1x10^6km^2$) in the southern hemisphere.
C) Urban areas now occupy 0.03 of Earth, thus $3.8x10^6km^2$, and are very sparsely vegetated, therefore they are powerful solar air heaters. Unfortunately, since they are mostly situated on coastlines, they create damaging HACs which dry out the inner continental lands. The continental reforestations of point A and B require stopping the HACs; thus, as many trees and green canopies as possible are necessary in urban areas for a successful global climate restoration. Revegetation density in urban areas can be no more than ¼ of that of rainforests, thus it would be equivalent to only $1x10^6km^2$ of full plantation, and it is situated mainly in the northern hemisphere. This would balance revegetation to $7x10^6km^2$ in each hemisphere, resulting in a total global revegetation requirement of $14x10^6km^2$.

7.3.5.1 The Carbon Budgets

When revegetating the intertropical and subtropical lands, a restructuring of land use should take place, according to the suggestions in Section 7.2.7, that would bring great benefits as increased land productivity, financial performance and human health benefits. As result, most areas should be planted with dense useful trees, while vegetable croplands would have tree lines. Cereal crop fields and pastures should be greatly minimised, as their products are inefficient and unhealthy.

The result would be a distribution of land use categories as depicted in Figure 81. The useful tree plantations in the figure include the fruit orchards, which would provide a significant part of food for people and for animal meat providers. The rest tree areas are for Natural Reserves (chapter 5).

Since the focus will be to revegetate the intertropical area, the carbon requirement figures would be those of rainforests, but halved because even the densest orchards or other useful tree plantations cannot store as much carbon as natural rainforests.

From multiple sources, it results that natural rainforest carbon storage capacity is about 60,000t/km^2 (35,000t above ground, 10,000t underground and 15,000t in the soil). Therefore, the most a dense plantation could achieve at maturity (200 years) would be around 30,000t/km^2 because of canopy difference and it does not have time to accumulate much carbon in the soil.

FIGURE 81: Revegetation of the tropics

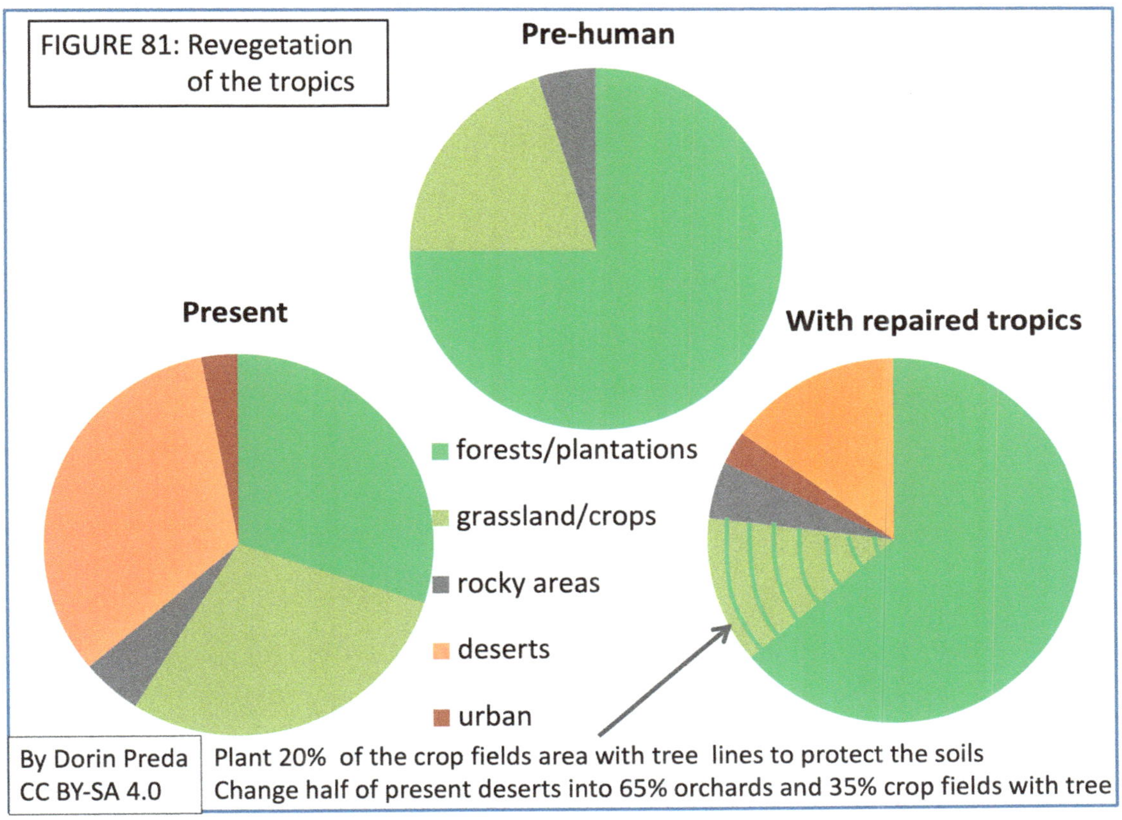

Additionally, the vegetable fields and minimal grain fields with tree lines cannot store as much carbon as a plantation, but they would occupy a significant part of the land. We could assume that they decrease the average carbon sequestration capacity to around 22,000t/km^2. The global requirement of carbon for the next 200 years becomes 22,000t/km^2 x 14x10^{6}km^2 = 300Gt. <u>Therefore, around 300Gt of carbon would be required for Earth revegetation in 200 years, but only 150Gt in 100 years, or around 100Gt during this century.</u>

7.3.5.2 Total Reforestation to bring the Earth to how it was before humans is unattainable because:

- As estimated in section 3.6, approximately 1,500Gt of carbon has been lost through deforestation and soil erosion over the past tens of thousands of years. The atmosphere now only holds 800Gt, hence there is not enough carbon available for full revegetation.
- Large areas of the land are occupied by settlements, and other needed developments.
- From the astronomical point of view, the Earth is now entering a glaciation, thus total revegetation would probably trigger a glaciation [115]. Half of the Northern Hemisphere would be lost to glaciers, perennial snow and extreme cold. However, the vast water content of oceanic areas would prevent glaciation in the southern hemisphere.

7.3.5.3 The Carbon Action Strategy

Approximately 2Gt of carbon corresponds to 1ppm of CO_2 in the atmosphere, which now has a concentration of 420ppm of carbon dioxide. It is known that at least 300ppm of CO_2 must remain in the atmosphere for plants and life to continue [38]. This means that there would be about 120ppm of CO_2 or 240Gt of carbon available in the atmosphere for reforestation.

As seen in above Section, 300Gt of carbon are needed for revegetation with fully matured trees in 200 years, but only 100Gt is needed this century. This means that accelerated revegetation can immediately start to reduce the carbon from the atmosphere, and the inevitable carbon emissions in the next century will provide the difference of around 60Gt.

The plans for Net Zero already consider reforestation in most developed countries, that is underway in North Europe and China, which are not included in the reforestation for climate restoration planned in this book. In fact, reforestations in the Earth mid-latitudes enhances climate change by taking more of the humidity needed at low latitudes as explained above. Those reforestations attract more humidity, overheating the mid latitudes and even the Arctic. The result is increased flooding in those areas, that is already apparent in China, which has large oceanic humidity input. This effect is less significant in Northern Europe, that is land-locked from the East, while the westerlies bring less humidity because of the cold North Sea. Therefore, developed high latitude countries should fulfil their Net Zero plans by funding revegetation in the intertropical area.

As seen in Figure 22, fossil fuel burning produces about 9Gt of carbon per year. The ocean absorbs 2Gt, but half of that compensates for the CO_2 emissions due to ongoing deforestation, so there still are 8Gt of emissions per year that accumulate in the atmosphere. The increase in carbon emissions should be stopped, to limit the annual input to 8Gt of carbon that could gradually be taken up by revegetation. Revegetation can be very effective, as suggested in Figure 82.

FIGURE 82 Tropical Revegetation Scenarios

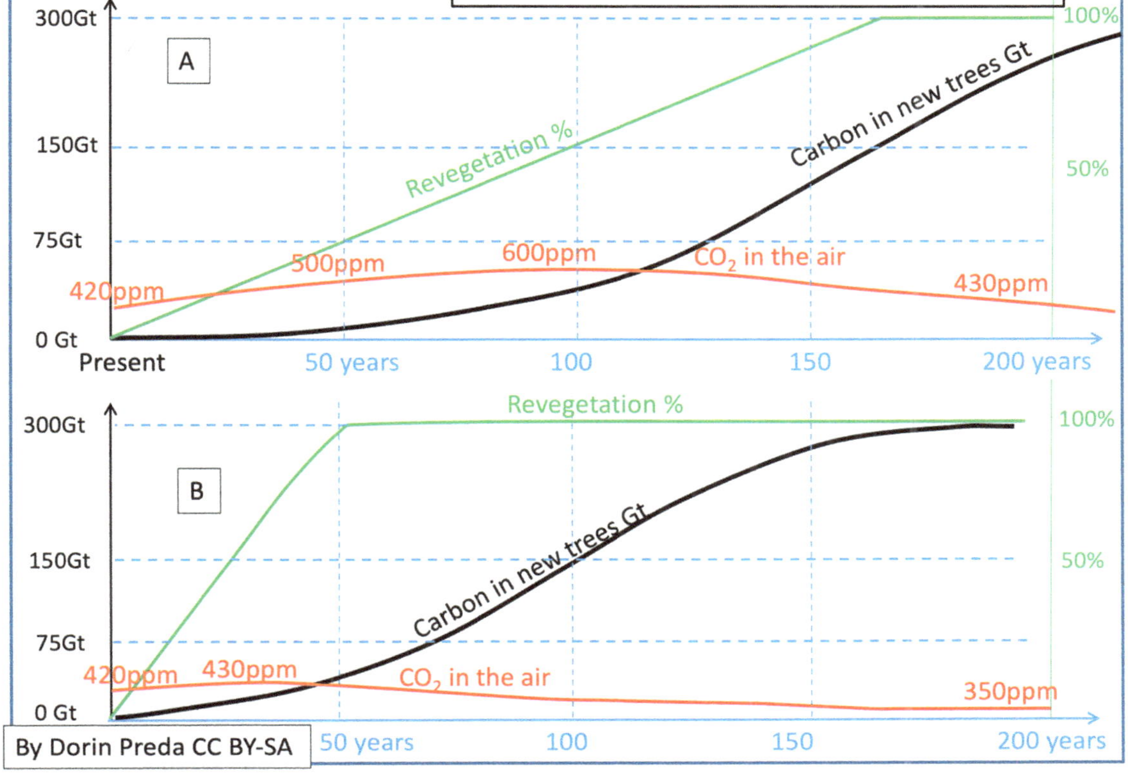

Initially, with the 240 Gt handicap in the atmosphere and the lower carbon initial uptake by young trees, the CO_2 concentration in the atmosphere will continue to rise. However, it would then begin to decrease as trees grow and their absorption capacity increases.

Simultaneously, there should be transition to renewable energy, while the revegetation program should not increase the fossil fuels consumption, which is possible by reducing energy waste.

As shown in the top A version of Figure 82, if the increasing in carbon emissions is stopped, but reforestation is slow, too high levels of CO_2 in the atmosphere would be reached in less than 100 years. Instantaneous switching to 100% renewables is not feasible because there is not yet the know-how, and manufacturing those renewable energy devices would double fossil fuel use and emissions. Though, transition to renewables must be sustained at the maximum possible pace.

The most realistic solution is to revegetate with useful trees the tropical areas within 50 years (bottom B version), while transitioning to renewable energy without increasing fossil fuel consumption. This revegetation and the transition to renewable energy would be powered by an increase in energy efficiency and by eliminating unnecessary activities.

More than 200 years in the future, when the oldest planted trees would begin dying, to avoid their carbon from returning to the atmosphere, they could be transformed into biochar to create new soils in the vast desert areas that have an almost unlimited capacity to receive them.

To achieve this goal, there should be scientific accord on the above principles, followed by proper engineering planning and ideally international cooperation. However, it should be noted, that the benefits produced by the suggestions at point 7.2.7 would allow some countries pursue by themselves the revegetation with useful trees, even if the international accord is not reached.

7.3.5.4 Revegetation of the Tropical Regions

Intertropical revegetation with useful trees as suggested in Section 7.3.5 would achieve:

- More rain on tropical inland areas due to the elimination of HAC, solving the drought problems.
- Preventing intertropical heat from overheating and flooding the mid-latitudes.
- Halting the permafrost melting, thus preventing methane release.
- Reduction of overall heat received by Earth through increased cloud reflection at low latitudes
- Reflective and evaporative cooling of the low latitude lands, especially the desserts.
- CO_2 sequestration, thus decrease of the greenhouse effect.
- Global cooling because of decreasing temperatures both at high and at low latitudes.
- The tropical oceans will receive less solar radiation (due to increased cloud cover), hence will evaporate less water and decrease the water vapours partial pressure, limiting their transport.
- Cooling of settlements by reducing UHI, diminishing the rain imbalance
- Stopping soil erosion through tree roots stabilisation, stopping landslides.
- Stopping the tornadoes in the fields by turning them into tree areas.
- When AGC reaches Western Sahara, the hurricanes that damage the USA will be diminished.
- Stabilisation of the climate by reducing extreme climate events (heatwaves, storms and floods).
- Rehabilitation of deserts, tripled food production, new settlements and Nature conservation.

- Substantially improving human health and wellbeing through better diet.
- Resolving the issue of African migration to Europe and the social instability in the Sahel.
- Resolving the piracy problem in Somalia by providing local work and income.
- Resolving the international economic growth issues (since this a very large project).

Therefore, revegetating half of the Earth deserts would halve the 3^0C of warming produced through their initial deforestation (Section 3.6.2), and will compensate for the present excess of 1.5^0C.

All of the above are possible if each country finances the sequestration of all its carbon emissions in new trees in the intertropical area.

The future continuous need for 2Gt of carbon annually to compensate for the CO_2 naturally absorbed by the ocean could be provided by the biochar production process from timber.

Reforestation has been previously suggested by other authors, but this 2020 justification in [52] and [90] is the first to address the HAC and ASH issues, identify tropical deforestation as a major cause of climate change and propose viable solutions.

7.3.5.5 Conclusions on the Action Strategy:

- Reforestation at mid-latitudes amplifies climate change.
- Developed mid-latitude and high latitude countries should fulfil their Net Zero plans by funding revegetation with useful trees in the intertropical area.
- A balance can be achieved between carbon availability and climatic goals through tropical revegetation with useful trees within 50 years and transitioning to renewable energy.
- Many social, economic and environmental benefits can be attained while stabilizing the climate, preferably through international cooperation.
- Thus, planting useful trees is the solution to solving all major climate issues.
- Geoengineering should be banned both as research and testing.

7.3.6 AGC Suggestions

There are different requirements for each location where the AGC would be implemented, so the solutions and general suggestions outlined here will vary. However, there are common characteristics. AGC can be achieved as capabilities and resources by most nations and is within reach for well - organised and well - funded consortiums.

The greatest expense for AGC is the URBs, as the working population requires accommodation. It would be inefficient to first setup traditional villages. Instead, URB construction workers would live in mobile accommodation or relocatable camps. In the second stage, after the URB is established, the actual AGC work will begin when most workers arrive and are housed in the URB. Worker requirements are accessible to developing nations:

- Basic experience in earth works for water reservoirs and terracing
- Basic knowledge of hydraulic operations for canals, pipes and pump stations
- Understanding of soil preparation using inorganic and organic materials
- Basic skills in nursery work, digging and planting,
- Basic transportation skills including trucking
- Skills in agriculture, and horticulture
- Higher skills are necessary for URBs construction, but they are still usually accessible.

7.3.6.1 Indonesia

According to the information presented in this study to since 2020, Indonesia could take measures to mitigate climate change in its proposed new capital in Borneo, which shares a similar geographic position with Somalia.

7.3.6.1.1 Somalia

Figure 83 demonstrates the significant impact of vegetation changes on coastal and inland areas. Both coasts lie on the Equator and face the same direction (SE), experiencing easterly oceanic winds and monsoons. However, in Somalia, these winds are blocked from moving inland by the strong coastal HAC. Deforestation in Madagascar created a powerful HAC, reducing Equatorial Africa precipitation and redirecting marine humidity to North, towards the Horn of Africa (see Figure 52).

In the past, before human intervention, moist marine air could penetrate the coast above a densely vegetated Somalia, and reach the highlands of Ethiopia, leading to abundant rainfall due to orogenic lift. This rain also fed the rivers that flowed back into the Somalian rainforest as shown in Figure 83.

FIGURE 83 Vegetation Effect

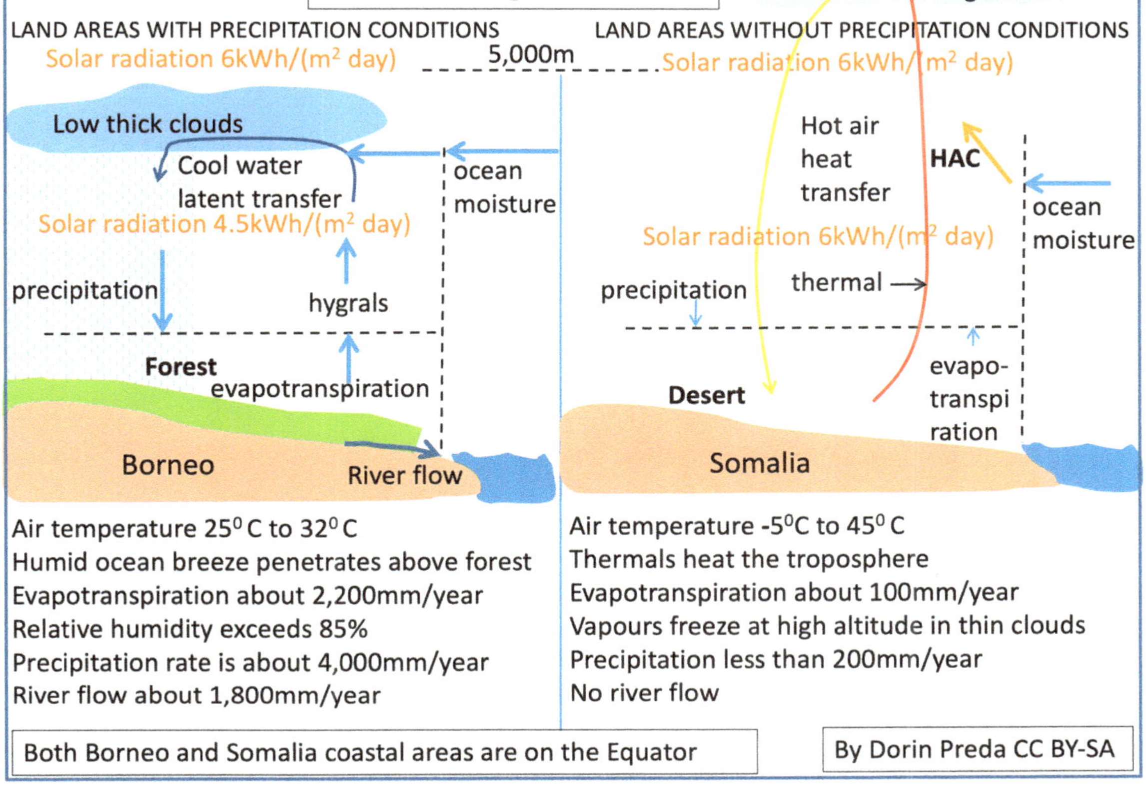

The robust Somalian HAC diverts the humid ocean breeze northwards; only a small part of it is heated and rises, creating high altitude thin clouds that produce a greenhouse effect, but no rainfall. The entire troposphere above the Horn of Africa is hot and dry, but has a high greenhouse effect due to significant absolute humidity.

The primary causes of this situation in Somalia are deforestation and desertification due to the traditional charcoal production and overgrazing with goats (https://en.wikipedia.org/wiki/Geography_of_Somalia). A few hundred years ago, Somalia was not a desert, and a thousand years ago it had lush vegetation (https://www.earth.columbia.edu/articles/view/3262).

The aberrant deforestation of Somalia, including the export of charcoal, has led to desertification, human suffering, and piracy, with global social and economic consequences resulting from human-induced local climate change.

7.3.6.1.2 Borneo

A contrasting situation to Somalia is found on the identical coastline of Borneo, as depicted in Figure 83. The dense vegetation covering Borneo produces high evapotranspiration, which keeps the temperature low and generates slow lifting air currents I call "hygrals" (fig 69). These hygrals raise the relative humidity and dew point of the incoming marine air. The high dew point leads to the formation of low-altitude clouds, with rain occurring between the treetops in some highland areas (https://www.shodor.org/os411/courses/_master/tools/calculators/cloudbaseheight/). Without a HAC in natural Borneo, marine humidity reaches the highlands, combining with local evapotranspiration to produce abundant rainfall that sustains the forest and rivers.

Temperatures in Borneo are lower than in Somalia due to the frequent low thick clouds that reflect solar radiation, resulting in lower solar radiation levels (16MJ/m^2*day) and higher relative humidity. Most solar radiation is used for water evapotranspiration, with no heating of the air. There are no hot thermals, but the cold and slow hygrals lift the water vapours to the clouds.

Borneo is a healthy Earth with rainforests, while Somalia is locked in a self-destructive HAC despite having rainforests in the past. If deforested, Borneo would initially transform into a landscape similar to present-day Madagascar, and eventually resemble Somalia.

In 2020, Indonesia planned to relocate its capital from the overcrowded Jakarta, affected by flooding, to Borneo. The proposed capital would involve clearing approximately 600km^2 to create a "capital in the forest ". The concern lies in the extensive vegetation clearance, especially if nearby towns like Balikpapan and Samarinda expand and connect with the new capital, deforesting a large strip and leading to a powerful HAC that could dry out the inland areas of Borneo. The reason is that, unlike the small Indonesian islands with ocean-dominated climates, Borneo is large enough to create its own climate. The outcome, whether Borneo remains humid or dries out, will depend on its land management practices.

7.3.6.1.3 Calculation of heat should not differentiate between a natural Borneo Forest and a tall palm oil plantation, as both exhibit similar evapotranspiration characteristics. Our interest here is climate protection rather than biodiversity conservation. Therefore, a palm oil plantation is deemed acceptable, while total devegetation (no vegetation) for an urban development is not.

Initially, a forest or plantation would emit 15% in infrared, store 15%, reflect 15%, photosynthesize 5% and evaporate the difference of 50%, with no air heating due to the equivalent temperature with the ocean. In contrast, an urban capital would emit 10%, store 15%, reflect 25%, have negligible photosynthesis if the city has the typical vegetation, evaporate 5%, and the rest is air heating of 45%.

Following the construction of the 600km^2 capital, solar insolation would increase to 18MJ/m^2*day, equivalent to 6,570MJ/m^2*year, or 4x10^{12}MJ/capital*year.

The solar heating resulting from replacing the forest/plantation with the capital would amount to 45% x 4x10^{12}MJ/capital*year = 1.8x10^{12}MJ/capital*year. In average power plants (see 4.7) this would mean 1.8x10^{12}MJ/year / 19x10^{9}MJ/year = 95 power plants-equivalent of additional solar heat due to deforestation.

Therefore, the solar heating due to replacing a forest or a tall tree plantation with the 600 km^2 classical capital would produce as much additional solar heating as 95 average power plants.

This substantial heat generation could lead to a potent HAC, hindering marine humidity from reaching inland areas and ultimately causing drought in Borneo. Initially, there will be a diminishing of river flow due to HAC and to urban extraction of water for the city, resulting in less water available for vegetation, leading to avalanche devegetation as shown at point 4.4. The reduced evapotranspiration of the forest/plantation would result in increased solar heating; thus, an ASH will start drying out the natural reservations in the highlands.

To avoid creating HAC and ASH, URBs should be used for the capital and all the land between them should be forest or tree plantations, as suggested in Figure 84. URBs require less land area and release heat as water vapours, mimicking vegetated areas and avoiding solar heating.

The robust URBs would last for many generations, surrounded by natural or plantation vegetation, creating a pleasant environment. Each URB could be dedicated to another government department, thus offices and apartments of their clerks would be in the same URB. Indonesia has the opportunity to establish the world's first modern capital, setting an example for others to follow, and achieving a remarkable feat.

If URBs would not be used for the capital, other designs like the green rooftops could be employed, and provisions for evapotranspiration of at least 1.5m of water per year should be made to prevent the formation of a HAC. The solutions outlined in section 7.3 could mitigate these risks.

FIGURE 84: Suggestion for the new Indonesian capital

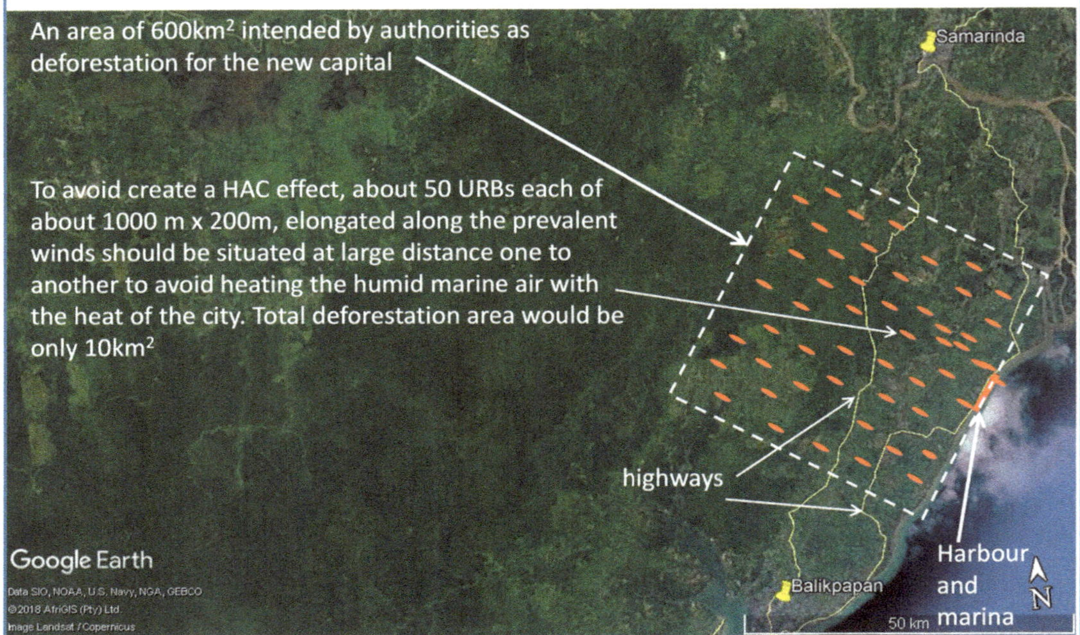

By Dorin Preda CC BY-SA

7.3.6.2 Africa

There is an unfortunate tradition in very populous countries with recent poor land to practice intensive pastoralism, when it is known that meat is far less land efficient than cereals or fruit, and that goat pastoralism is a major cause of desertification.

7.3.6.2.1 Central Africa

The Sahara Desert is rapidly advancing, posing a threat to Central Africa with droughts and desertification. Nearby Somalia proves that an equatorial desert is possible even on Easterly coasts.

There is an accelerated humanitarian program in Africa involving low-depth drilling for safe drinking water, that unfortunately started to be used for other, more demanding purposes. The water extracted should only be used for drinking water by people and animals, as extracting large amounts of water lower the groundwater level leading to the drying up of trees and forest fires (see section 3.6). This can result in a drier climate and the gradual loss of the equatorial Central Africa Forest, similar to Sahel and Somalia.

Central Africa receives less humidity from the Indian Ocean due to the deforestation of Madagascar. Additionally, it receives little humidity from the Atlantic due to the rotation of the Earth. This puts Central Africa at risk of becoming a desert if deforestation continues and its forests are dried out through groundwater extraction.

7.3.6.2.2 Madagascar

Madagascar plays an important role in allowing Indian Ocean marine humidity to reach Central Africa, but the recent HAC of deforestation of the island has disrupted this process. This has not only caused the island to slowly transform into a desert, but to divert the humid marine air away as seen in Figure 52, affecting the humidity levels in Central Africa. Urgent action is needed to stop deforestation in Madagascar and fully revegetate the island with useful trees or with original forest.

7.3.6.2.3 Somalia

Somalia was deforested a long time ago as mentioned in section 7.3.6.1.1, transformed into a desert and the resulting HAC and ASH prevent the Indian Ocean humidity from reaching the South-East side of the Sahara Desert. Revegetation of Somalia should be one of the first steps in the restoration of North Africa. Directing international resources to this country would:

- Provide enough food for its population to give up goat herding and stop desertification
- Provide revegetation work for people to renounce piracy.
- Increase the country's agricultural production, stopping its dependence on foreign aid.
- Allow ocean humidity to provide rainfall on the Ethiopian Plateau, and water for the Nile
- Allow oceanic humidity to reach SE Sahara again after the Ethiopian plateau is revegetated.

7.3.6.2.4 Sahara

The Great Green Wall is a massive tree plantation and land rehabilitation project on the southern edge of the Sahara launched in 2007 through international cooperation with the aim of halting the desert's advance to the south and creating a stable economic, social and political zone under the umbrella of the United Nations: https://www.unccd.int/our-work/ggwi .

Unfortunately, the discovery of the HAC and ASH mechanisms published in 2020 in [52] was ignored by climatologists. Therefore, in 2021 an Accelerator to the Great Green Wall program was launched, without properly considering the issue of average Sahara winds.

Since the dominant wind blows to the South [80], as illustrated in Figure 85, the ASH downloads desert heat and sand on top of the Great Green Wall, which will somehow slow down the advance of the desert's advance. However, as described in section 4.8, it will not be able to reduce the quantity of heat or the volume of sand; it will only capture the sand between its trees which in the process will be buried. The ASH cannot be stopped at its front, because it is much more powerful than human capability, as calculated in section 4.9.

Fortunately, less than 1/3 of the Great Green Wall of Africa was completed, hence it is not too late to redirect resources and start a proper AGC to really regreen the Sahara from the north to the south.

As shown in the lower panel of Figure 85, the ASH can only be stopped at its source (upwind) and gradually starve it of energy (stop the solar heating) by covering the desert with vegetation in a downwind direction. Sahara revegetation is only possible through sustained international cooperation. If the world community would mobilise, the ASH could be stopped in 50 - 100 years, before it totally covers Sudan, Chad Niger and Mali with sand.

The proper high vegetation in the coastal AGC depicted in the lower panel of Figure 85 would not heat, but further humidify the incoming marine air, bringing it close to the dew point. There is a special manoeuvre of generating a sharp increase in air humidity by suddenly turning to maximum the EC, generating for a few tens of minutes at night a very humid mass of air. When this misty or cloudy air is pushed by the dominant wind over the adjacent desert areas, it will be propelled upwards by the desert HAC. In this way, it will reach abrupt condensation conditions and produce rainfall over that desert strip, preparing it for planting with young trees from the nurseries in the coastal area and elsewhere.

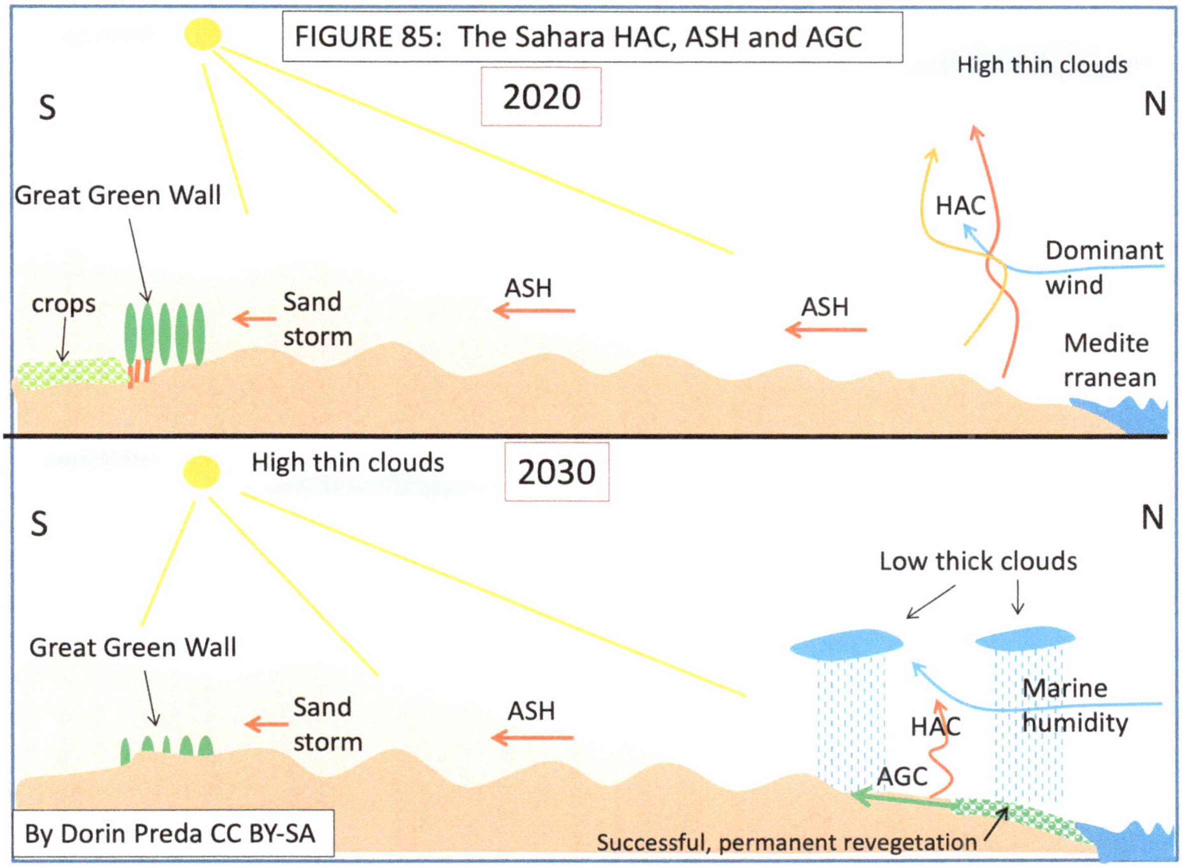

Sahel people will be needed on the Southern shores of the Mediterranean to work for the North Sahara revegetation program. Their relocation to North Sahara would stop present migration to Europe and they could permanently settle in the recovered Sahara areas. The general AGC program is described in section 7.3.2, uses the tools of section 7.3.1. and has the framework:

a) Extract all large feral herbivores from all territories envisaged for the future AGC.
b) Stop all human activities in non-AGC regions with less than 500mm water input annually.
c) Setup rational Natural Reserves in the territory as per section 5.5.
d) Water reservoirs, boreholes and pumping networks stop water from returning to the sea.
e) Turn pastures and grain fields into dense orchards.
f) Vegetable crop fields should be planted with 20% lines of trees.
g) Forests, useful tree plantations and orchards should block at least 90% of solar radiation
h) All grazing animals must not be allowed outside of their barns.
i) All possible trees and green canopies should be planted in urban and rural areas
j) Water supply and air humidity can be increased as in section 7.3.1.1.
k) EC should compensate everywhere evapotranspiration to a minimum of 1500mm per year.

Stage 1 of the Sahara revegetation, as depicted in Figure 86, should focus on Morocco, Algeria and Tunisia, to create a marine humidity corridor above the Atlas Mountains, which are now a barrier.

Figure 86: Stages of Sahara Revegetation

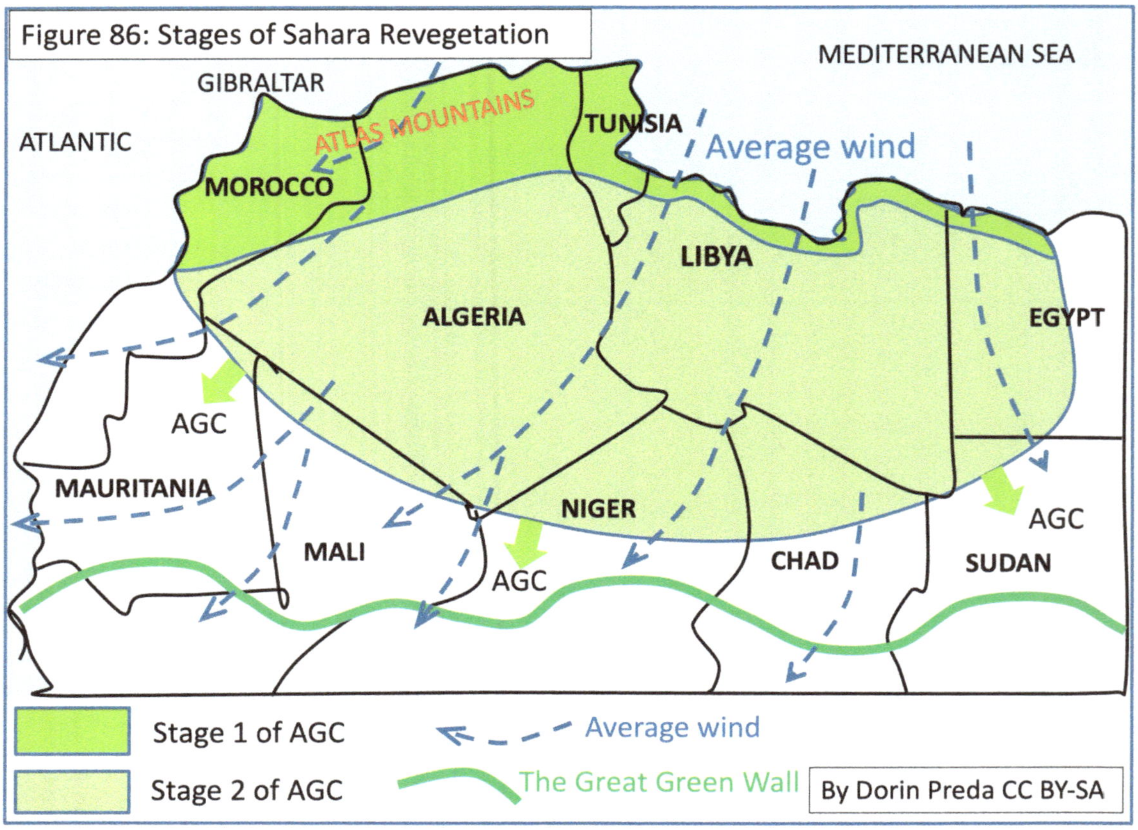

Proper AGC vegetation planting (high density) will cancel solar heating and cool the air. Through concentrated AGC measures, local humidity will increase on the seaside slopes of the Atlas Mountains, while marine humidity pushed by wind towards the crests will achieve condensation through orogenic lift. The resulting rainwater will help extend the AGC vegetation on the entire Northern flanks of the mountains. In about 10 years vegetation on those slopes will start to mature and increase water precipitation conditions, such that rain will also form frequently on the Southern slopes of the mountains, allowing the expansion of the AGC downhill towards the desert. A smaller number of water reservoirs will be needed on these slopes than on the Northern slopes.

About 20 years would be needed for greening the entire Atlas Mountains, which will revitalise the ancient creeks flowing southwards into the desert. During this time, the rest of the coast to the Nile Delta should be prepared with a first 20km wide band of vegetation and URBs every 10km along the shore to house the future working settlers. Water reservoirs should be built deep inland in advance. In 20 years, palm trees would grow tall enough to allow a second level of vegetation underneath, for example nurseries for the step 2 of AGC.

As shown above, the revegetation of Somalia would also help bring Indian Ocean humidity over the Sahara through a repaired double-monsoon climate of Somalia.

Stage 2: After greening all the slopes of the Atlas Mountains to allow marine humidity pass above and ensuring consistent coastal evapotranspiration, the AGC can extend to the flatter areas of the Sahara and start advancing from North to South as in Figure 86. The fossil lake beds in the region will gradually form permanent lakes that will create a humid environment. When the entire coast is cool and humid, precipitation conditions would become frequent, and a natural humidity pump as in Figure 77 will start by itself, regularly discarding humidity on the land. The climate will rapidly improve, with fewer high thin clouds and more low thick clouds forming, reducing ground level solar radiation and therefore the above ground temperature. Humid conditions of the Sahara will then accelerate at the pace at which people can provide compost and saplings. If this is an international effort and saplings and compost are freighted in from around the world, the Sahara could be greened in up to 100 years.

Additionally, if a similar effort is done at the same pace in the Southern Hemisphere, in Australia, by year 50 of the AGC, the global climate will start to improve and substantial rains will form in tropical areas, especially the AGC ones, further easing the revegetation of these problem areas. Global temperatures will begin to decrease and the Arctic will freeze again in winter. Without a doubt, the first AGC successes will inspire people in other drought-stricken areas like the Middle East to start local AGCs, and this way a global greening avalanche will restore Earth's vitality. The forward progress will be also propelled by the great economic advantages produced by the increased agricultural production (section 7.2.7).

Other colonists from different parts of the world could help, because the AGC areas can become very attractive. Once changed back to vegetated and productive lands, AGC areas could be a more pleasant place to live than the cold and misty higher latitude areas, which as AGC progresses, will cool back to their pre-warming temperatures. Rational settlements like the URBs can offer civilised and comfortable living conditions and have a smaller environmental footprint than the present sprawling amalgamations of concrete and asphalt, and will attract new settlers.

Hurricanes form usually in the Atlantic, in front of the Sahara coast, where the prevailing Easterly winds bring the hot air produced by Sahara above the ocean. As calculated in section 4.9, that heat is 15 times larger than all human produced energy annually. It is obvious that the heat of the Sahara enhances Atlantic water evaporation because the hurricanes start just in front of the Sahara coast.

Without that heat, it might be possible some hurricanes would not even start. It is certain that if western Sahara is greened up and cooled down, the entire wind pattern and air temperature in the area would change. Sahara revegetation would reduce the frequency of hurricanes in the Atlantic, with great economic and social benefits for the Gulf of Mexico Islands and the South-east USA.

The Sahara is a global problem that must be addressed through international cooperation.

7.3.6.2.5 Egypt

Egypt was the largest ancient civilisation and is the most populous country in North Africa due to the Nile water, which now could help start the revegetation of the entire Sahara. Ideally, this AGC would integrate with the Sahara AGC international project, or Egypt could proceed with this AGC independently by combining its two former projects as below:

- The Qattara Depression Flooding project is costly due to the tunnel/channel, has a limited life span, would produce less atmospheric humidity and electric power in the long run because brine does not evaporate easily, and would contaminate the Sahara groundwater with brine if the depression is not waterproofed, which would be prohibitively expensive. There is not enough Nile water available to compensate for the evaporation in the Qattara depression.
- The Ras El Hekma City is small, its green areas are located behind the HAC of the city, thus they will require Nile water, it is expensive and would not be sustainable due to lack of natural water, resulting in large running costs. Therefore, neither of these projects is viable in their current form in the long term.

A much larger benefit would be to combine the above two old projects with an AGC project that would produce its own water and would gradually expand to the SW, regreening and bringing development in the future to the entire Egypt. The great advantage of Egypt is that it has Nile water for the start-up of the AGC before the revegetation will produce its own water. Egypt also has a larger workforce than the other North Africa countries, so it is another option from where the regreening of the entire Sahara could start.

The outcome would be a more extensive and sustainable AGC that naturally expands over time through the benefits it generates. The funding for the initial Ras El Hekma City project could cover the AGC start along the entire coast to El Hamam, gradually expanding to Qattara, while also accomplishing the coastal city Ras el Hekma in an improved design.

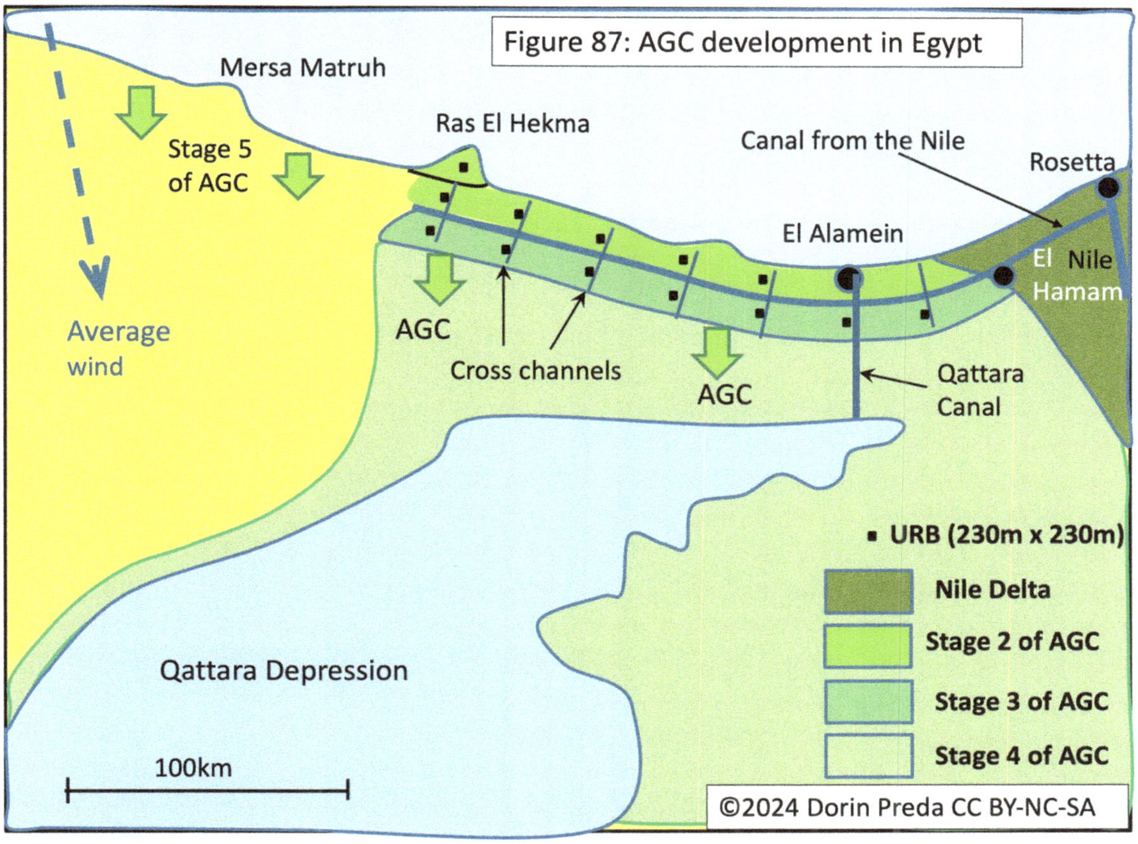

The general AGC program is outlined in section 7.3.2, utilising the tools from section 7.3.1. and follows the framework:

a) Establish rational Natural Reserves in the territory as outlined in section 5.5.
b) Restrict goats and camels from leaving their barns.
c) Plant all possible trees and green canopies in existing urban and rural areas.
d) Compensate with EC the evapotranspiration to a minimum of 1500mm per year.
e) URBs in Egypt could have the traditional pyramidal shape, therefore rectangular as framework and easier to construct than an oval-shaped URB of chapter 6. An URB of the same size and shape as the Great Pyramid of Giza would have 45 levels and could comprise 3,500 apartments, 400 offices, a hospital, a primary and a secondary school, all having windows and terraces on the sides of the pyramid. It could house up to 10,000 people in terraced apartments (thus similar to Djoser's pyramid, but having more steps), while also including all necessary service areas. It would be extremely energy-efficient because of its compactness and its integrated services, as described in chapter 6. Its immediate surroundings would be occupied by gardens, parks and sports fields, while further there would be orchards, crop fields with tree lines and other tree plantations.

The AGC steps as illustrated in Figure 87 would be as follows:

Stage 1: Over two years, as many trees and green canopies as possible should be planted in the existing settlements on the shoreline between El Hamam and Ras el Hekma to stop solar heating and loss of marine humidity through HAC. Agricultural workers would complete the greening of the coast with useful trees 1km from the sea using Nile water and other properly adapted AGC measures. Large nurseries would prepare seedlings and saplings for the AGC. Construction workers would build:

- a large canal 10km from the coast to bring Nile water from Rosetta to Ras el Hekma,
- runoff retaining structures and water reservoirs in the 10km strip from the shore.
- five URBs around 5 km inland, and approximately 25km apart along the coast to Ras el Hekma. One additional URB would be constructed near the harbour on the Ras El Hekma peninsula.

Stage 2: Over the next 10 years, agriculture workers would plant the entire strip up to 10km from the shore with two species of trees of different heights in alternating rows, work on field crops, establish more plant nurseries and complete planting in the 10km strip between the coast and the canal to increase local rainfall.

During this time, the construction workers would install runoff retaining structures and water reservoirs in the 10 to 20km strip from the shore, would complete the first 6 URBs and start work at the second row of 7 URBs at 15km from the shore.

It is essential that stage 2 is fully implemented with enough vegetation density before proceeding to the nest stage. Nurseries have to prepare saplings and young trees in advance, giving the long time they need to grow.

Stage 3 will start when rainfall will increase, requiring less Nile water for vegetation in the shoreline to 10km strip of stage 2, allowing agriculture workers to manage more easily the technical plantations, crop fields and orchards. As rainfall increases close to the shore and all excess is retained in the reservoirs, Nile water can be used in the 10 to 20km strip of land for planting permanent trees and establishing new nurseries. Some rain is also expected in the 20 to 30km strip in mild wind conditions; this advanced rain front would allow the tallest tree species to be planted in advance if dripping irrigation is assured for them when needed, even if the AGC front will reach them a few years later.

Construction workers would finalize the second line of URBs and new population will occupy them and complete the planting work in the 10 to 20km strip. Subsequently, they would prepare with runoff retention works and water reservoirs the next strip 20 to 30km. This expansion would progress southward using natural rainfall water and income from agriculture. Residents in URBs would be self-sufficient, also providing community services, as described in chapter 6. Medium-heavy industry parks could be developed if needed, providing they are covered with roofs which do not produce solar heating.

A few years after the third stage is fully implemented the local climate will become permanently wet and the rain fronts will advance faster inland than the vegetation front planed by people. There will be the temptation of trying to catch up with the rain front by planting in low density, but that would be a mistake. Depending on the capacity of the plant nurseries and if the AGC is internationally driven or only relying on Egypt resources, stage 3 will end when the rain will begin to fall on the Qattara Depression, before the first URBs are built on the third line from the sea.

Stage 4 will be marked when the AGC expansion would eventually reach the Qattara depression, gradually forming a freshwater lake with AGC excess rainwater, greening the surrounding land. It will take a long time to replenish the groundwater and have a permanent lake in Qattara. Nevertheless, the vegetation front will extend at pace with assuring housing for the communities of workers, with the rate of growth of the trees and with the maturity of the agriculture production system. It is essential that the AGC properly encompasses all the territory between Qattara Depression and the Nile Valley, even the hilly areas. There should be no bare land or rockface left without vegetation cover, because they would otherwise dry their surroundings. It would be advisable to place small trees in large pots in the Qattara Depression before there is permanent water there because vegetation evapotranspiration is needed to decrease solar heating. As the water level rises, those mobile trees would be withdrawn. Similarly, on the western shore of the Qattara it is necessary to establish a permanent thick and tall vegetation cover.

Step 5. As soon as the NE side of the Qattara depression starts to have some permanent water, a new AGC front should be started westward on the coast, beginning to create a new 10km green belt from the shoreline to improve the regional coastal climate because 10 km is the height of the troposphere and would create a new AGC climatic prism. Excess water from the first AGC will serve this one, like Nile water was used for the former. This process should start forming soils, in a long process in which no water runoff and erosion is allowed.

This way, Egypt could become self-sufficient in food production, reducing demographic pressure in the large cities and in the Nile Valley, and enhancing living conditions with URBs, which require a significant initial effort, but will last for hundreds of years. The process would very much be eased if it is part of an international cooperation that regreens the entire Sahara, as shown above.

7.3.6.3 Suggestions for the USA

As shown since 2020 in [52] and in Section 3.8 above, the main causes for the California forest fires are HAC, forest thinning and groundwater extraction, to which the global warming (air heating and CO_2) adds up to unfortunate ignition.

American Indians might have preferred to have more grasslands with bison than forests with wolves. Thus, there could be a human explanation for the very few woodlands on the Great Plains. The white settlers were known to have completed that work, and the result is the "dust-bowl" as shown in Section 4.4 above. The effect compared to the original forested land, is an increase in solar heating by about 30%, when crops are fully grown, or by about 40% after ploughing (despite that also increases evaporation for a few days).

The considerable resulting solar heating creates thick slabs of hot air on top of the ground, which further increases the atmospheric thermal and density instability. In conditions of wind shear, these hot slabs may sharply rise and create small tornadoes. When a cold front happens from the North, it creates a cold air wedge that instils under the hot air slabs, promoting their rapid ascent. The result is a large tornado that has ample kinetic power and produces damages, including soil erosion through strong wind and heavy rain.

The solution against tornadoes could be planting long rows of useful trees everywhere in the Great Plains, as in Figure 79. This way, the evapotranspiration of the crop fields can be increased, the soil could be stabilised not lost in dust storms, solar radiation and solar heating would be decreased. The useful tree rows would also break the wind and prevent the incipient tornadoes. An additional advantage of planting rows of trees would be a statistical increase in rainfall on the Great Plains.

The present climate tendency is to increase desertification and air temperatures on the continents, therefore in cities. As shown at 7.4.7, Memphis will soon have the same climate as Baghdad because they are both land-locked, are situated at the same latitude, and both have a river flowing through them. The difference is that Iraq harboured successive intensive civilisations over more than 5,000 years, therefore it is surrounded by empty desert, while the USA had this type of civilisation for less than 500 years. The solution to the rapid desertification is improved land management.

With the revegetation (trees) of the Great Plains, the humid air entering the region would not be solar-air-heated like at present, but would be kept closer to the ground, where it would have larger chances of producing rain when it combines with the vegetated ground evapotranspiration. Possibly this is how more rain would penetrate to the arid Western areas.

7.3.6.4 Suggestions for Australia

Besides a lot of solar radiation, Australia has 34,000 km of coastline, which could make it the wettest and richest continent, not the driest. This large humidity potential means easier climate recovery than all the other continents. Other advantages of Australia are its large territory, small population and short modern development, resulting in coastlines not yet entirely lost to dense city sprawling.

Although humans were very efficient at changing the initially green continent into the present desert with scarce coastal vegetation, the redevelopment using the high coastal humidity potential would bring great economic benefits by doubling production per hectare (7.2.7) and making most land productive. If done rapidly and with international support, it would make a significant contribution (with the other revegetation projects of Section 7.3.5) to restoring Earth's climate.

7.3.6.4.1 Pre-human Conditions

The wind and humidity regime derives from Australia being an island surrounded by oceans.

On the eastern side it had abundant and regular humidity input from the warm Pacific Ocean, which in summer sometimes sent a tropical cyclone over Northern Queensland, as depicted in Figure 89, where the dotted wind arrows show how the pre-human wind used to penetrate the low-lying coastline without changing its direction, bringing humidity to Eastern Australia.

The influence of the Pacific is strong up to the Great Dividing Range and the Australian Alps range which create an almost north-south barrier in the eastern quarter of the continent, thus less Pacific humidity used to pass over them to Central Australia, but much more than at present. The Great Dividing Range is the reason Central Australia never was a continuous rainforest like Amazonia.

On the north-western side Australia had a double monsoon (now extinct) with the warm and humid Indian Ocean that used to regularly bring humidity to Central Australia and occasional cyclones that used to penetrate over the western side of the continent.

FIGURE 88: Average Queensland wind and the suggested AGC

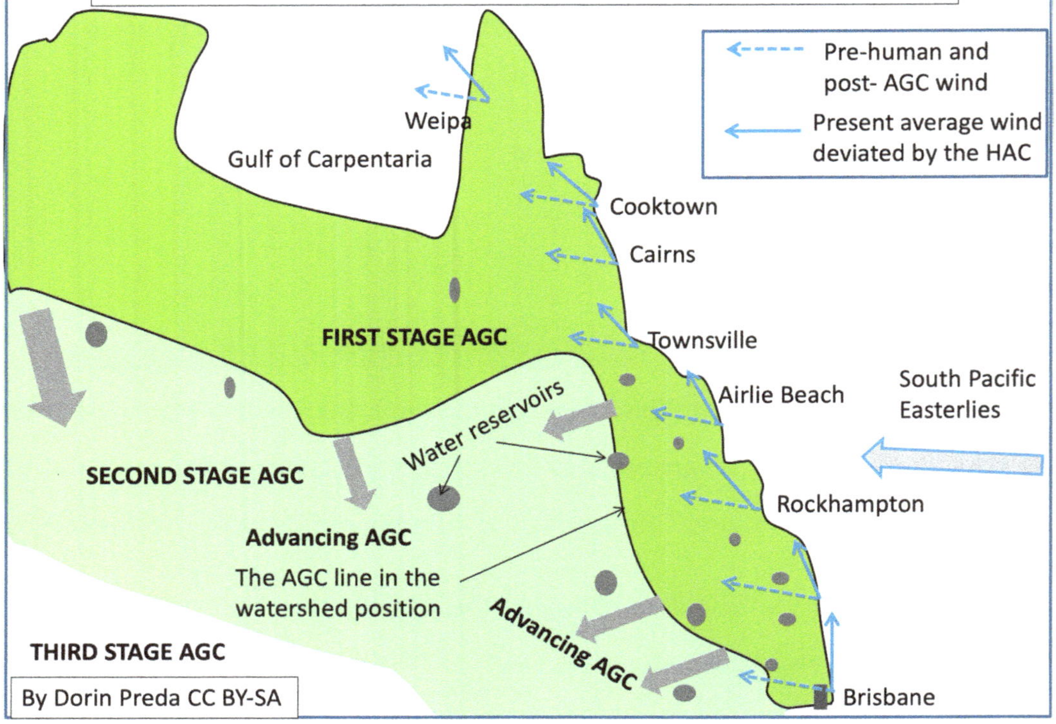

On the southern and south-eastern sides of Australia and in Tasmania there was the powerful influence of the cold Southern Ocean, that penetrated up into the central parts of the continent, while in SE it has sometimes a reverse circulation with the Australian Alps.

The wind and weather pattern were substantially changed by the deforestation HAC (Section 4.5.1).

The vegetation cover that once created a blanket over the entire continent was in equilibrium and inter-dependence with the wind and humidity regime (Section 2.2.2.5). In its strive for life, vegetation had covered all territory due to the high adaptability of plants and to their capacity of attracting precipitation (Section 2.2.2.1). As result, the forests compensated the high solar radiation by evaporating water that enhanced the reflective cloud cover and cooled evaporatively (Section 2.2.2.3), and by establishing humidity pumps (Section 2.2.2.4) that brought very large quantities of water from the ocean (Section 2.2.2.2), similar to the Amazonian Forest.

There are many scientific studies on the former vegetation that covered the continent before humans: "Dating back more than 100 million years, Australia has pockets of rainforests that are more than ten times older than the Amazon. These pockets still contain ferns, conifers and angiosperms that once covered the Gondwana supercontinent" [20]. The ferns and conifers need a much wetter environment than at present, so their present existence in deep canyons and ravines in the centre of the continent proves that Australia was rather recently much wetter than now.

Since the warm and humid climate plants survived in Australia for hundreds of millions of years (much longer than the Ice Ages), it means that the cold weather of the glaciations was not significant in the Southern Hemisphere, and that the many interglaciations were never too dry.

The central-south regions of the continent had a cool climate and vegetation of tall trees, similar to present time Tasmania due to the cool influence of the Southern Ocean. On the eastern third of the continent including the Murray-Darling basin there were tall forests of Redwood, Cedar, and Mahogany, while the top third of the continent was covered by intertropical rainforests.

There were no large fires before humans because of the too humid climate. Lightning certainly burned one-two trees at a time or a patch of dry grassland, but there could not have been large fires because of the wet green vegetation and damp ground carrying a thick cover of moist litter. Humans would not have found large forests of very old trees if large forest fires would be natural.

The fauna included large mammals, some adapted to living in large trees, like the tree kangaroo. Skeletons of animals weighing hundreds or even thousands of kilograms ant trunks of very large trees were preserved to present all over the continent, indicating the past great vegetation cover.

7.3.6.4.2 The Human-influenced Period

The Aboriginal people arrived around 60,000 years ago in Australia and used fire as their main tool of subduing the original jungle and its animals. Like in other parts of the world, the first people who arrived in a pristine environment were confronted by large vegetation amassed in impenetrable forests, while meadows were easier to use as a hunting ground for the herbivore animals. It was assumed that the Aboriginal people did not use bows and arrows because they had to deal with the large initial animals of Australia, against which they had to develop efficient spears and spear throwers, that they highly mastered [77].

Fire is a very efficient (though wasteful) method of both hunting and transforming the land from unusable forests to manageable grassland. Grasses quickly grew back after a fire attracting new herbivores which could be hunted. The Aboriginal peoples are known to have developed a fire culture in land management that suited their needs at that moment and in that environment. There was a complex relationship between the land to which they thought they belong and their life practices. That was the longest surviving culture and the closest as connection to the land, but people still used fire to their outmost advantage, as all the other human populations did.

There is no natural explanation for Central Australia becoming a desert in the past tens of thousands of years, during the last glaciation, when global climate was colder and wetter. The dried-up giant trunks in the Australian deserts are only a few thousand years old, not millions of years as for the times when the continent was located at different latitudes. Unfortunately, the loss of vegetation cover was followed by the powerful solar radiation breaking down the soil and transforming it into dust that was (and still is) easily and totally eroded by dust storms, leaving the barren rock or dunes of sand in place of the once rainforests.

The tropical geographical position of Central Australia makes this continent prone to drought and desertification, once vegetation is diminished and periodically cleared with repeated fires.

Central rainforests have disappeared together with their fauna which included the tree kangaroo which obviously needs large trees [20]. In Australia we see a consistent, repetitive, burning of the forests. Captain Cook in 1770 called Australia "the continent of smoke" because of the smoke plumes rising everywhere from the patches of forests which at that time occupied all coastal area (those forests were since cut down by the white colonists).

In the Central and western parts of the continent the fires were especially damaging due to the smaller natural humidity input induced by the Great Dividing Range in the east. Without that mountainous ridge, Australia would have had a climate and vegetation similar to the Amazon. The humidity brought by the Easterlies can be much larger than that brought by the Westerlies winds. I assert that the Indian Ocean western monsoon was a vegetation-induced weather phenomenon, favoured by and overlapping with the humidity pumps created by the former Central Australian forests with the Pacific Ocean. As result, the vegetation in those regions was more susceptible to fire, thus when often set alight, central forests irreversibly burned down. The land overheated and stopped the Indian Ocean monsoon about 40,000 years ago, and the seasonal humid air did not penetrate any more in the centre of the continent [24]. Without water, eventually the grasslands also disappeared and soil erosion left behind bare rock on the western half of the continent.

The wetter region in the Murray Darling River basin had rainforests of tall dense trees with patches of grassland where the first humans used and perfected their "fire stick" hunting method. By setting fire to the forest meadows, they not only hunted the animals but burned the bushes and the tree saplings, stopping forest regeneration. This way the frequent fires before the European colonisation changed the inland rainforest to the paradoxical fire prone and fire resistant but less dense Eucalyptus trees, which allow more solar ground heating and create a drier local climate.

On the eastern and south-eastern coasts, the high natural humidity input from the Pacific Ocean and the Southern Ocean maintained rainforests with tall, dense, old trees so wet that it was impossible to start large forest fires. The grass fires could not reach the branches or penetrate the thick bark of those large trees that frequently had more than 1 metre in diameter.

Before the arrival of Europeans, the entire Pacific coast of Australia was an almost continuous green canopy of a gigantic forest, even though Aboriginal people regularly burnt the grass and small bushes in the forest clearances. It is known that they created a system of land management through regular fire to suit their needs, adapting the new flora and fauna to it. The large trees on the eastern coastline remained, but the small flora and fauna was changed by the regular burning.

On the Eastern side of the continent the Aboriginal people reached a balance with the natural environment, where most of the large trees were preserved in a huge, continuous canopy which is visible in the first aerial photographs from the beginning of the 20-th century.

However, those large, old trees of Redwood, Cedar, and Mahogany were cut down by the European settlers for their valuable timber and to make room for sheep grazing that were up to 100 years ago the most lucrative trades. Since then, where pastures were not maintained after the clearance of the old trees, Eucalyptus trees took over or were even planted to create today's forests.

The largest clearing as area of the coastal forests took place in the past 100 years for cattle pastures, agriculture, mining, natural gas drilling and settlements. Conservationists still try to protect the last remaining forests of native large trees in Tasmania from the pressure of the timber industry.

In the past 50 years large land reserves of the relatively new Eucalyptus growth were put aside around rivers, creeks, mountains and some hills, but they are under assault from the recent forest mega-fires. There also is an increase of the preventive and of the cultural grass fires.

After the great natural forests were replaced by the new sparse forest, there is a gradual transition from new forest to smaller and drier bushland because of recent human activity. The drought-stricken patches of forests, bushes and pastures further decayed through periodical, extensive small and large fires that intensify with deforestation and threaten to transform the entire Australian continent into a human-made desert. Without any vegetation cover, large areas of bare ground are burned by the intense sunshine and the soil is destroyed and entrained in large dust storms which engulf the coastal cities.

7.3.6.4.3 Present Conditions Influencing Climate

a) Australia has a very favourable oceanic wind regime. The Eastern coast is dominated by the Easterlies Pacific wind which is warm and moist and the most constant wind as direction because of the rotation of the Earth, but it is stopped by the Great Dividing Range. The Southern Ocean wind is powerful and can penetrate deeply on the continent from the South-East with cold air of rather small absolute humidity, but its bursts are irregular at present and frequently reversed by the circulation with the Australian Alps. Australia also receives warm humid air from the Indian Ocean over its north-west coastline, but that is irregular.

b) The first settlements and clearings on the Eastern coast were small, but in the past 100 years the population pressure has totally replaced the small vegetated townships with city asphalt, concrete and tiles, which are the best solar air heaters. The resulting powerful HACs prevent marine humidity from entering deep inside the continent by deflecting the oceanic Easterlies towards the North. In summer the oceanic breeze is deflected towards the South, showing that the cause of deflection is the HAC, not the low-lying coast. Figure 89 shows how Queensland coastal HACs increase in power from north to south, in direct proportion to the increase in local deforestation. In Central Queensland the average wind is deviated by about 40^0 (Cato Island and Gladstone Airport B.O.M. data comparison). However, the power of the HAC increases towards the largest deforested area (Brisbane), where the average wind deviation is about 70^0 (Point Lookout and Brisbane Airport B.O.M. data comparison).

   Within about 10 km from the sea, there is still intermittent coastal precipitation because of the intense marine wind and the lifting of the marine air by the underneath thermals. However, most of the coastal rainwater ends up in drains, creeks and ocean because cities have waterproofed most of their coastal land.

   The further inland 10-50km band has also become a solar air heater due to halving of its evapotranspiration (through change from forest to pasture or similar). It generates a HAC (section 4.5.1) which intensifies, as deforestation and complete land covering progress inland.

   The power of the human-made hot air curtain (HAC) on the Eastern coast of Australia is thousands of times larger than the energy used by our society, as calculated in section 4.5.2 and illustrated in Figure 53.

During strong wind days, the HAC is blown inland by the incoming wind and becomes an avalanche solar air heater (ASH) that overheats and dries out the ground and vegetation as explained in section 4.5.3. It increases in efficiency and power as explained in section 4.5.4 and depicted in Figures 54 and 55. That solar energy collected from the ground substantially reduces relative humidity (air becomes dry, although it has significant absolute humidity). The part of that air which remains in contact with the ground blows onto the inland vegetation and dries it out, while most of the air flows up due to its smaller density, and rises in the atmosphere to create convective currents, haze and thin high clouds that increase global warming.

c) The summer storms sometimes succeed to humidify the ground a few kilometres inland, and by doing this, they stop the HAC. Above-land atmospheric pressure decreases through cooling, and the place of the condensed clouds is taken by further humidity coming from the sea. These marine clouds cool the ground by allowing only 10-20% of solar radiation through, while at night they cool by radiating their heat to the cosmic space, and then condense to produce night rain. That rain extends more inland until it creates a cool blanket on top of which more ocean humidity penetrates deeper into the continent for a few days, or longer, depending on the Southern Oscillating Index and may produce floods through a conveyor belt of humidity as explained in Section 2.2.2.4. An additional problem in this changing climate is that once the storms and rain stop, and the sunshine returns for a few days, it quickly re-creates a HAC, which for months afterwards reduces the penetration of ocean humidity on the land and creates drought. Moreover, these flooding events are local, as they happen on a sector of coast only, while the rest is still in drought. The result is a spatial and temporal alternation of extremes: flood to drought and vice-versa.

d) The 50 to 300km inland band is still subject to strong ASH, which becomes more robust by the day, gathering the heat from the drying pastures in that area which replaced the original forests. Few patches of forests have remained, and small settlements have encroached into them and produced the present catastrophic fires by further drying that forest, as explained in section 3.8. The drought that has accentuated gradually is just the result of this ASH drying the land. The solar heat collected by the deforested areas heats the ground wind which dries further the inland regions. The drier the ground becomes, the less vegetation grows; thus, there is less evapotranspiration to decease the solar heating. The result is the avalanche desertification depicted in section 4.4.

e) The tornadoes that became more frequent especially in New South Wales are the result of the increased deforestation which transforms large ground areas in solar air heaters. As explained before, only the air in contact with the ground is heated, while the layers above are not (as the atmosphere is transparent). The result is a temperature and density inversion (the layers near the ground are lighter than the air layers above). When a large slab of hot ground air swiftly rises, it creates the twister underneath, which can cause damage to houses.

f) The dryness produced by local deforestation as detailed in section 3.8 and through the regional HAC (described in section 4.5.1), and the ASH (section 4.5.3) is further increased by the El Niño episodes and creates conditions for the forest fires which also destroy the small encroached settlements, as described in section 4.6. The multiple simultaneous ignition points at the worst time of the dry season results in mega-fires. Interestingly, the 1939 fire has burned 69 sawmills.

The 2019-2020 mega-fires have been particularly devastating, increasing hundreds of times the anthropogenic global warming for tens of years until new trees grow to cool the land and form clouds that would reflect most of solar radiation.

The rain usually follows forest fires because the ash particles in the smoke become centres of condensation for the water vapours. Storm rainwater erodes the ash layers on the ground as well as the soil that is not protected any longer by the burnt vegetation. The ash is mostly hydrophobic and bare ground has a smaller water retention capacity than vegetated land; therefore, only a small part of ash nutrients and runoff water seep into the ground, while most of them are lost to the sea, where they damage the marine ecosystem.

The Aboriginal people who preferred grasslands to forests, did land management with fire in a way in which they ignited the grass and the fallen branches, while unfortunately also burning the tree saplings to prevent the natural return of the forests. Claims that preventive burning would stop larger fires and that it returns the nutrients to the soil are incorrect partial truths. <u>The preventive, the controlled and the cultural vegetation burning are wrong because:</u>

- Decreasing ground fuel does not deter the well-known cause of forest fire ignition.
- Large fires are started and propagate through crown fire, not through ground fire.
- Controlled burning can lose control and start a large fire (which has happened).
- The nutrients are already in their best place (in the living creatures); no fire is needed.

- In a fire, most nutrients go up in smoke, do not stay in the ground ash.
- Most ash is taken up by wind or washes to the ocean, damaging the marine ecosystem.
- Through fire the local water in vegetation and top soil is lost as vapours taken by wind.
- The rain which follows a fire erodes the ash and the soil now unprotected by plants.
- Burning the small animals and insects is mass-biocide with no equivalent in Nature.
- Burning the grass exposes the ground to more solar radiation until it grows back.
- Burning the top soil turns it into dust easily taken by the wind caused by bare ground.
- Trying to rebuild with parts of previous destruction always results in something smaller.

The Aboriginal peoples have an interesting culture and beautiful traditions connecting them to the Dreamtime. <u>However, their "fire stick" is not an environmentally friendly practice, but the same mistake people have made all over the world when they have put fire to the vegetated land.</u>

g) Similar to North Africa, Arabia, or the Horn of Africa, the present Australian goat farm fashion means that in a short time all vegetation will disappear, followed by loss of agricultural value, followed by total soil loss, as shown in section 3.4.1. Somalia has similar coastline orientation and latitude to Northern Queensland and used to have a rainforest before humans due to the monsoon climate, but 1000 years after deforestation it now has an impenetrable HAC, bare ground lands, and is one of the driest places on Earth, having lost its monsoon.

The humidity of Central and Northern Queensland will decrease due to the present coastal forest clearings for cattle farms, natural gas drilling and partly due to sugarcane, and HAC will strengthen due to the decrease of evapotranspiration in the coastal areas, so less humid oceanic wind will penetrate the coast. <u>Thus, the Murray Darling basin will receive less and less water and will progress further towards desertification.</u>

Deforestation in all its forms continues in increasing scale. The soils are extremely precious because, without them, nothing will grow. In the past 200 years the mass of the NSW and Victorian forests was reduced by at least three quarters, thus they evaporate three times less water and heat the air, leading to desertification. The current line of events means that the Eastern and Southern Australian coasts are on the verge of becoming deserts, completing desertification of the entire Australia.

There is substantial misunderstanding of the realities of vegetation, climate, fire, social and economic interests that lead to the decisions resulting in the present unfortunate situation.

7.3.6.4.4 Restoring Australian Land and Climate

There are many projects and schemes aiming to rehabilitate Australia, from terraforming projects like the old and the new John Bradfield schemes, to the small adaptative Natural Supply Management schemes. Each project has its requirements, achievements and merits. They cannot be discussed here, but hopefully this suggestion will join them in the overall endeavour because this project has multiple advantages for agriculture, natural resources, food industry, finance, land development, construction, transport, etc. To these there will add societal benefits and solutions for protection of the natural environment and climate.

7.3.6.4.5 The understanding needed to improve the productivity and climate of Australia:

- Solar radiation is the source of energy on Earth, but global climate would not allow life without cloud and vegetation shade and evapotranspiration to diminish solar heating.
- Global climate is the resultant of regional climates, which are results of local climates.
- The largest local climatic change occurs when deforestation replaces forest with desert.
- The main causes of climate change are the ongoing deforestations and the repeated fires which devegetated the land and create barren lands that lead to desertification.
- Fragmentation of forests and deforestation for encroaching small settlements has dried out the forest and conduced to violent fires which destroy both forests and settlements.
- Deforestation of coasts has replaced evaporative cooling with solar heating that created HACs which in turn prevent ocean humidity from entering the continent.
- Vegetation and especially large and dense trees are directly interlinked with water.
- There would be no water on land if oceanic humidity would not condense above land.
- All climate change forms: global warming, drought, sea level rise, flood, extreme storm, heatwave, are due to changes in the water cycle, and aided by large CO_2 variations.

- Vegetation and especially trees are the most powerful and least expensive water gatherers, storers, regulators and distributors, thus revegetation can restore climate.
- The best revegetation is with tall and densely planted, useful trees in private ownership.
- Massive revegetation requires large scale, co-ordinated and consistent action.
- Proper revegetation can bring large climatic, economic, and social benefits.

7.3.6.4.6 Stop Climate Change:

- Stop net deforestations and plant cleared out areas with useful trees to protect the soils.
- Stop and prevent all forms of vegetation fire, "preventive burning", "back-burning", etc.
- Stop pastoralism (especially with goats, lamas and camels) in regions with less than 500mm rainfall per year. It could be replaced in a first step with free range kangaroo farms [78], while useful trees would be planted in a second step.
- Goats should only be kept in barns and all fodder should be brought to them
- Feral goats, camels and horses should be put in barns and fed there, or culled.
- Stop encroaching houses in forests. Stop the sprawling of existing settlements.
- Use forest groundwater only as drinking water for people and animals.

7.3.6.4.7 **Territorial planning and restructuring**

- There is an outstanding opportunity for Australian development in line with the international necessity to stop the climate change as presented in section 7.1.3 and revegetate the tropical and subtropical deserts as suggested in section 7.3.5 to reverse global warming and the changes in climate.
- A unitary optimum plan for long-term development is needed because the present inefficiencies (for example those shown in section 6.1 for the urban issue) can be resolved with the beneficial suggestions presented further.
- Australia is in a unique position in the world to achieve excellent development through large scale and long-term planning because: its humidity input is excellent, its territory

is relatively undeveloped while lightly developed areas can be improved, and because its educated and resourceful population can generate effective solutions to the very diverse challenges presented by continental developmental reshaping.

- It is necessary to create a unitary development vision that provides for future needs.
- Decide upon the optimum technical and economic methods of achieving that vision.
- Properly inform population about the general plan and the methods of achieving it.
- Organise public debates to achieve a common vision plan and refine methods of achieving it. Improvements are expected after public debate, which nevertheless have to be in line with the objective requirements of proper development.
- Make a detailed plan of action and of methods to achieve it.
- Allocate suitable land to large Nature reserves as at section 5.5. for both flora and fauna, and this way free the rest of the land for development purposes.
- Private ownership of the land by the settlers of those areas is important in assuring safety and proper contractual development and maintenance of the land.
- Aeroponics in urban, rural and bare land (desert) areas is a valuable tropical solution of increasing evapotranspiration and productivity of that land. Architectural codes for new buildings and engineering improvements would allow increased loads on the roofs.
- Start training programs in the trade specialities needed for the rehabilitation work.
- Setup/adapt the infrastructure for providing the needed materials and energy.
- Decide strategic sites for future compact settlements, in the URB style according to requirements for workforce and distribution of population in the territory.

7.3.6.4.8 **The AGC**

- The AGC would aim to gradually revegetate Australia with useful trees that create economic value and restore the climate through the water cycle, starting from the coast and progressing inland according to the created water availability. As shown in Section 7.3.2.9, the condition is not to allow any freshwater flow into the ocean, but use it to advance the AGC towards the inlands; this is achieved with increased ground storage and large reservoirs and pumping stations able during floods to send all surplus uphill and store it.

- **For the first stage of the AGC**, the strategy should rely mainly on the dependable humid air brought by the Pacific Easterlies to revegetate the northern and eastern sides of the continent to the watershed of the Great Diving Range as depicted in Figure 88 and in Figure 89. In the northern part of Queensland there are corridors of lower land altitude that would allow AGC penetrate deeper inland. It is necessary to cancel the strong HAC created especially on the eastern coast by deforestations and developed areas. For this, it is necessary to increase the water reservoirs and create a larger water stock, and to plant as many trees and green canopies and setup compensating EC as per Section 7.3.2. It is necessary to overcome the powerful HAC on the entire eastern side of the continent.

FIGURE 89: Restoring Australian Climate with AGC

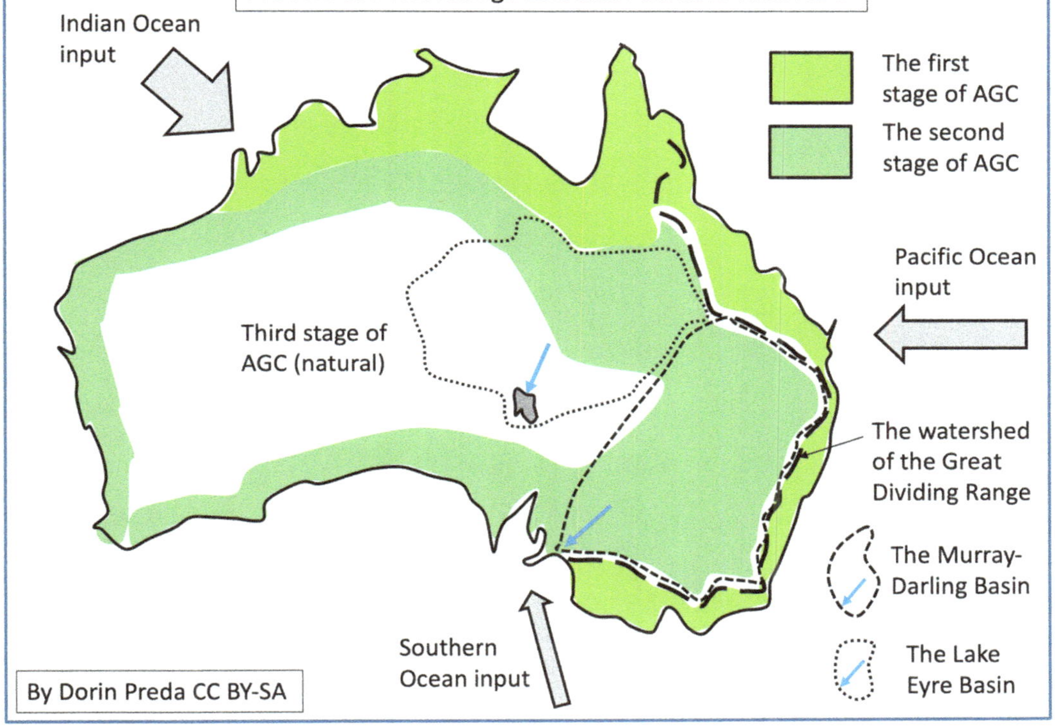

By Dorin Preda CC BY-SA

- All around the coast of Australia there would be the opportunity to start AGC from the shore and setup dense tree plantations of useful trees to allow the marine wind penetrate the coast, while keeping it cool. The humid wind has to be kept at low altitude by cancelling all ground air heating with evapotranspiration and cloud shade. The useful trees will bring income that would fund the extension of the AGC inland. Gradually, the humidified air will achieve frequent condensation and precipitation conditions to regularly produce small rainfalls, not rare deluges as at present.

- **The second stage of the AGC** can start when sufficient humidity starts to fall on the western side of the Great Dividing Range. By stopping the HAC with increased evapotranspiration on the coast, the marine humid wind will rotate in a cross-coastline direction, as shown in Figure 88 and penetrate deep inland. As the dominant Easterlies are easily pushed above the cool and humid green land, they suffer orogenic lift above the Great Dividing Range and produce rainfall on the westward slopes, providing humidity to the Murray Darling Basin. Since the hills are not too tall in the northern and central part of the coast, a significant part of the oceanic humidity will begin discharging onto the Eastern part of the Eyre Lake Basin, while the planting should keep up with the maximum available humidity.

  - Even if a land area is not downwind from an AGC connected to the ocean, it can increase cloud cover and precipitation if it is larger than around 25km^2 of densely planted tall trees and it evaporates enough water. The increase in precipitation would originate in the general atmospheric circulation, not in an oceanic humidity pump, and would produce significantly less rainfall than a humidity pump, thus less than it would consume. Therefore, that land area must have a sufficient artificial source of water that allows it not heat the air with solar energy. Obviously, if the isolated land area is small, the increase in cloud shade and precipitation will not occur directly above it. Parts of the Murray-Darling basin could be such an example of isolated land area if cancelling the HAC of the East Australia coast would be only partially successful. This is why AGC should develop downwind as a continuous strip of dense vegetation from the ocean and progress inland without allowing any significant non-evaporating area.

- The advancing speed of AGC will be mostly dependent on the rate of tree planting; without international co-operation it will be slower than the rate of saturation of the groundwater with the excess rain, that is in itself a guarantee of success.

- Towards the end of the second stage, when trees are close to maturity, it is expected that the increase in humidity allows that the entire land area of the artificial AGC as shown in Figure 89 receives more and more rain, possibly to the stage in which the entire Murray-Darling permanently has an excess of water, while the Lake Eyre becomes perennial and extends much in size.

- **The third stage** would start by itself and manifest as irregular monsoonal pulses. If left without artificial water or planting, the vegetation will gradually increase in size and slowly in density. Ideally, not Eucalyptus, but the original gymnosperms would be seeded initially and allowed to spread by themselves as the climate will become more humid. This avalanche effect (feedback) would increase to the point in which the monsoon with the Indian Ocean would restart. I assume that the contact of its warm humid air with the cold air of the Southern Ocean would generate inland storms with abundant rain. Lake Eyre could become a large, permanent wetland. Gradually, the wet and naturally vegetated Central Australian area would expand westwards until it would re-green the entire continent. The high humidity potential of Australia as a relatively small continent surrounded by vast areas of warm ocean means that enough rainfall is available for maintaining a lush green cover over the entire continent, if that green cover is desired and allowed to exist. Subsequent to the greening, there would be no need of EC on the coasts. The central and western sides of Australia, naturally regreened in the 3-rd stage are too remote from the coast, thus could be left as a great Nature reserve.
- Unlike the old and the new Bradfield schemes, this scheme when established does not rely on expensive artificially pumped river water, but on much larger quantities of oceanic humidity brought by natural humidity pumps directly to Central Australia.

7.3.6.5 Suggestions for South America

Of all inhabited continents, the north-western part of South America is in the least damaged condition due to the less invasive way of life of the indigenous South Americans who managed to live in harmony with their natural environment and did not cause significant damage. As a result, we have witnessed the Amazonian Forest, and understand how the rest of the world initially was. However, the eastern, central and southern part of the continent was lost to agriculture due to European settlers, and significant deforestation is underway in the central Amazonian Basin as seen in Figure 23A.

Since the northern coast of South America was not developed due to the short time of modern development, it does not yet have a strong coastal HAC, thus humid marine winds still enter the continent as shown in Figure 52 and bring some rainfall to Amazonia. Unfortunately, rapid deforestation is taking place on a line connecting Boa Vista to Matupa. Figure 90 shows how the decrease in precipitation follows the advancement of deforestation on this line.

Another cause of dryness is the thousands of forest clearings for timber, field crops or pastures which generally lead to warm, dry conditions. As calculated in section 4.7, around ten square kilometres of deforestation are equivalent to a power plant's heating, meaning that the current deforestations in the Amazonian Basin are equivalent to the heating by thousands of large power plants, inducing drying of the surrounding forest and creating a fire hazard.

Only the southernmost Magellanic subpolar forests remain of the forests of the southern part of the continent, therefore the humidity pump with the Atlantic has ceased working in the centre of the continent, with average precipitation being only 1/3 of that in Amazonia. Bursts of the local HACs and flooding of the deforested lands like in Southern Brazil intensify due to the build-up of ocean humidity near the coast, meeting regular humid air from the Northern forests.

Figure 90: South American Deforestation and Precipitation

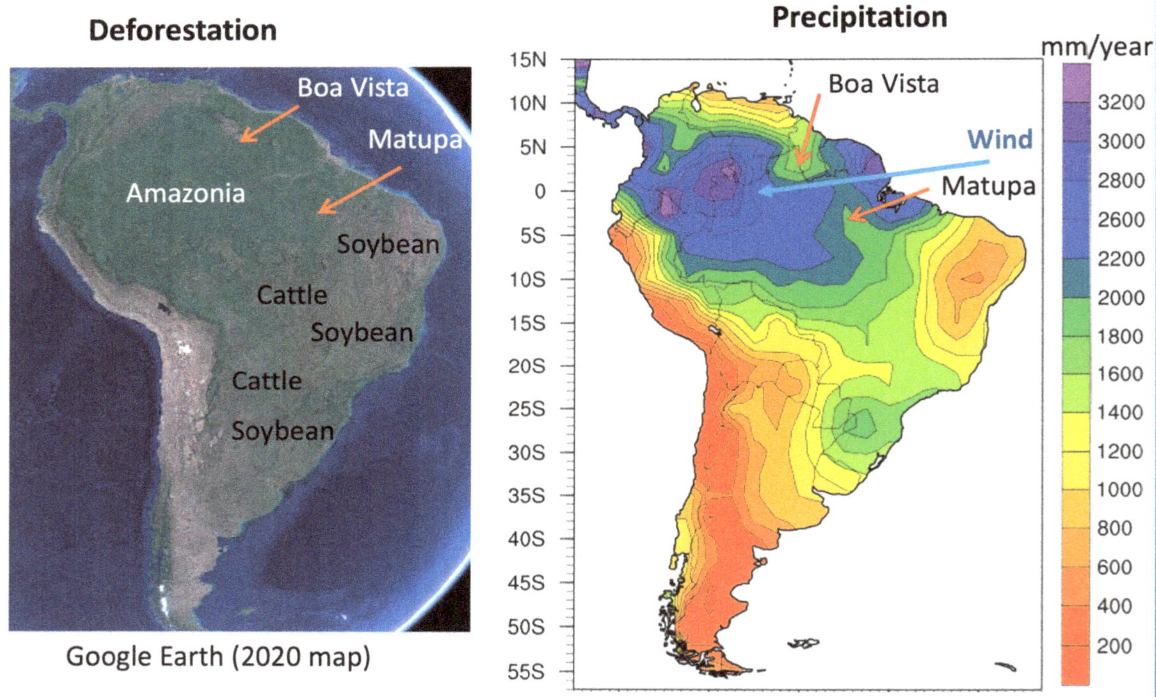

https://en.m.wikipedia.org/wiki/File:South_America_Precipitation_Map_%28data_from_1976-2009%29.gif

The problem in the North of the continent is that the Boa Vista-Matupa line crosses the direction of the average wind that comes from the Atlantic Ocean and brings vital humidity to the Amazonian Forest. The ongoing deforestation on this line has already decreased the flow of oceanic humidity, dried the forest and increased the intensity of forest fires in the clearing areas.

Unfortunately, climatology has correlated deforestation with cooling (A4.4 in [16]) and with increased local precipitation [A4.6 in 16], instead of the true decrease in overall precipitation. I explain it in section 4.6 similarly to the urban heat island rain, as due to the dry thermals of deforestation lifting humid forest air to its precipitation altitude. While forgotten studies in published letters like [114], [115] state the detrimental effect of deforestation, including decrease in precipitation, the most detailed and expensive climatology study in the Amazon [101] states that deforestation increases precipitation, thus favouring deforestation. Recently, mainstream climatology accepted, though, that large scale deforestation could decrease in the future 80 years rainforests precipitation by only 2 – 3% in the Amazon and by 8 -10% in the Congo [81]. However, we know that deforestations that produced current deserts resulted in around 85% precipitation loss. Thus, the above little percentages of precipitation are incorrect and support the short-term economic benefit of deforestation versus alleged small precipitation loss in the future.

Deforestation has accelerated in Amazonia with political help, and threatens to totally stop the tree humidity pump. The rate of waterflow of the Amazon River has significantly decreased, while climatology cannot explain it through the raise in global temperatures and in El Nino effect, that would contrary, produce increased precipitation. The true reason is that firstly, most of the time the HAC thermals produced by deforestations like those on the Boa Vista-Matupa line interrupt the inflow of oceanic humidity towards Amazonia (stop the tree humidity pump). The second reason is that local deforestation patches increase solar air heating, and that dry air blown by the wind onto the surrounding forests dries them out and increases their flammability.

The rare and violent floods are caused by bursts through the HAC of the increased ocean humidity pressure, and are the last convulsions before the total shut down of the humidity pumps.

It is crucial for the South American peoples to understand that if Amazonian Forest deforestation continues, and especially if it completes the Boa Vista-Matupa line, a HAC will develop on that line and will suddenly interrupt the oceanic wind and the humidity pump feeding the Amazonia.

The rapidly ensuing forest drying after the setup of the HAC will result in unstoppable continental megafires that will quickly consume the entire Amazonia.

The rare but powerful floods that will follow the mega-fires will erode the soil, and the climate will soon become drier, allowing soybean cultivation only close to the coast in 5 years, while inland goats or llamas will replace cattle. Not far later, the Amazon River will disappear and dust storms will render the continental inlands useless. In a few tens of years Brazil can resemble the geographically similar Somalia.

The loss of the South American clouds, and of the Amazonian humidity and heat sink will have profound repercussions on the global climate. As calculated in section 2.2.2, the same rate of net precipitation (continuous 200,000m^3/s), or even larger due to increased ocean temperature, will have to fall on other regions on Earth, producing violent floods that will erode soils from unforested lands and will lead to famines. The excessive floods have already begun, but they will increase. In the long term, the excessive heating of the high latitudes will melt the Siberian permafrost and release methane, which will significantly raise global temperatures, possibly in a positive feedback effect.

<u>Therefore, the Amazonian and Central African deforestation is an urgent international problem that should be managed by the United Nations.</u>

<u>Recommendations for South America:</u>

- Do not clear out coastal forests or large, compact areas of the interior; any clearing should be immediately replanted with dense, tall, useful trees.
- Replace the croplands and pastures with orchards and plant lines of useful trees in the remaining few crop-fields and pastures.
- Do not practice forest thinning, slash and burn, or preventive burning of vegetation. All vegetable residue has to be used as mulch for soil increase.
- Take all social measures necessary to categorically prevent forest fires.
- Logging (if really necessary) should only be done with lighter-than-air airships with water ballast which extract vertically no more than one tree per hectare per century.
- Do not deforest the coastlines for settlements. In the existing settlements, as many trees as possible should be planted, and roofs should be covered with green canopies or aeroponics if structurally possible.
- Stop deforestations for usual settlements and consider either the method at 3.8 or URBs.
- Do not extract forest groundwater for anything else than drinking for people and animals.

7.4 Earth's Water Budget

Despite mainstream science states that water cannot escape Earth's gravitational field, water vapours exist both on Earth and in the cosmic space as were first observed with aircraft and balloons [98], raising the question whether there is exchange of water between the two systems. The origin of Earth water is disputed, and hypothesised to be brought as comet ice.

Mainstream science speaks about water in space as ice, and about comet tails disintegrating under Sun's radiation, but does not discuss water vapour that are separate molecules which cannot freeze. In my assumption, they exist as vapours, even though we cannot yet detect them individually. They should absorb the faint light of distant stars and re-emit that energy in far infrared, which is not detected by our radiometers, thus explaining Olber's paradox.

However, cosmic water vapours should be collected by Earth's atmosphere and gravitational field during its travel through space, and fall on Earth especially at high latitudes. Conversely, at small latitudes, because of Earth's rotation and the tangential push of cosmic radiation, there would be increased conditions for water vapours to evade Earth into space. The rate between these two phenomena would give the Earth's net rate of water loss or gain.

If there is an increase of water vapours at the top of the atmosphere because of higher Earth temperature and sublimation of high thin clouds, the rate of water escaping Earth would exceed the intake. Mars has lost its water although it is cooler and it receives less solar radiation and solar wind than Earth. Therefore, it would be plausible that, with the increase in temperature, high altitude convection and high thin clouds, significant water quantities make their way into the top layers of the atmosphere and they are lost to space, helped by the solar wind.

Earth's water budget seems to be in deficit, as we have lost a significant amount of water from above the ground through deforestation and desertification and from the aquifers through pumping [76], while the smaller rise in sea level only corresponds to the melting of land ice.

Another mechanism of water loss would be through the known water vapours dissociation into oxygen and hydrogen in the ionosphere, with the hydrogen being lost to space (see the top part of the H concentration curve in Figure 4). If there is no loss of oxygen to space, its atmospheric concentration should increase. Recent pictures of a new type of Aurora borealis suggest an increased oxygen concentration in the ionosphere [44].

As mentioned in [52], I question whether Earth loses significant amounts of water to space.

7.5 Meteorology Suggestions

Perhaps the reason the HAC mechanism was not known until 2020 [52] is the lack of instruments for measuring vertical wind. To understand atmospheric processes accurately, we need to measure vertical air currents. Monitoring the 3D airflow and temperature will enhance our understanding of the true dynamics of the troposphere and lead to successful computer models.

However, we should not delay large scale revegetation in favour of first conducting 3D wind measurements and similar meteorological research. The calculations in the previous chapters clearly demonstrate the severity of the solar heating issue resulting from deforestation. Therefore, we should immediately begin Earth revegetation, while continuing climatic research concurrently.

7.6 Advantages of Revegetation over Geoengineering:

We have become addicted to making changes to Nature, and now our artificial alterations have led us to the point of depleting the vitality of the Earth.

There have been numerous proposals for large scale artificial interventions in climate, creating an artificial and dangerously unpredictable Earth. Some of these proposals are supported by certain climatologists, who highlight the "beneficial" cooling effect of sulphur dioxide (SO_2) in their reports. The prominent use of large capital font titles in IPCC reports such as "OBSERVED WARMING IS DRIVEN BY EMISSIONS FROM HUMAN ACTIVITIES, WITH GREENHOUSE GAS WARMING PARTLY MASKED BY AEROSOL COOLING" [102] indicates the importance they put on such a climate mitigation measure.

In the latest IPCC document AR6 of 2021 the highly polluting scenario SSP3 7.0 in Figure SPM4(b) is depicted as capable of reducing global warming by 0.5^0C through deforestation (land use) and aerosols (SO_2 emissions). Both of these methods are inaccurate and detrimental, casting doubt on the intentions of the IPCC. For instance, one proposed method involves injecting fine dust into the stratosphere to produce SO_2 (similar to that from fossil fuel pollution) to absorb solar radiation. This dust will produce highly corrosive sulfuric acid with the moisture in the atmosphere. The proponents of this method acknowledge that the dust will eventually fall into the troposphere, leading to acid rains that can damage all vegetation and acidify the oceans. Continuous dispersal of new quantities into the stratosphere would be necessary, resulting in a perpetual acid rain that burns the Earth below. The aluminium of the airplanes will also be attacked by the sulphuric acid and weaken their structure, leading to possible accidents. The quantities needed would amount to tens of thousands of tonnes per year, meaning an entire fleet of planes carrying it continuously to the stratosphere at large expense and use of fossil fuels. Regrettably, this fundamentally flawed and ethically unacceptable research is promoted by top universities, like Harvard: https://geoengineering.environment.harvard.edu/.

Artificial geoengineering methods are not suitable because we aim to repair the climate, not further damage it. Therefore, we should return to the path we knew and trusted. The current situation has arisen due to deforestation of tropical lands, which can no longer retain and evaporate sufficient water, leading to their dry overheating, while intertropical ocean vapours precipitate, overheat and flood mid-latitude areas. The simple way back is to restore the tall vegetation cover of the land, especially tropical regions, which will cool the air and absorb excess atmospheric carbon dioxide.

Advantages of revegetation over artificial geoengineering:

- Plants (trees) are proven to be effective, unlike artificial contraptions.
- Once established, plants grow and propagate naturally, unlike artificial systems
- No need for airplanes, artificial materials or energy, just basic gardening work.
- Forests/plantations can sustainably provide fruit, timber, and game, unlike geoengineering.
- Trees do not pollute, but diminish atmospheric and waterways pollution.
- While geoengineering is expensive, revegetation with useful trees is very beneficial.
- Restoring trees does not pose a risk of further destabilizing Earth, but returns it to original state.

At present, the known official opposition to water and vegetation in land and Earth management leads to an increase in the regional and global warming and to an exacerbation of the climatic disturbances. This greatly amplifies the effects of the carbon gasses emissions and pushes the Earth into a climatic crisis. It would be catastrophic for climate, Nature and society to introduce (impose) geoengineering "protection" that would further destabilize Earth. Especially, the use of sulphates would burn vegetation and literally would endanger life on Earth.

Throughout our history we have demonstrated that we do not understand Earth well enough to venture into more changes. **Thus, we simply need to replant (useful) trees.**

7.7 The Need for Unity

Climate change is a complex, multi-parameter issue, that affects everyone on Earth.

As illustrated in Figure 91, the main climatic parameters are CO_2 and equally important, water (H_2O). Water is intricately linked to vegetation, which is its main attracter, storer and disperser.

FIGURE 91: Perceived Interests

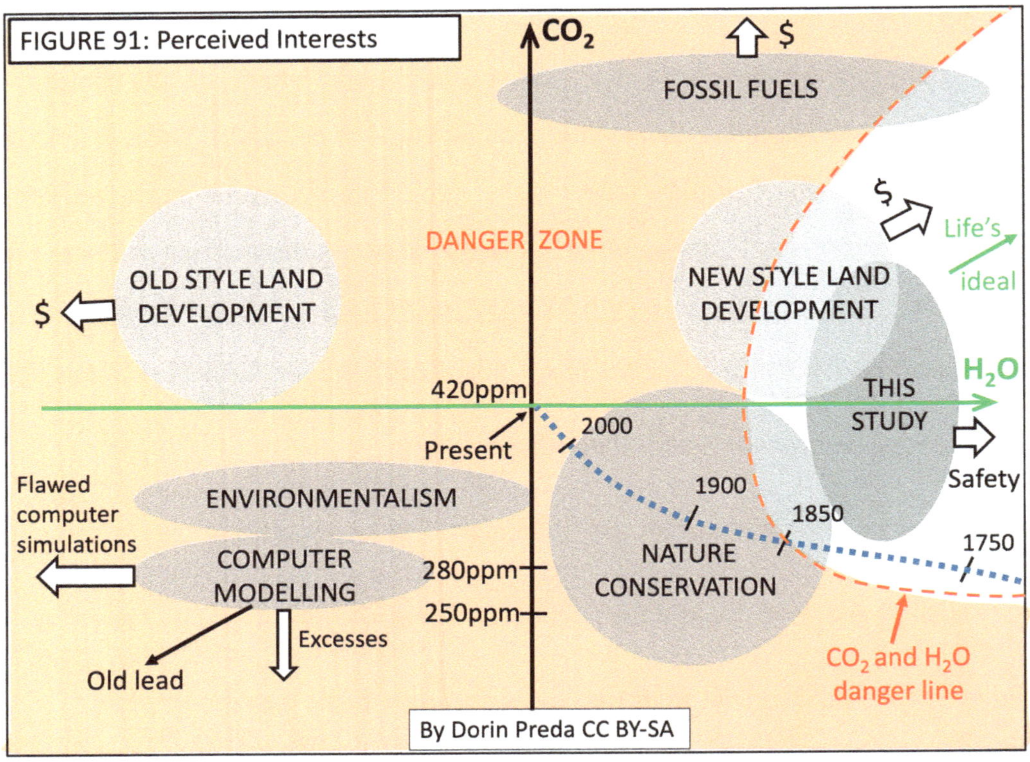

Earth's climate requires cooling, and Earth's life depends on large vegetation, both of which necessitate water on land. This study suggests that the climate tipping point was probably reached in 1850, when high economic development and deforestation began to be supported by railways and steam timber mills. This added to millennia of deforestation and desertification, leading to excessive land dryness and heat. The heat generated by barren lands, or too small vegetation is amplified by increased CO_2 levels, resulting in climate change (refer to Section 7.1.3).

It is also argued that plants can thrive in much higher CO_2 concentrations than present levels, as evidenced by agriculture and the geological past, provided there is sufficient water on land, thus air humidity around plants. Despite significant water evaporation from the ocean, it now condenses without reaching the land plants locked behind HACs, as demonstrated in [52]. Unfortunately, present mainstream climatology promotes flawed computer simulations that portray water vapours and vegetation as undesirable climate warmers (as in Section 2.1). Efforts to reduce CO_2

starve plants, which are the climate solution (Section 7.2.3). Mainstream climatology influences the environmentalists, while conservationists aim to the landscapes from the 1900s' first photographs.

Climate change and climate repair have proven to be very divisive issues, and the result is our inability of solving this problem. People, society and organisations have segregated in the three categories shown in Figure 92, which separates the assumed "climate destroyers" from the "climate friends", each pulling the Earth in a "tug-of-war" that threatens not only to damage the planet, but the social and sometimes personal relationships. It will be shown below that this abnormal situation can be corrected with rational solutions that resolve all problems, while all parties gain benefits.

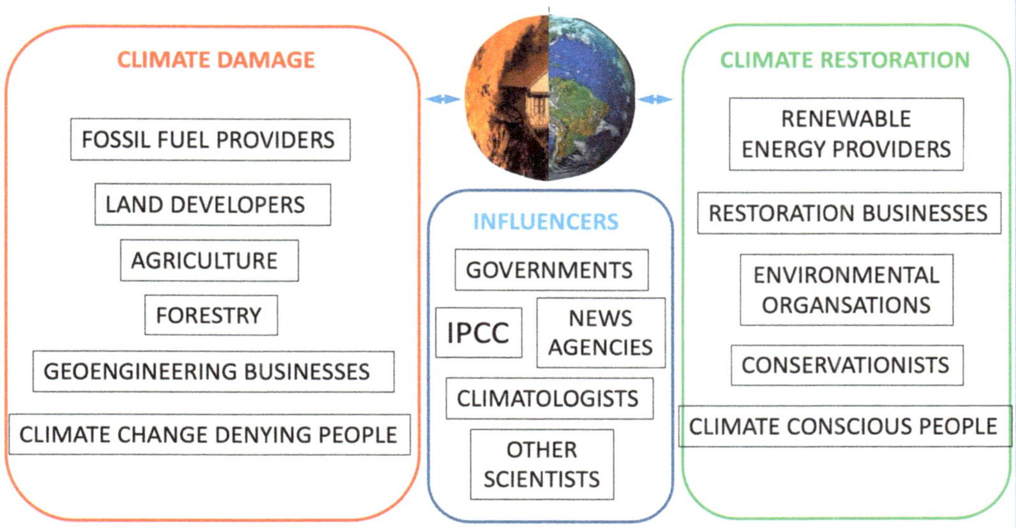

FIGURE 92: The Unreal, Divisive Picture

By Dorin Preda CC BY-SA

There is a significant increase in climate change effects, which proves that Earth's water cycle is destabilised, as shown in the previous chapters. The most acute is the decrease in the annual rate of flow of the Amazon River, indicating that the tree humidity pump is shutting down due to HAC and deforestation solar heating as shown in Section 7.3.6.5. The situation is critical, because deforested Amazonia and Central Africa will dry out and overheat, and their thermals will generate more high thin clouds (Section 2.2.3), significantly increasing the global greenhouse effect. The intertropical land is forced to lose its precipitations, creating an excessively hot dry belt at low latitudes, and an excessively wet, warm belt at high latitudes including Arctic permafrost melting. Therefore, there is a global emergency that requires urgent climate restoration, which is now hampered by the much smaller conflicting interests of diverse parties. A few conflicting issues:

- Energy: reduce fossil fuels that are also needed to manufacture renewable energy devices
- Development: high interests locked in "business as usual models" ignore new opportunities
- Agriculture: traditional malpractices, dependence on consumer irrationality.
- Economy: Increased waste from excess consumerism, wars, uneconomical economy.
- Climatology: excessive reliance on computer simulations versus conceptual simulations.
- Scientific policy: stress on carbon emissions reductions, while in fact opposing reforestation.

However, if we renounce the disjunctive parts and rely on the common ones, we observe that both solar heating of deforestation and atmospheric carbon gases act in tandem to disrupt the climate through the water cycle, hence they both have to be reduced as soon as possible to restore climate. Fortunately, we can resolve both these problems by planting useful trees in the intertropical zone.

General benefits for people and climate:

- Cool the low latitudes with cloud shade and restored evapotranspiration
- Increase precipitation in intertropical lands, transforming deserts in useful lands
- Cool the mid-latitudes through reduced heating and decreased flooding
- Stop the melting of the Arctic and of the permafrost
- Improve people diet with more fruit than cereals and more white meat than red meat.

Benefits for the groups and parties of climate change in Figure 92:

a) In conflict resolution, the weakest must be protected first. We have an ethical duty to preserve the essential parts of the biological Nature, this being the benefit for the **conservationist organisations**. Since global population is rapidly growing with demands for increased consumption, we have to urgently setup Natural Reserves as suggested in above Chapter 5, before all large land areas are fragmented and habitats are altered or destroyed.

b) Subsidies to all energy providers should be stopped to eliminate all bias and assure free competition between **energy providers**, and let consumers decide the balance.

c) Education and support in implementation of energy efficiency and rational consumption.

d) **The forestry businesses** would have access to the forests of unsuitable trees like the fire-prone Eucalyptus, as they are transformed into orchards or useful tree plantations after putting aside the Natural Reserves for koalas. The large orchards that would replace most pastures and crop fields will provide, after maturity, a large future timber source.

e) Restructuring existing land allocations will create large benefits for **land developers**, who will also benefit from the creation through AGC of new usable lands from deserts. This will transform into an economic revival for the entire society through the new opportunities.

f) **Agriculture** would receive very large benefits from changing most of present pastures and crop fields into orchards and vegetable beds, resulting in a three-fold increase in production. The intensive and extensive changes in agriculture would be nothing less than a new revolution in agriculture, with wide economic, social and health implications for everybody.

g) Besides the health and wellbeing from improved diet and inexpensive living conditions in URBs, **people** would be able to enjoy the improved world of reconstituted forests, parks.

h) The **renewable energy businesses** are set to take over from the **fossil fuel businesses**, that will have to continue, but diminished, because of inevitable ocean carbon sequestration.

i) **The environmental organisations** would receive satisfaction from participation in the setup of Natural Reserves, and to observe transition to renewables and climate restoration.

j) **The restoration businesses** would receive orders for AGC implementation, constructing URBs and other measures that would bring not only financial benefit, but also social status.

k) **The geoengineering businesses** are very young, at the beginning of their activity, thus it would be easy for them to change into prestigious climate restoration businesses.

l) **Climate conscious people** will not only observe, but participate in the climate restoration process and enjoy the benefits of owning, caring for and benefitting from their useful trees.

m) **Climatologists** will monitor and fine-tune the changes induced in the local and global climate by large scale revegetation, and will use that information in their climate models.

n) It is beyond the scope of this study to address **governments**, but they will have a capital role in correct climate restoration, by reorganising and improving everything, including the **IPCC.**

o) The **news agencies** always had a great role in conveying unbiased information to the public, and correctly mobilising public opinion. Therefore, they could have merit in creating a positive image about the efforts of climate restoration. In turn, the information about climate actions would provide hundreds of years of news for their broadcasting.

Besides multiple economic benefits from restoring Earth, we shall be rewarded with its beauty.

7.8 Suggested Urgent Actions

1) This study requires assessment by physicists, Earth scientists, thermal engineers, climatologists.
2) Policy makers and the public should be clearly informed that deforestation is a major cause of climate damage, while intertropical revegetation can cool and restore global climate.
3) Earth management should be guided by conclusions of land use and water cycle studies using conceptual modelling based on observations and measurements.
4) Geoengineering solar radiation modification and its research should be banned.
5) Natural Reserves should be established before there are no more large areas available.
6) Intertropical net deforestation should be stopped, and former forested areas should be planted with useful trees or crop fields with useful tree lines.
7) Large AGC programs at least in North Africa and Australia could start global climate restoration.

EPILOGUE

- The higher the value of the object under action, the more correct understanding is needed.
- There would be more benefit from a revitalised Earth, than from a deteriorated Earth.
- We were not banished from the Garden; we are banishing Life out of it.
- Plants help us repair without the ultimate mistake: geoengineering.
- After restoring Earth, we can confidently look at other planets.
- Earth is our home; we have it as we manage it.
- This is not Earth's rehabilitation, but ours.
- Earth simply reflects and follows us.

REFERENCES

[1] Palle E, Goode PR, Montañes-Rodriguez P, Koonin SE. Can Earth's Albedo and Surface Temperature Increase Together? Eos. 2006;87(4).

[2] Baffled Scientists Say Less Sunlight Reaching Earth. LiveScience. 2006 Jan 24 [cited 2019 Nov 29]. Available from: https://www.livescience.com

[3] Menne MJ. The Global Historical Climatology Network Monthly Temperature Dataset, Version 4. J Clim. 2018 Dec 15;31(24):9835-9854. doi: 10.1175/JCLI-D-18-0094.1.

[4] Stocker TF, Qin D, Plattner GK, Tignor M, Allen SK, Boschung J, et al., editors. Summary for Policymakers. In: Climate Change 2013: The Physical Science Basis. Contribution of Working Group I to the Fifth Assessment Report of the IPCC. Cambridge, United Kingdom and New York, NY, USA: Cambridge University Press; 2013.

[5] Imbrie J, Imbrie JZ. Modeling the Climatic Response to Orbital Variations. Science. 1980;207(4434):943-953. doi: 10.1126/science.207.4434.943.

[6] Carbon dioxide in Earth's atmosphere. Wikipedia [Internet]. Available from: https://en.wikipedia.org/wiki/Carbon_dioxide_in_Earth%27s_atmosphere

[7] Water cycle. Wikipedia [Internet]. Available from: https://en.wikipedia.org/wiki/Water_cycle

[8] Water distribution on Earth. Wikipedia [Internet]. Available from: https://en.wikipedia.org/wiki/Water_distribution_on_Earth

[9] Atmospheric escape. Wikipedia [Internet]. Available from: https://en.wikipedia.org/wiki/Atmospheric_escape

[10] Chen C. Determining the Leaf Emissivity of Three Crops by Infrared Thermometry. Sensors. 2015 May 15;15(5):11387-11396. doi: 10.3390/s150511387.

[11] The Numbers Behind the Value of Tropical Forests in the Climate Change Equation. World Resources Institute [Internet]. 2018 Oct 18. Available from: https://www.wri.org/blog/2018/10/numbers-value-tropical-forests-climate-change-equation

[12] Deforestation facts and statistics. The World Counts [Internet]. Available from: https://www.theworldcounts.com/stories/deforestation-facts-and-statistics

[13] Forest transition theory. Wikipedia [Internet]. Available from: https://en.wikipedia.org/wiki/Deforestation#Forest_transition_theory

[14] Gavashelishvili A, Tarkhnishvili D. Biomes and human distribution during the last ice age. Glob Ecol Biogeogr. 2016 May;25(5):563-574. doi: 10.1111/geb.12437.

[15] Carbon Cycle. NASA Earth Observatory [Internet]. Available from: https://earthobservatory.nasa.gov/features/CarbonCycle

[16] Shukla PR, Skea J, Calvo Buendia E, Masson-Delmotte V, Pörtner HO, Roberts DC, et al., editors. Summary for Policymakers. In: IPCC Special Report on Climate Change and Land. In press.

[17] Shukla PR, Skea J, Slade R, van Diemen R, Haughey E, Malley J, Pathak M, et al., editors. Technical Summary. In: IPCC Special Report on Climate Change and Land. In press.

[18] Stocker TF, Qin D, Plattner GK, Alexander LV, Allen SK, Bindoff NL, et al., editors. Technical Summary. In: Climate Change 2013: The Physical Science Basis. Contribution of Working Group I to the Fifth Assessment Report of the IPCC. Cambridge, United Kingdom and New York, NY, USA: Cambridge University Press; 2013.

[19] Albedo. Wikipedia [Internet]. Available from: https://en.wikipedia.org/wiki/Albedo

[20] Convict Creations. Convictcreations.com [Internet]. Available from: http://www.convictcreations.com/aborigines/rainforest.html

[21] Pörtner HO, Roberts DC, Masson-Delmotte V, Zhai P, Tignor M, Poloczanska E, et al., editors. Summary for Policymakers. In: IPCC Special Report on the Ocean and Cryosphere in a Changing Climate. In press.

[22] Houghton JT, Ding Y, Griggs DJ, Noguer M, van der Linden PJ, Dai X, et al., editors. Climate Change 2001: The Scientific Basis. Contribution of Working Group I to the Third Assessment Report of the IPCC. Cambridge, United Kingdom and New York, NY, USA: Cambridge University Press; 2001.

[23] Bushfire ignition sources. ABC News [Internet]. Available from: https://www.abc.net.au/news/science/2019-11-20/bushfire-ignition-source-how-we-know/11701132

[24] Notaro M, Wyrwoll K, Chen G. Did Aboriginal vegetation burning impact the Australian summer monsoon? Geophys Res Lett. 2011;38(11). doi: 10.1029/2011GL047774.

[25] Ukkola AM, Pitman AJ, Decker M, De Kauwe MG, Abramowitz G, Kala J, et al. Modelling evapotranspiration during precipitation deficits: Identifying critical processes in a land surface model. Hydrol Earth Syst Sci. 2016;20(6):2403-2419. doi: 10.5194/hess-20-2403-2016.

[26] Rickards L, Howden SM. Transformational adaptation: agriculture and climate change. Crop Pasture Sci. 2012;63(3):240-250. doi: 10.1071/CP11172.

[27] Tarim mummies. Wikipedia [Internet]. Available from: https://en.wikipedia.org/wiki/Tarim_mummies

[28] Ruddiman WF. The Anthropogenic Greenhouse Era Began Thousands of Years Ago. Climatic Change. 2003;61(3):261-293.

[29] Mesopotamia. Wikipedia [Internet]. Available from: https://en.wikipedia.org/wiki/Mesopotamia

[30] Liu W, Wu J, Fan H, Duan H, Li Q, Yuan Y, et al. Estimations of evapotranspiration in an age sequence of Eucalyptus plantations in subtropical China. PLoS One. 2017;12(4). doi: 10.1371/journal.pone.0174208.

[31] Foldawheel Electric Wheelchair. Available from: http://bit.ly/2WalyKy

[32] KD Smart Electric Wheelchair. Available from: http://bit.ly/2EaBDek

[33] van Geel B, van der Plicht J, Kilian MR. The sharp rise of 14C ca. 800 cal BP: Possible causes, related climatic teleconnections, and the impact on human environments. Radiocarbon. 1998;40(1):535-550.

[34] Mousa HA-L. Prevention and treatment of influenza, influenza-like illness, and common cold by herbal, complementary, and natural therapies. J Evid Based Complementary Altern Med. 2017;22(1):166-174. doi: 10.1177/2156587216641831.

[35] Sugawara H, Takamura T. Surface albedo in cities: Case study in Sapporo and Tokyo, Japan. Boundary-Layer Meteorol. 2014;153(3):539-553.

[36] Prevedello JA, Winck GR, Weber MM, Nichols E, Sinervo B. Impacts of forestation and deforestation on local temperature across the globe. PLoS One. 2019;14(3).

[37] Liu X, et al. Tree species richness increases ecosystem carbon storage in subtropical forests. Proc R Soc B Biol Sci. 2018; doi: 10.1098/rspb.2018.1240.

[38] Moore P. The Positive Impact of Human CO_2 Emissions on the Survival of Life on Earth. 2019. ISBN: 978-0-9686404-1-8.

[39] Wada Y, Beek LV, Kempen C, Reckman J, Bierkens MF. Global depletion of groundwater resources. Geophys Res Lett. 2010; doi: 10.1029/2010GL044571.

[40] Creed IF, Jones JA, Archer E, Claassen M, Ellison D, McNulty S, et al. Managing forests for both downstream and downwind water. Front For Glob Change. 2019;2:64. doi: 10.3389/ffgc.2019.00064.

[41] Ellison D, Futter MN, Bishop K. Trees, forests and water: Cool insights for a hot world. Glob Environ Change. 2017;43:51-61.

[42] Trabucco A, Zomer RJ, Bossio DA, van Straaten O, Verchot LV. Climate change mitigation through afforestation/reforestation: A global analysis of hydrologic impacts with four case studies. Agric Ecosyst Environ. 2008;126(1-2):81-97. doi: 10.1016/j.agee.2008.01.015.

[43] Syktus JI, McAlpine CA. More than carbon sequestration: Biophysical climate benefits of restored savanna woodlands. Sci Rep. 2016;6:29194. doi: 10.1038/srep29194.

[44] Saplakoglu Y. They turned out to be a new kind of aurora. LiveScience. 2020 Jan 29. Available from: https://www.livescience.com/citizen-scientists-discover-new-dunes-aurora.html

[45] Titov DV, Bullock MA, Crisp D, Renno NO, Taylor FW, Zasova LV. Radiation in the atmosphere of Venus. Space Sci Rev. 2013; doi: 10.1007/s11214-012-9911-5.

[46] Wright DK. Humans as agents in the termination of the African Humid Period. Front Earth Sci. 2017;5:4. doi: 10.3389/feart.2017.00004.

[47] Bushfire ignition source. BBC News. Available from: https://www.bbc.com/news/world-australia-50400851

[48] Nentwig W. Human environmental impact in the Paleolithic and Neolithic. In: Handbook of Paleoanthropology. Berlin, Heidelberg: Springer; 2007. doi: 10.1007/978-3-540-33761-4_62.

[49] Lemmen C. World distribution of land cover changes during pre- and protohistoric times and estimation of induced carbon releases. Géomorphologie. 2009; doi: 10.4000/geomorphologie.7756.

[50] Kaplan JO, Pfeiffer M, Kolen JC, Davis BA. Large-scale anthropogenic reduction of forest cover in Last Glacial Maximum Europe. PLoS One. 2016;11(11). doi: 10.1371/journal.pone.0166726.

[51] Ventresca Miller AR, Spengler RN, Haruda A, Miller B, Wilkin S, Robinson S, et al. Ecosystem engineering among ancient pastoralists in Northern Central Asia. Front Earth Sci. 2020;8:168. doi: 10.3389/feart.2020.00168.

[52] Preda D. Home – Calculations for Hope. Q Solar Pty. Ltd.; 2020. ISBN: 978-0-6487764-1-3.

[53] Gascoin S, Etchevers P, Lhermitte S, Marty C. Sensitivity of bare soil albedo to surface soil moisture on the moraine of the Zongo glacier (Bolivia). Geophys Res Lett. 2009;36(3). doi: 10.1029/2008GL036377.

[54] Maurellis A. The climatic effects of water vapour. Phys World [Internet]. 2003 May 1. Available from: https://physicsworld.com/a/the-climatic-effects-of-water-vapour/

[55] Bastin JF, Garcia FG, Mollicone D, Rezende M, Routh D, Zohner M, et al. The global tree restoration potential. Science. 2019;365(6448):76-79. doi: 10.1126/science.aax0848.

[56] Boucher O, Myhre G, Myhre A. Direct human influence of irrigation on atmospheric water vapour and climate. Clim Dyn. 2004;22(6-7):597-603. doi: 10.1007/s00382-004-0402-4.

[57] Otterman J. Anthropogenic impact on the albedo of the earth. Clim Change. 1977;1(2):137-155. doi: 10.1007/BF01884408.

[58] Sagan C, Toole OB, Pollack JB. Anthropogenic albedo changes and the earth's climate. Science. 1979;206(4425):1363-1365. doi: 10.1126/science.206.4425.1363.

[59] Voigt A, Albern N, Ceppi P, Grise KM, Li Y, Medeiros B. Clouds, radiation, and atmospheric circulation in the present-day climate and under climate change. WIREs Clim Change. 2021;12(4)

. doi: 10.1002/wcc.694.

[60] Stephens GL, O'Brien D, Webster PJ, Pilewski P, Kato S, Li JL. The albedo of Earth. Rev Geophys. 2015;53(1):141-163. doi: 10.1002/2014RG000449.

[61] Wang L, Wang L, Li Y, Wang J. A century-long analysis of global warming and earth temperature using a random walk with drift approach. Decis Anal J. 2023;7:100237. doi: 10.1016/j.dajour.2023.100237.

[62] Abbass K, Qasim MZ, Song H, et al. A review of the global climate change impacts, adaptation, and sustainable mitigation measures. Environ Sci Pollut Res Int. 2022;29(30):42539-42559. doi: 10.1007/s11356-022-197

[63] Masson-Delmotte V, Zhai P, Pirani A, Connors SL, Péan C, Berger S, et al., editors. Summary for Policymakers. In: Climate Change 2021: The Physical Science Basis. Contribution of Working Group I to the Sixth Assessment Report of the IPCC. Cambridge, United Kingdom and New York, NY, USA: Cambridge University Press; 2021.

[64] Penner JE. Carbonaceous Aerosols Influencing Atmospheric Radiation: Black and Organic Carbon. Livermore, CA: Lawrence Livermore National Laboratory; 1994.

[65] Teuling A, Taylor C, Meirink J, et al. Observational evidence for cloud cover enhancement over western European forests. Nat Commun. 2017;8:14065. doi: 10.1038/ncomms14065.

[66] Pachauri RK, Reisinger A, editors. Climate Change 2007: Synthesis Report. Contribution of Working Groups I, II and III to the Fourth Assessment Report of the IPCC. Geneva, Switzerland: IPCC; 2007.

[67] Spencer RW, Braswell WD. How Dry is the Tropical Free Troposphere? Implications for Global Warming Theory. Bull Am Meteorol Soc. 1997;78(6):1097-1106.

[68] Inamdar AK, Ramanathan V. Satellite observations of the water vapor greenhouse effect and column longwave cooling rates: Relative roles of the continuum and vibration-rotation to pure rotation. J Geophys Res. 2004;109(D6). doi: 10.1029/2003JD003980.

[69] Pöschl U, Martin ST, Sinha B, Chen Q, Gunthe SS, Huffman JA, et al. Rainforest Aerosols as Biogenic Nuclei of Clouds and Precipitation in the Amazon. Science. 2010;329(5998):1513-1516. doi: 10.1126/science.1191056.

[70] Mahmood R, Pielke RA Sr, et al. Impacts of Land Use/Land Cover Change on Climate and Future Research Priorities. Bull Am Meteorol Soc. 2010;91(1):37-46. doi: 10.1175/2009BAMS2769.1.

[71] Kaplan JO, Pfeiffer M, Kolen JCA, Davis BAS. Large Scale Anthropogenic Reduction of Forest Cover in Last Glacial Maximum Europe. PLoS One. 2016 Nov 30;11(11)

. doi: 10.1371/journal.pone.0166726.

[72] Robertson J, Watson A. Thermal skin effect of the surface ocean and its implications for CO2 uptake. Nature. 1992;358(6389):738-740. doi: 10.1038/358738a0.

[73] Cotton WR, Pielke RA. Human impacts on weather and climate. New York, NY: Cambridge University Press; 1995.

[74] Peterson TC, Gallo KP, Lawrimore J, Owen TW, Huang A, McKittrick DA. Global rural temperature trends. Geophys Res Lett. 1999;26(3):329-332. doi: 10.1029/1998GL900322.

[75] Numata I, Khand K, Kjaersgaard J, Cochrane MA, Silva SS. Forest evapotranspiration dynamics over a fragmented forest landscape under drought in southwestern Amazonia. Agric For Meteorol. 2021;306:108446. doi: 10.1016/j.agrformet.2021.108446.

[76] Poeter E, Fan Y, Cherry J, Wood W, Mackay D. Groundwater in our water cycle – getting to know Earth's most important fresh water source. The Groundwater Project, Guelph, Ontario, Canada. 2020. doi: 10.21083/978-1-7770541-1-3.

[77] Habgood PJ, Franklin NR. The revolution that didn't arrive: A review of Pleistocene Sahul. J Hum Evol. 2008;55(4):187-198. doi: 10.1016/j.jhevol.2008.06.001.

[78] Kangaroos: The solution to lingering livestock gas challenge. ANU Reporter [Internet]. Available from: https://reporter.anu.edu.au/all-stories/kangaroos-the-solution-to-lingering-livestock-gas-challenge

[79] Adkins J, de Menocal P, Eshel G. The African humid period and the record of marine upwelling from excess 230Th in Ocean Drilling Program Hole 658C. Paleoceanography. 2006;21(4). doi: 10.1029/2005PA001200.

[80] Shi W, Dong Z, Chen G, Bai Z, Ma F. Spatial and Temporal Variation of the Near-Surface Wind Environment in the Sahara Desert, North Africa. Front Earth Sci. 2021;9:789800. doi: 10.3389/feart.2021.789800.

[81] Smith C, Baker JCA, Spracklen DV. Tropical deforestation causes large reductions in observed precipitation. Nature. 2023;615(7950):270-275. doi: 10.1038/s41586-022-05690-6.

[82] Makarieva AM, Gorshkov VG, Li BL. Conservation of water cycle on land via restoration of natural closed-canopy forests: implications for regional landscape planning. Ecol Res. 2006;21(6):897-906.

[83] Makarieva AM, Gorshkov VG. Biotic pump of atmospheric moisture as driver of the hydrological cycle on land. Hydrol Earth Syst Sci. 2007;11(2):1013-1033.

[84] Happer W. CO2, The Gas of Life [Internet]. YouTube. 2023 Oct 3. Available from: https://www.youtube.com/watch?v=tXJ7UZjFDHU

[85] Pickering BJ, Duff TJ, Baillie C, Cawson JG. Darker, cooler, wetter: forest understories influence surface fuel moisture. Agric For Meteorol. 2021;300:108311. doi: 10.1016/j.agrformet.2020.108311.

[86] McKitrick RR, Michaels PJ. Quantifying the influence of anthropogenic surface processes and inhomogeneities on gridded global climate data. J Geophys Res Atmos. 2007;112(D24). doi: 10.1029/2007JD008465.

[87] Peterson TC. Assessment of Urban Versus Rural in Situ Surface Temperatures in the Contiguous United States: No Difference Found. J Clim. 2003;16(18):2941-2959.

[88] Hasler N, Williams CA, Denney VC, et al. Accounting for albedo change to identify climate-positive tree cover restoration. Nat Commun. 2024;15(1):2275. doi: 10.1038/s41467-024-46577-1.

[89] Mohajerani A, Bakaric J, Jeffrey-Bailey T. The urban heat island effect, its causes, and mitigation, with reference to the thermal properties of asphalt concrete. J Environ Manage. 2017;197:522-538. doi: 10.1016/j.jenvman.2017.03.095.

[90] Preda D. Earth – Calculations for Climate. Q Solar Pty. Ltd.; 2021. ISBN: 978-0-6487764-2-0.

[91] Preda D. Improved Covers for Solar Collectors. Australia: AU-2002300258; 2002 Jul 25.

[92] Preda D. Advanced Solar Air Heaters. Australia: AU-2007901243; 2007 Mar 12.

[93] NASA Earth Observatory. Global Warming [Internet]. Available from: https://earthobservatory.nasa.gov/features/GlobalWarming

[94] Gerber PJ, Steinfeld H, Henderson B, Mottet A, Opio C, Dijkman J, et al. Tackling Climate Change Through Livestock: A Global Assessment of Emissions and Mitigation Opportunities. Rome: FAO; 2013.

[95] Bouvard V, Loomis D, Guyton KZ, Grosse Y, El Ghissassi F, Benbrahim-Tallaa L, et al. Carcinogenicity of consumption of red and processed meat. Lancet Oncol. 2015;16(16):1599-1600. doi: 10.1016/S1470-2045(15)00444-1.

[96] Taft CT, Hamilton EG, Leviyah X, et al. Animal Consumption Associated with Higher Intimate Partner Aggression. J Fam Viol. 2023. doi: 10.1007/s10896-023-00556-0.

[97] Williams CA, Gu H, Allen CD, Law BE, Meddens AJH, Hicke JA, et al. Climate impacts of U.S. Forest loss span net warming to net cooling. Sci Adv. 2021;7(7). doi: 10.1126/sciadv.aax8859.

[98] European Space Agency (ESA). Earth Observation Brief [Internet]. Available from: https://www.esa.int/esapub/bulletin/bullet91/b91brie.htm

[99] Cess RD, Zhang MH, Minnis P, Corsetti L, Dutton EG, Forgan BW, et al. Absorption of Solar Radiation by Clouds: Observations Versus Models. Science. 1995 Jan 27;267(5197):496-499.

[100] Herman GF, Wu ML, Johnson WT. The Effect of Clouds on the Earth's Solar and Infrared Radiation Budgets. J Atmos Sci. 1980;37(6):1251-1261. doi: 10.1175/1520-0469(1980)037<1251

>2.0.CO;2.

[101] Silva Dias MAF, Rutledge S, Kabat P, et al. Cloud and rain processes in a biosphere-atmosphere interaction context in the Amazon Region. J Geophys Res Atmos. 2002;107(D20). doi: 10.1029/2001JD000335.

[102] Matthews JBR, Möller V, van Diemen R, Fuglestvedt JS, Masson-Delmotte V, Méndez C, et al., editors. Annex VII: Glossary. In: Climate Change 2021: The Physical Science Basis. Contribution of Working Group I to the Sixth Assessment Report of the IPCC. Cambridge, United Kingdom and New York, NY, USA: Cambridge University Press; 2021.

[103] von Schuckmann K, Palmer M, Trenberth K, Cazenave A, Chambers D, Champollion N, et al. An imperative to monitor Earth's energy imbalance. Nat Clim Chang. 2016;6(2):138-144. doi: 10.1038/nclimate2876.

[104] Preda D. Home – Earth Calculations [Internet]. Available from: https://dpreda.net/earth/

[105] Solomon S, Qin D, Manning M, Chen Z, Marquis M, Averyt KB, et al., editors. Climate Change 2007: The Physical Science Basis. Contribution of Working Group I to the Fourth Assessment Report of the IPCC. Cambridge, United Kingdom and New York, NY, USA: Cambridge University Press; 2007.

[106] Arias PA, Bellouin N, Coppola E, Jones RG, Krinner G, et al. Technical Summary. In: Climate Change 2021: The Physical Science Basis. Contribution of Working Group I to the Sixth Assessment Report of the IPCC. Cambridge, United Kingdom and New York, NY, USA: Cambridge University Press; 2021.

[107] Watanabe Y, Abe-Ouchi A, Saito F, et al. Astronomical forcing shaped the timing of early Pleistocene glacial cycles. Commun Earth Environ. 2023;4(1):113. doi: 10.1038/s43247-023-00765-x.

[108] Matthews JBR, Möller V, van Diemen R, et al. Annex VII: Glossary. In: Climate Change 2021: The Physical Science Basis. Contribution of Working Group I to the Sixth Assessment Report of the IPCC. Cambridge, United Kingdom and New York, NY, USA: Cambridge University Press; 2021.

[109] Wall CJ, Hartmann DL. Balanced cloud radiative effects across a range of dynamical conditions over the tropical West Pacific. Geophys Res Lett. 2018;45(21):11490-11498. doi: 10.1029/2018GL080046.

[110] Dickinson RE. Satellite systems and models for future climate change. In: World Survey of Climatology. 1995;16:245-279.

[111] Ouyang Z, Sciusco P, Jiao T, et al. Albedo changes caused by future urbanization contribute to global warming. Nat Commun. 2022;13(1):3800. doi: 10.1038/s41467-022-31558-z.

[112] Wei PS, Hsieh YC, Chiu HH, Yen DL, Lee C, Tsai YC, et al. Absorption coefficient of carbon dioxide across atmospheric troposphere layer. Heliyon. 2018;4(10)

. doi: 10.1016/j.heliyon.2018.e00785.

[113] Wei PS, Chiu HH, Hsieh YC, Yen DL, Tsai YC, Ting TC. Absorption coefficient of water vapor across atmospheric troposphere layer. Heliyon. 2019;5(1)

. doi: 10.1016/j.heliyon.2019.e01145.

[114] Aragão LEO. The rainforest's water pump. Nature. 2012 Sep;489(7415):217-218. doi: 10.1038/nature11485.

[115] Feulner G. Formation of most of our coal brought Earth close to global glaciation. Proc Natl Acad Sci USA. 2017;114(43):11333-11337. doi: 10.1073/pnas.1712062114.

[116] Oliveira LJC, Costa MH, Soares-Filho BS, Coe MT. Large-scale expansion of agriculture in Amazonia may be a no-win scenario. Environ Res Lett. 2013;8(2):024021. doi: 10.1088/1748-9326/8/2/024021.

[117] Spiridonova EA. Pollen-based reconstruction of Holocene vegetation and climate in the Eurasian Steppe. Quat Sci Rev. 2002;21(1-3):193-206.

[118] Hunt CD, et al. Impact of pastoral nomadism on Holocene vegetation and soil development in the Eurasian Steppe. The Holocene. 2011;21(4):493-503.

[119] Kates RW, et al. Sustainability science. Science. 2001;292(5517):641-642.

[120] Hulme M. Why We Disagree About Climate Change: Understanding Controversy, Inaction, and Opportunity. Cambridge University Press; 2009.

[121] Trenberth KE, et al. The changing character of precipitation. Bull Am Meteorol Soc. 2003;84(9):1205-1217.

[122] Ellison D, Futter MN, Bishop K. On the forest cover–water yield debate: from demand-to supply-side thinking. Glob Change Biol. 2012;18(3):806-820.

[123] Robock A, et al. Benefits, risks, and costs of stratospheric geoengineering. Geophys Res Lett. 2009;36(19).

[124] Pielke RA, et al. Unresolved issues with the assessment of climate impacts on land-use change. Glob Environ Change. 2007;17(3-4):400-410.

[125] Gentine P, et al. Coupling between the land surface and atmospheric boundary layer in convective conditions: A review. J Adv Model Earth Syst. 2019;11(6):1833-1855.

[126] Bonan GB. Forests and climate change: forcings, feedbacks, and the climate benefits of forests. Science. 2008;320(5882):1444-1449.

[127] Bright RM, et al. Carbon and radiative forcing of different land uses in continental Norway. Glob Change Biol. 2013;19(9):3106-3119.

[128] Veldman JW, et al. Where tree planting and forest expansion are bad for biodiversity and ecosystem services. BioScience. 2015;65(10):1011-1018.

[129] Sherwood SC, et al. An assessment of Earth's climate sensitivity using multiple lines of evidence. Rev Geophys. 2020;58(4).

[130] Held IM, Soden BJ. Water vapor feedback and global warming. Annu Rev Energy Environ. 2000;25(1):441-475.

[131] Lawrence DM, et al. The Community Land Model version 5: Description of new features, benchmarking, and impact of forcing uncertainty. J Adv Model Earth Syst. 2019;11(4):4245-4287.

[132] Stephens GL, et al. The global character of the flux of downward longwave radiation. J Clim. 2015;28(4):1203-1214.

[133] Goode PR, Palle E. Short-term variations in the Earth's reflectance. Eos Trans AGU. 2007;88(22):245-250.

[134] Bala G, Caldeira K, Wickett M, Phillips TJ, Lobell DB, Delire C, et al. Combined climate and carbon-cycle effects of large-scale deforestation. Proc Natl Acad Sci USA. 2007;104(16):6550-6555.

[135] Trenberth KE, Fasullo JT. Tracking Earth's energy: From El Niño to global warming. Surv Geophys. 2010;33(3-4):413-426.

[136] Lacis AA, Schmidt GA, Rind D, Ruedy RA. Atmospheric CO2: Principal control knob governing Earth's temperature. Science. 2010;330(6002):356-359.

[137] Thurairajah B, Shaw JA. Cloud Statistics Measured with the Infrared Cloud Imager (ICI). IEEE Trans Geosci Remote Sens. 2005;43(9):2000-2007.

[138] Pielke RA, et al. The influence of land-use changes on climate. Clim Change. 2002;52(1-2):5-23.

[139] Norris JR, Evan AT. Empirical evidence for a recent reduction in Earth's albedo derived from satellite observations. Geophys Res Lett. 2015;42(10):3749-3756.

[140] Bowman DM, Balch JK, Artaxo P, et al. Fire in the Earth system. Science. 2009;324(5926):481-484.

[141] Kaplan JO, Krumhardt KM, Zimmermann N. The prehistoric and preindustrial deforestation of Europe. Quat Sci Rev. 2009;28(27-28):3016-3034.

[142] Weiss H, Bradley RS. What drives societal collapse? Science. 2001;291(5504):609-610.

[143] Ellison D, Futter MN, Bishop K. On the forest cover–water yield debate: from demand-to supply-side thinking. Glob Change Biol. 2012;18(3):806-820.

[144] Geist HJ, Lambin EF. Dynamic causal patterns of desertification. BioScience. 2004;54(9):817-829.

[145] Kremenetski CV, Boettger T, MacDonald GM, Vaschalova T, Sulerzhitsky L. Climate change and the formation of steppe landscapes in Eastern Europe and Central Asia. J Quat Sci. 1998;13(2):149-157.

[146] Diamond JM. Collapse: How Societies Choose to Fail or Succeed. Viking Press; 2005. ISBN: 978-0670033379.

[147] Archer S. Woody plant encroachment into southwestern grasslands and savannas: Rates, patterns and proximate causes. Ecol Monogr. 1994;64(1):111-127.

[148] Houghton RA. Revised estimates of the annual net flux of carbon to the atmosphere from changes in land use and land management 1850–2000. Tellus B. 2003;55(2):378-390.

[149] Wilkinson TJ. Environmental fluctuations, agricultural production and collapse: A view from Bronze Age Upper Mesopotamia. Antiquity. 1997;71(273):183-189.

[150] Butzer KW. Early Hydraulic Civilization in Egypt: A Study in Cultural Ecology. Chicago: University of Chicago Press; 1976.

[151] D'Odorico P, Bhattachan A, Davis KF, Ravi S, Runyan CW. Global desertification: Drivers and feedbacks. Adv Water Resour. 2013;51:326-344.

[152] Dong SK, Wen L, Liu SL, Zhang XQ, Lassoie JP, Yi SL, et al. Vulnerability of worldwide pastoralism to global changes and interdisciplinary strategies for sustainable pastoralism. Ecol Soc. 2011;16(2).

[153] Manzano P, White SR. Intensifying pastoralism may not reduce rangeland degradation: Ecological insights from a simulation model of a Somali pastoral system. J Arid Environ. 2019;166:84-92.

[154] Lambin EF, Geist HJ. Land-use and land-cover change: Local processes and global impacts. Springer; 2006.

[155] Pimentel D, Pimentel M. Food, energy, and society. CRC Press; 2008.

[156] Houghton RA. Revised estimates of the annual net flux of carbon to the atmosphere from changes in land use and land management 1850–2000. Tellus B. 2003;55(2):378-390.

[157] Pan Y, Birdsey RA, Fang J, Houghton R, Kauppi PE, Kurz WA, et al. A large and persistent carbon sink in the world's forests. Science. 2011;333(6045):988-993.

[158] Betts RA. Offset of the potential carbon sink from boreal forestation by decreases in surface albedo. Nature. 2000;408(6809):187-190.

[159] Scheffer M, Holmgren M, Brovkin V, Claussen M. Synergy between small- and large-scale feedbacks of vegetation dynamics in the climate system. Glob Change Biol. 2005;11(6):1003-1012.

[160] Foley JA, Levis S, Costa MH, Cramer W, Pollard D. Incorporating dynamic vegetation cover within global climate models. Ecol Appl. 2000;10(6):1620-1632.

[161] Steffen W, Richardson K, Rockström J, Cornell SE, Fetzer I, Bennett EM, et al. Planetary boundaries: Guiding human development on a changing planet. Science. 2015;347(6223):1259855.

[162] The energetic basis of the urban heat island. Q J R Meteorol Soc. 1982;108(455):1-24.

www.ingramcontent.com/pod-product-compliance
Ingram Content Group UK Ltd.
Pitfield, Milton Keynes, MK11 3LW, UK
UKHW061846041026
15591UKWH00015B/929

9 780648 776444